Praise for *Modern Business Analytics*

Modern Business Analytics is an essential guide for anyone looking to enhance data-driven decision-making. This book not only bridges theory with practical application in Python and R, but also empowers both beginners and experienced practitioners to unlock the full potential of their data—from foundational concepts to advanced models and visualizations. A must-have for business analysts and data scientists alike.

—*Balaji Dhamodharan, Global Data Science Leader, NXP Semiconductors*

This book does an excellent job of exploring predictive and prescriptive analytics through data modeling using R and Python. If you are new to data science or if you just need a refresher, this guide is an excellent place to start!

—*Christopher Gardner, Lead Business Intelligence Analyst, University of Michigan*

Modern Business Analytics expands the horizons of business analytics for anyone interested in the myriad world of data. With its focus on the power of Python and R, this book simplifies complex concepts and empowers readers to unlock the exponential value of their data. It's a can't-miss resource for anyone striving to master today's rapidly evolving analytics landscape!

—*Lipi Deepaakshi Patnaik, Senior Software Engineer, Zeta*

I don't get excited about too many books, but I am about this one. I published several books about data science applications in the business domain and have been a practitioner my whole professional life. So I'm aware of the need for a book on business analytics that stresses, explains, and illustrates both R and Python for analyzing business data. This book satisfies this need and then some. The chapters on advanced data visualization and working with modern data types alone take these languages into territory all business data scientists need to be aware of. I highly recommend this book to increase your background in data science.

—Walter R. Paczkowski, Ph.D., President, Data Analytics Corp.

Modern Business Analytics

Increasing the Value of Your Data with Python and R

Deanne Larson

Modern Business Analytics

by Deanne Larson

Published by O'Reilly Media, Inc., 1005 Gravenstein Highway North, Sebastopol, CA 95472.

O'Reilly books may be purchased for educational, business, or sales promotional use. Online editions are also available for most titles (*http://oreilly.com*). For more information, contact our corporate/institutional sales department: 800-998-9938 or *corporate@oreilly.com*.

Acquisitions Editor: Michelle Smith
Development Editor: Jill Leonard
Production Editor: Katherine Tozer
Copyeditor: nSight, Inc.
Proofreader: Kim Wimpsett
Indexer: BIM Creatives, LLC
Interior Designer: David Futato
Cover Designer: Karen Montgomery
Illustrator: Kate Dullea

December 2024: First Edition

Revision History for the First Edition

2024-12-10: First Release

See *http://oreilly.com/catalog/errata.csp?isbn=9781098140717* for release details.

978-1-098-14071-7

[LSI]

Table of Contents

Preface

Business analytics is evolving rapidly, and the use of powerful tools like Python and R is revolutionizing how organizations harness the power of their data. I'm excited to share this book, *Modern Business Analytics*, which is designed to help you enhance your data-driven decision-making skills using these tools. The ability to leverage Python and R can increase the value of your data exponentially, whether you're performing statistical analysis, building models, or creating advanced visualizations. What's especially exciting is how accessible these languages are for business analysts and data professionals alike.

This book will guide you step-by-step, from basic concepts to advanced applications, allowing you to unlock the full potential of your data.

Who Should Read This Book

This book is ideal for business analysts, data scientists, and anyone interested in enhancing their data analytics skills using Python and R. Whether you are an experienced professional looking to expand your toolkit or a beginner eager to learn the fundamentals, this book will provide valuable insights and practical applications. A basic understanding of statistics and general data concepts is recommended, but you don't need to be an expert programmer to benefit from the content. This book aims to make complex analytics more approachable while helping you improve the efficiency and impact of your data projects.

Why I Wrote This Book

The world of business analytics has changed significantly over the past decade, and it's clear that tools like Python and R are becoming essential for any serious data professional. I wrote this book because I saw a need for a comprehensive resource that blends the theoretical foundations of analytics with practical, hands-on coding in these two languages. This book fills the gap by not only teaching you how to perform analytics but also showing you how to apply those skills to real-world business problems. By the end, I hope you'll feel more confident in your ability to extract meaningful insights from data and contribute more effectively to your organization's decision-making processes.

Navigating This Book

This book is organized into 11 chapters, each focusing on a key aspect of business analytics:

Chapter 1, "The Role of Business Analyst and Analytics"
: This chapter covers the evolving role of business analysts and how analytics plays a critical part in modern organizations.

Chapter 2, "Methodologies for the Business Analyst and Analytics Projects"
: Learn about the methodologies that drive successful business analytics projects, including agile frameworks and the CRISP-DM model.

Chapter 3, "Introduction to R and Python"
: Get an introduction to both programming languages, with a focus on their role in data analysis and their unique strengths.

Chapter 4, "Statistical Analysis with R and Python"
: Discover how to perform key statistical analyses using Python and R to draw insights from your data.

Chapter 5, "Exploratory Data Analysis with R and Python"
: This chapter will teach you how to perform exploratory data analysis to identify patterns, relationships, and anomalies in datasets.

Chapter 6, "Application and Evaluation of Modeling in R and Python"
: Learn how to apply various modeling techniques and evaluate their effectiveness using both languages.

Chapter 7, "Modeling and Algorithm Choice"
: This chapter discusses how to select appropriate modeling algorithms based on the nature of your data and business needs.

Chapter 8, "Model Operations"
: Explore the operational aspects of maintaining and scaling models, from deployment to monitoring their performance.

Chapter 9, "Advanced Visualization"
: Dive into creating advanced visualizations that communicate your data insights effectively using both Python and R.

Chapter 10, "Working with Modern Data Types in Analytics"
: This chapter covers the challenges and techniques for working with modern data types such as text, social media, image, and video data.

Chapter 11, "Measuring Business Value from Analytics and the Role of AI"
: Finally, you'll learn how to measure the impact of your analytics work and explore the growing role of AI in business analytics.

Additional resources, including the code examples used in this book, are available on the book's GitHub page (*https://github.com/deannelarson/modernBA*).

Conventions Used in This Book

The following typographical conventions are used in this book:

Italic
: Indicates new terms, URLs, email addresses, filenames, and file extensions.

`Constant width`
: Used for program listings, as well as within paragraphs to refer to program elements such as variable or function names, databases, data types, environment variables, statements, and keywords.

`Constant width bold`
: Shows commands or other text that should be typed literally by the user.

`Constant width italic`
: Shows text that should be replaced with user-supplied values or by values determined by context.

This element indicates a warning or caution.

Using Code Examples

Supplemental material (code examples, exercises, etc.) is available for download at *https://github.com/deannelarson/modernBA*.

If you have a technical question or a problem using the code examples, please send email to *support@oreilly.com*.

This book is here to help you get your job done. In general, if example code is offered with this book, you may use it in your programs and documentation. You do not need to contact us for permission unless you're reproducing a significant portion of the code. For example, writing a program that uses several chunks of code from this book does not require permission. Selling or distributing examples from O'Reilly books does require permission. Answering a question by citing this book and quoting example code does not require permission. Incorporating a significant amount of example code from this book into your product's documentation does require permission.

We appreciate, but generally do not require, attribution. An attribution usually includes the title, author, publisher, and ISBN. For example: "*Modern Business Analytics* by Deanne Larson (O'Reilly). Copyright 2025 Larson & Associates LLC, 978-1-098-14071-7."

If you feel your use of code examples falls outside fair use or the permission given above, feel free to contact us at *permissions@oreilly.com*.

O'Reilly Online Learning

For more than 40 years, *O'Reilly Media* has provided technology and business training, knowledge, and insight to help companies succeed.

Our unique network of experts and innovators share their knowledge and expertise through books, articles, and our online learning platform. O'Reilly's online learning platform gives you on-demand access to live training courses, in-depth learning paths, interactive coding environments, and a vast collection of text and video from O'Reilly and 200+ other publishers. For more information, visit *https://oreilly.com*.

How to Contact Us

Please address comments and questions concerning this book to the publisher:

O'Reilly Media, Inc.
1005 Gravenstein Highway North
Sebastopol, CA 95472
800-889-8969 (in the United States or Canada)
707-827-7019 (international or local)
707-829-0104 (fax)
support@oreilly.com
https://oreilly.com/about/contact.html

We have a web page for this book, where we list errata, examples, and any additional information. You can access this page at *https://oreil.ly/ModernBusinessAnalytics*.

For news and information about our books and courses, visit *https://oreilly.com*.

Find us on LinkedIn: *https://linkedin.com/company/oreilly-media*.

Watch us on YouTube: *https://youtube.com/oreillymedia*.

Acknowledgments

Writing this book has been an incredible journey, and I am deeply grateful to everyone who contributed to making it a reality. I would like to express my sincere thanks to my colleagues and peers, who provided invaluable feedback throughout the writing process. Their insights helped refine the content, ensuring that it is both practical and relevant for readers in the ever evolving field of business analytics.

A special thanks to the technical reviewers, whose keen attention to detail and commitment to accuracy were essential in shaping the book's technical integrity. I am particularly grateful to Milind Chaudhari, Balaji Dhamodharan, Christopher Gardner, Aditya Goel, Andreas Kaltenbrunner, George Mount, Walter R. Paczkowski, Lipi Deepaakshi Patnaik, and Tobias Zwingmann. Your thoughtful and meticulous feedback ensured that every example and concept was clear, accurate, and applicable. Your expertise was invaluable in strengthening the technical aspects of this book.

I would also like to acknowledge the many discussions I've had with industry professionals and fellow educators; these conversations have greatly influenced the direction of the book and its focus on blending theory with real-world application.

To my editor, Jill Leonard, I extend my heartfelt gratitude. Jill, your guidance, patience, and meticulous attention to detail were instrumental in bringing this project to life. From helping shape the initial concept to the final draft, your editorial expertise has been invaluable. Your support made the challenging process of writing a book much smoother, and I couldn't have asked for a better collaborator throughout this journey.

Finally, I want to thank my family, whose unwavering support allowed me the time and space to dedicate myself to this project. To my spouse and children, thank you for your patience and understanding during the long hours spent researching and writing. Your encouragement kept me motivated and focused. I am also deeply appreciative of my friends and mentors, whose advice and support have been a constant source of inspiration.

Thank you to all who played a part in making *Modern Business Analytics* possible. This book would not have been the same without your contributions, and I am truly grateful for your help and encouragement along the way.

CHAPTER 1

The Role of Business Analyst and Analytics

If you are new to business analytics or considering a career in this space, this chapter will be particularly useful to get started or transition from a business analyst into analytics. If you are already familiar with the fundamentals or currently working in the industry, you may find it helpful to jump ahead to the next chapter, where we start with analytics application.

The role of a business analyst has been around for some time but has grown in importance as the need to strategically leverage data has increased. Organizations want to leverage their data as an asset and improve decision making at all levels. While better decision making can be defined in multiple ways, ultimately it means the right answer was chosen the first time, which results in improved efficiency, effectiveness, and a better bottom line. There are many roles that leverage data, but a business analyst is the one role that provides the necessary context to create business value.

If you research what knowledge and skills a business analyst has, you will likely come across what business analysts do but not the business value of the results achieved. This chapter defines what the business analyst's role is in analytics and the critical part it plays in attaining value from analytics. It also addresses the business problems that analytics solve and the project life cycle that analytics projects progress through. Let's get started with understanding the role of a business analyst and types of analysts.

What Is the Role of a Business Analyst?

As a business analyst, you leverage your analysis and industry skills to decompose, or break down, problems and find root causes resulting in continuous improvement and successful strategy execution. This isn't always easy, because the business landscape of an organization can be complex. It may include multiple information systems, business processes, and business departments (both people and structure), resulting in the analyst working across all of these functional areas. Ultimately as a business analyst, you have subject matter expertise about your organization and its industry, and you are key to solving problems and providing solutions.

Due to the diversity of knowledge a business analyst has, one of the primary areas that business analysts focus on is *requirements management*. Requirements management is the process of defining, documenting, and analyzing requirements to address the scope of business needs. Requirements are determined by understanding the problem and identifying the best solutions. Analytical projects often start with understanding a problem's symptoms. For instance, let's say you work for a telecom company and it appears customer attrition (percentage of customers no longer using a company's products or services) has increased. An analytical project might start with research to determine the root cause of the problem. Requirements analysis would continue based on how clear the problem is and where analytics might be used to solve the problem. Increased customer attrition is a problem, but it is not yet clear how analytics might help solve it. This is why things aren't quite as clear cut as they may seem. Requirements and how analysts help solve problems will be covered more in a later chapter.

As a business analyst, you will always be addressing new requirements due to the constantly changing business environment. Because of this, the best way to understand your role is to explore the skills and responsibilities and to see examples of the different types of analyst roles that may exist in an organization.

Skills

Business analysts are quite versatile, drawing upon multiple skills to produce results. As a business analyst, you research new approaches to address current problems, which means searching for new processes, systems, and other options to be able to tackle challenges your organization may be facing. You use your industry knowledge and technical skills to evaluate the results of your research and then determine the applicability of each option. To do this, business analysts apply statistics to determine if new options will produce the desired results and communicate findings to peers and senior leadership. Central to each of these examples is leveraging your ability to analyze—which means examining details and applying these details to the big picture of problem solving. Table 1-1 highlights many of the skills a business analyst can be expected to have.

Business analysts can have many skills that are applicable in different work environments and scenarios as well as industries. It is quite common for business analysts to transfer skills across domains and industries throughout their careers. The skills you choose to focus on and develop will help shape your future projects, responsibilities, and day-to-day activities as a business analyst.

Table 1-1. Business analyst skills

Skill	Definition	Example
Research	Systematic identification of materials and sources to fact-find and establish a conclusion	Reviewing different vendor offerings to determine if these offerings meet requirements
Communication	Exchanging of information to leverage effective methods	Presenting to an executive audience possible solutions to a problem
Problem solving	The process of identifying solutions to complex issues	Completing a root cause analysis effort, identifying the cause, and applying resolution
Industry expertise	Knowledge and skills specific to an industry group and ability to apply this knowledge	Having knowledge of the manufacturing industry and using this knowledge to complete problem solving
Statistics	The practice of collecting and analyzing data to infer results	Collecting and analyzing data on a process with the intent of improvement
Technical	Specialized knowledge and expertise required to perform technical activities	Using a programming language to complete data analysis
Analysis	Examining complex components or processes to gain knowledge of the nature and features	Examining system interfaces to understand how points interact

Responsibilities

Responsibilities are activities a business analyst is expected to perform, and depending on the focus of the business analyst, these can vary. As a business analyst, you might be focused on marketing, or you could have a focus on information technology (IT) projects—both common areas of specialization. Here are some of the primary responsibilities of business analysts, but this is not an exhaustive list:

Managing requirements
: Gathering, preparing, and validating requirements for projects

Identifying problems and opportunities
: Performing analysis on potential causes, sorting through the noise, and identifying root contributors and new opportunities

Determining and proposing solutions
: Building on identifying problems and opportunities, taking the analysis further to identify and recommend the next course of action

Budgeting and forecasting
: Creating a budget for business expenses and revenue for short-term planning, then projecting business outcomes for the future

Planning and monitoring
: Scheduling activities, allocating resources, identifying milestones, and determining progress made to the plan

Process modeling
: Creating a graphical representation of business processes or workflows for analysis

Ongoing analysis
: Continuous gathering of data and review to check results for a specific purpose

Testing
: Measuring the overall quality, functionality, performance, and reliability before use

Given the breadth of the various responsibilities that a business analyst can have, analysts can reside in many different organizational departments and support different efforts. As an analyst, you can specialize in an industry or functional area. The next section explores different types of analysts and their areas of focus.

Types of Analysts

Because the business landscape is broad and multiple functions support each business model, the types of analyst roles can vary. Any team that uses data may be able to use a business analyst, and all of these roles leverage the skills and responsibilities we just discussed while focusing on the needs of their part of the business. Some common types of analysts include marketing, finance, functional, system, and data analyst. Again, this is not an exhaustive list as it is possible to specialize in several areas; however, the roles we're about to cover provide a good foundation of the diversity that comes with being a business analyst. The different roles outlined can also collaborate on projects in the organization.

Marketing analyst

Marketing efforts are directly linked with organizational strategy such as increasing revenue or the number of customers. Marketing has grown due to the digitization of how businesses interact with their customers. Marketing has shifted into digital marketing, also known as online marketing, because most marketing occurs online. Communications via social media, websites, and mobile devices enable businesses to connect directly with customers and are the primary way of developing customer relationships. Being able to analyze the data points created through these touchpoints is critically important to gain insight about how to best engage customers. The marketing analyst role doesn't stop there. Since digital marketing is measurable, the need to analyze the results is a priority.

Marketing analysts focus on gathering and cleaning data from sources such as surveys and campaign results. Campaigns that include offers and promotions need analysis such as measuring reach, engagement, and lift (revenue increase) to determine what is working and what needs to be improved. Marketing analysts take the results and determine how to adapt and drive the digital marketing strategy. They also leverage analytics to predict a customer's choice or use segmentation to target and personalize marketing campaigns.

Marketing analysts can also focus on research such as understanding competitors, reviewing trends in markets, and price analysis. They support product development and other business teams as marketing is a support organization and each of these areas leverage different types of analytics.

Financial analyst

Finance is another support organization like marketing. All organizations have to have financial health to stay operational. Finance supports the organizational strategy by providing insight into financial investment decisions. A financial analyst will primarily focus on activities that support these decisions by reviewing economic trends, industry direction, and competitive analysis. Analytics is a common skill set leveraged by financial analysts.

One key area a financial analyst works on is analyzing historical results and completing forecasts and predictions. Financial analysts create different financial models that could involve optimization and simulation scenarios to determine risk or the optimal price for a product. Knowing Excel and predictive analytics tools is often necessary as they support the creation of different financial models. These models start with financial statements, but can include discounted cash flows, mergers and acquisition analysis, and the impact of decisions on the organization's stock price. Financial analysts are involved in risk analysis based on investment decisions and produce written reports related to financial status.

Both financial and marketing analysts are examples of business-focused analysts. Analysts can also be specialized when working in IT, where they may be a functional, system, or data analyst.

Functional analyst

Functional analysts can specialize in areas such as manufacturing, supply chain, or specific applications. Functional analysts are subject matter experts in their areas and leverage this expertise to find improvement opportunities. The primary goal of functional analysts is to improve the productivity of an area by focusing on the functional requirements and reviewing the systems. An example of this is reviewing the supply chain process for a company and identifying opportunities to improve efficiencies and then recommending improvements in the system to the operations team. Functional analysts also focus on reviewing existing systems and coordinating updates to keep the technology current. This type of analyst tends to be part of the IT department, but it is possible this role may exist in the department that the functional analyst has expertise in. A functional analyst is often the liaison between the business and technical departments in an organization and can be a part of either department.

As the previous examples show, functional analysts have a deeper technical skill set than other analyst types. They have a broad understanding of technology including networking, databases, and applications, and they leverage process and technical knowledge to identify opportunities for improvement and recommend technical solutions. Technical analysts, like functional analysts, leverage software development methodologies including agile or Scrum and may model data, prepare design diagrams, complete testing, and train users. This role can be visible within the organization and requires strong communication and leadership skills. A functional analyst may lead a portion of a technical project and need project management skills as well.

Functional analysts contribute to the requirements for analytical projects by identifying measures, metrics, indicators, and other data needed for improvement opportunities. Process improvement relies on analysis and measurement of goal progress. Analysis also focuses on checking position on progress and making decisions about next steps. Many of the skills needed to be a functional analyst is used by a system analyst. We will review the system analyst next.

System analyst

System analysts are considered technology professionals who define requirements, assist in design, and support the deployment of information systems in an organization. There are some parallels with a functional analyst, but a systems analyst is more technical. Like a functional analyst, a system analyst can research problems, recommend solutions, and work with stakeholders on developing requirements. The difference between the functional analyst and the system analyst is that the system analyst is familiar with operating systems, application configurations, hardware platforms, cloud platforms, and programming languages. System analysts are often involved from the analysis stage of a project to post implementation to ensure system stability.

System analysts are often in a liaison between the business stakeholders and the technology teams to translate requirements into technical design. System analysts also focus on integration of technologies to solve business problems linking different platforms, protocols, networking, and software together. System and functional analysts can collaborate on solutions. System analysts are typically part of the IT department.

System analysts can also contribute to the requirements for analytical projects, by identifying data points required for decision making. Data analysts, which we will review next, work as team members on analytical projects.

Data analyst

Data analysts work with data to find insights that can be leveraged for business value. Business value can be a good decision, the discovery of a problem, the solution to a problem, or new trends and patterns that can be leveraged for new opportunities. Data analysts take insights and formulate data stories to communicate to leadership.

Data analysts can be in a technology or business department and share some of the skills of a functional analyst where knowledge of a business industry is important. Data analysts need to have a strong foundation in descriptive and inferential statistics and are often part of project teams that support larger analytical goals. One area where data analysts differ from the other types of analysts mentioned is the hands-on involvement working with the data.

As a data analyst, you will leverage technical skills for data collection and analysis, and report insights using data storytelling. A data analyst might acquire data from the source system, merge and blend the data with other sources, and use different analytical techniques to search for insights. For example, a data analyst might look at customers to determine if different clusters or segments exist so they can be treated differently. Data analysts can also cleanse and transform data to apply business rules or increase data quality.

While working with data, data analysts can partner with other analysts such as the system analyst. Since a system analyst has knowledge of the business applications, a data analyst may partner with them to better understand how to acquire the data or understand the structure of data in an application. In contrast, while system analysts focus on business applications to close business gaps, data analysts leverage analysis skills on diverse datasets to support improved decision making versus improving processes.

There are fine lines between a business analyst and a data analyst, with the primary difference being data analysts focus more on the technical aspect of wrangling and analyzing the data. Due to the overlap of multiple types of analysts, it is impossible to highlight precisely what a business analyst will focus on in an organization. Practically, a business analyst could leverage skill sets to work in multiple areas of an organization. If you review each type of analyst, you will see that data plays an important role in each, which begins to outline why business analysts need to understand analytics, which we will address in the next section.

Why Does a Business Analyst Need to Know Analytics?

Organizations are continuing to invest in analytics as it focuses on managing data and leveraging it to improve decision making, business processes, shaping strategy, and driving strategy. Analytics is used to discover and manage business risks and opportunities (*https://oreil.ly/oLqkr*). As outlined in the prior section, a business analyst, regardless of the primary role played, works with data, focuses on problem solving, and recommends solutions. This is the sweet spot where analytics is used.

Analytics will be explained in detail throughout this book, but to give it a definition, it is a capability organizations develop to support the analysis of data using statistics, math, and algorithmic models to improve decision making. Analytics can be descriptive, diagnostic, predictive, or prescriptive, according to Gartner (*https://oreil.ly/oLqkr*). These primary analytical types have been adopted industry-wide and categorize the different types of analysis that a business analyst can be involved in. To get a better understanding of why a business analyst needs to not only understand but apply analytics, the next few sections cover the explosion of data, the need to have business context for every analytical problem, and the role analytics plays in generating business value.

Data Explosion

The concept of digitization is the key to understanding the data explosion. Simply, the data explosion is the automated capturing of data points through technology. The term "big data" has been in use since the 1990s when John Mashey (*https://oreil.ly/pJ2TV*), a chief scientist from SGI, identified the start of digitalization with the growth of the internet. With data being captured that was not available before, the focus on using data combined with computing power, statistics, and math, the value of data started to evolve. Big data is mostly the norm now, but, the concept of the data explosion has continued with more data points (data captured as part of an event, transaction, or any interaction) being collected every second of the day.

Think about search engines, cookies on websites, global positioning systems, and each mobile application on your phone: these are all data collection points. Multiply the data collection points by the number of users and interactions; you get the picture (*https://oreil.ly/LNa48*). This is happening consistently and globally. Take an example of the automated toll collection systems that different states use for state highways. States on the east coast of the United States use EZPass, where the number of transactions for 19 different states comes to approximately 10 million transactions a day. Likely this information is already outdated, and the number is much more than quoted. This is an example of one organization capturing data for automation but also analysis. Traffic patterns, point-to-point travel, analysis of logistics, and road wear are examples of the types of analysis that can be completed with the data collected. YouTube, baseball, and analyzing the surface of the earth are all examples contributing to data generation and growth.

All organizations are investigating the use of data for cost reduction, improving efficiencies, and measuring results. But the challenge is the amount of data and understanding what to do with it.

So does this data have a use or value? This is one of the catalysts that is fueling the growth of analytics. The primary goal is to find insight that can be useful to an organization. This is where the role of the business analyst comes in. Business analysts are key to determining the value of data to an organization, and this comes with understanding the business.

Business Context

Data is not valuable without context. If you were handed a spreadsheet with columns and rows of numbers, but no column headers, could you do anything with it? You could argue that someone with experience could look at the numbers and guess the data contents, but not much more could be done with that data. Business context is the key to turning data into information.

Business context is about understanding where an organization fits in the business world. An example is understanding what industry an organization is in. Industries such as finance, manufacturing, high tech, telecommunications, shipping, or entertainment (not an exhaustive list) are examples of where a business may operate. An industry has a model for business operations that defines what a business does every day to service its customers. There are concepts about products and services in an organization that a business analyst knows. For example, in telecommunication a customer is called a "subscriber," and the average revenue per user, referred to as ARPU (pronounced R-PU), would be known to a business analyst working in this industry.

Data is generated by technology that can be segmented into systems and applications. Business analysts understand the primary business systems and applications in the organization they work for. Which systems are used in which business process? Which systems are the source of customer information? Which applications does a customer interface with when ordering a product or paying a bill? Knowing this information provides context to business data and insight into what the data represents and how it can be used.

Patterns and trends in data cannot be interpreted without context. It is not possible to understand root causes or provide valid recommendations to problems without context. It is impossible to determine if patterns or trends discovered are valuable without business context. Business analysts bring the experience and expertise needed to determine if data insights exist. This is why business analysts are a key part of the analytics process.

Analytics

Many terms exist that can be merged under the umbrella term of analytics. It has become an umbrella for business intelligence (BI) and in some cases has been identified as a specialized capability such as predictive analytics or marketing analytics. Many vendors use analytics to differentiate their products. Data and business analytics are often called out as different aspects of analytics, where data analytics focuses on the technical, statistical, and math aspects, and business focuses on the application of business expertise to the findings. Basically, the term analytics describes all activities and capabilities that an organization uses to exploit large datasets for insight.

Gartner (*https://oreil.ly/oLqkr*) has determined the primary analytical techniques that organizations use, and these were mentioned previously: descriptive, diagnostic, predictive, and prescriptive. Often another term that is thrown in is discovery analytics. Analytics can be segmented into five different techniques, which we explore next. Each technique can also be viewed as a step toward greater capability and maturity in analytics.

Descriptive

Most organizations have descriptive analytics capabilities as this area answers the question, What has happened? This capability stage uses business intelligence tools, dashboards, and scorecards to monitor and manage the business. Business questions include: What product category produced the most profit? How did customer care perform on average time supporting customers? What was the average production cycle for product X last month?

Diagnostic

Diagnostic analytics builds on descriptive analytics to answer the question, Why did it happen? This is the capability stage used to understand root cause analysis and find patterns and trends that lead to recommendations. Business questions include: Why did the profit margin decrease on product X? Why are sales dropping in the fourth quarter for product Y? Why did the manufacturing cycle double last month? Both descriptive and diagnostic analytics focus on hindsight.

Discovery

Discovery analytics focuses on answering the question, What else should I know? This is the stage where new capabilities start to emerge. Building on diagnostic analytics, discovery analytics is where a new direction is formed for analysis. This stage can involve incorporating new data sources and finding additional contributing factors to the root cause. Discovery analytics provides insight, which leads to foresight.

Predictive

Predictive analytics is the capability stage where most organizations want to be, as this is where the business value of analytics emerges. Predictive analytics focuses on answering the question, What is happening next? More specifically, predictive analytics focuses on answering these questions: What is the likelihood this customer will click on this ad? What is the probability of this customer churning? What is the likelihood that this marketing campaign will result in a certain lift in sales? Predictive analytics helps shape and drive the strategy of an organization. Predictive analytics relies on techniques such as classification, regression, and other machine learning approaches.

Prescriptive

Prescriptive analytics is the nirvana of analytics as it is used to drive outcomes. Prescriptive analytics focuses on automation and optimization. Rule-based approaches and operations management techniques are used to determine how best to automate decision making. Often combined with predictive analytics, prescriptive analytics includes the use of recommendation engines and automated decisioning such as insurance quotes and mortgage approvals. Prescriptive analytics starts to merge with AI once it matures.

At the heart of each of these analytical stages is the need to understand data and apply the results of the analytics to solving business problems. Investing in analytical technology and building large data repositories does not provide an organization with analytical capability or value. It is the people with the business expertise that are required to solve business problems.

Business Analyst Contributing to Analytics Value

Business analysts have the knowledge of their organization and focus on a primary function, which is to identify and validate the needs of the business. As a project becomes clear, the business analyst will develop an understanding and determine the requirements. Analytical projects are no different. There are many roles that are involved in analytical projects, including data engineers, data scientists, and other analysts, but none of these roles can validate the results of analytics or ensure that the results can be used correctly.

Business analysts are necessary to ensure analytical projects run efficiently. The technical resources can determine the technical direction and start to build software, but the business analyst provides information, answers questions, and assists in removing barriers to ensure the project moves forward as expected. Business analysts are involved in understanding the business problem to be solved (the most important step in the analytics life cycle) and how business needs can be met.

Business analysts also provide clarity if multiple stakeholders are involved in the business understanding of the project. Facilitating discussion, gathering facts, reviewing outcomes, and reconciling the feedback from the stakeholders is needed. Essentially, the business analyst is the business representative, advocating for stakeholders and ensuring issues are addressed throughout the analytics project.

Assisting with testing the analytics model is another aspect where business analysts become involved. Through testing, the business analyst determines if the model will support the requirements and if it addresses the original business problem. Testing of analytics is not specific to software testing but includes the testing of the outcomes and the results of decisions made from predictions.

Business analysts are really the key to analytics business value. Business analysts assist with change management as they are experts on how analytics will be used in the organization. If analytical models are created and they do not assist in decision making or problem solving, the model is not providing business value. To get a better understanding of business value, the next section looks at different business problems solved by analytics.

Business Problems Solved by Analytics

As mentioned, analytics without results would not provide business value to an organization. To provide more clarity on the value of analytics, we will explore some of the different business problems addressed by analytics. The following examples include better decision making, campaign optimization, and other examples from Marriott and UPS.

The goal of many organizations includes becoming data-driven and making better decisions. For example, a leader of a fintech organization wanted to increase the use of data by making business data available to departments. Historically, the IT department had control over the data, and this prevented collaboration between departments. Dashboards were created to make data consistent and shareable, which resulted in removing hours in report preparation by IT and data being available in real time for decision making. Prior to this, data was available only after preparation by the IT department.

An online business wanted to have insight into customer activity and the life cycle to optimize marketing campaigns and make quicker decisions on how to increase customer satisfaction and brand loyalty.

Tom Davenport of Harvard Business School cowrote a book called *Competing on Analytics* in 2006 and highlighted multiple examples of how analytics provides business value. In an abridged version, Davenport highlights several organizations that have gained value from incorporating analytics. Two examples that stand out are that of Marriott International and UPS.

Marriott International has leveraged analytics to determine the optimal price for rooms using an analytics process called revenue management. Marriott has leveraged analytics for total optimization. Marriott has used analytics to develop systems that optimize offerings to customers and determine customer churn, or what number of customers switch to their competitors. The analytics are in the hands of revenue managers to make the best decisions on pricing, which directly determines the profit margin for the company.

Logistics cannot be successful without the application of operations management and analytics. UPS uses analytics to track packages in real time and predict customer churn as well as root-cause analysis of problems. For example, UPS can accurately predict customer churn by reviewing patterns and complaints. If a customer is predicted to churn, a salesperson contacts the customer to resolve the problem before churn can happen.

Imagine being able to protect the profit margin and prevent customer churn. What is the value of both of these analytics-enabled capabilities to these organizations? Also consider how a business analyst would be needed for process and other business context to make these capabilities possible.

Collaboration with Other Teams

Organizations tend to have siloed data, which means data is not centralized or integrated to the point where it is possible to see an enterprise view. Additionally, business expertise can focus on a department or a particular area. Analytics is rarely focused on one data source or part of a process. Business analytics are necessary to bridge the different areas of expertise and understand the impact of analytics.

An example of this would be the use of a predictive model. A model can be created for predicting customer churn, for example, but there could be many different factors that impact a customer's experience. Marketing, sales, customer care, and product warranty could all be touching points experienced by a customer. How would an analytics project consider how these different viewpoints and processes impact churn? Business analysts help bridge these different areas and provide a horizontal view of the business versus a vertical one.

Skill Sets Used in Analytics

Math and statistics are some of the different skill sets mentioned in analytics. To expand on these further, business analysts become skilled at using different analytical techniques leveraged in the different analytical stages (descriptive, diagnostic, discovery, predictive, and prescriptive). Many of these techniques will be covered in chapters that follow, but analysts can be expected to understand descriptive and inferential statistics, applied math, some software development, and the analytical life cycle—all of which will be covered in the following sections.

Python and R

The current trend in analytics is to use open source tools: the leading tools are Python and R. Python is a full programming language that covers the programming capabilities to complete the data engineering as well as the machine learning algorithms to create models. R is used for statistics and analysis with strong visualization capabilities. Both tools are fueled by the different packages and libraries available.

Statisticians and academics developed R, and it currently has more than 12,000 packages available in its open source repository. Each package (synonymously used with library) contains many statistical capabilities needed by a business analyst. R surpasses Python with its output through visualization and the ability to publish findings in a document. The largest drawback to R is the ability to use the code in a production environment. While possible to do, R is hard to automate and use for operational processing.

Python has the same capabilities as R but excels at deploying and implementing large-scale analytics. Python code is easier to maintain and support. It was first a programming language before it was used in analytics. Today, most machine learning capabilities are available in Python first, then R. Python uses application programming interfaces (APIs) easily, and if you want to productionalize your machine learning code, Python is the simplest approach.

Analytics Project Life Cycle

Analytics professionals have adopted a life cycle for project delivery that is based on a data-mining approach used for some years. The life cycle is not sequential and is iterative as analytics projects are often about discovery. The life cycle provides the phases and steps that analytics professionals follow. Several of the large tech companies such as Microsoft and Amazon have incorporated these same phases and steps into their own methodologies. There are six primary phases to the analytics project life cycle: business understanding, data understanding, data preparation, modeling, evaluation, and deployment. Chapter 2 will explore the analytics project life cycle in detail, and the value of each of the phases will become clear.

Summary

This chapter focused on introducing the business analyst and the role of the business analyst in the analytics process. A business analyst role can take on several forms, but at the heart of the role, a business analyst focuses on understanding the requirements to address a business problem, working with data and providing business context to the data, reviewing the results of testing, and ensuring the business stakeholders are represented correctly in problem resolution. We also got introduced to the concept of analytics and how a business analyst gets engaged in the different analytics techniques. Business analysts are critical to analytics success due to business expertise and ensuring business value.

CHAPTER 2

Methodologies for the Business Analyst and Analytics Projects

Analytic projects are unique in that these projects produce working software. Within the industry, this software is referred to as a data product. Data products are those outcomes of analytics that are used by an organization as part of business operations. A common example of a data product would be a predictive model. For instance, this could be a model that predicts the likelihood of a customer clicking on a pop-up ad. The methodology that is the basis for most analytical projects today is called the Cross Industry Standard for Data Mining (CRISP-DM). While it is not referred to any longer due to its age and more modern methodologies being introduced, the modern methodologies still use the primary phases from CRISP-DM (*https://oreil.ly/ESg-0*). Each phase has a purpose and outcomes that business analysts often participate in. Figure 2-1 highlights the phases and possible iterations an analytics project might take.

Throughout this chapter, we'll dive into the six phases of an analytics project, which are business understanding, data exploration and preparation, modeling and evaluation, and deployment. We will also explore what happens to a data product once used as part of business operations. In the technology world, a data product used in business operations means that the software that produces the data is "in production." When software is in production, it requires support and monitoring, and this is referred to as model operations.

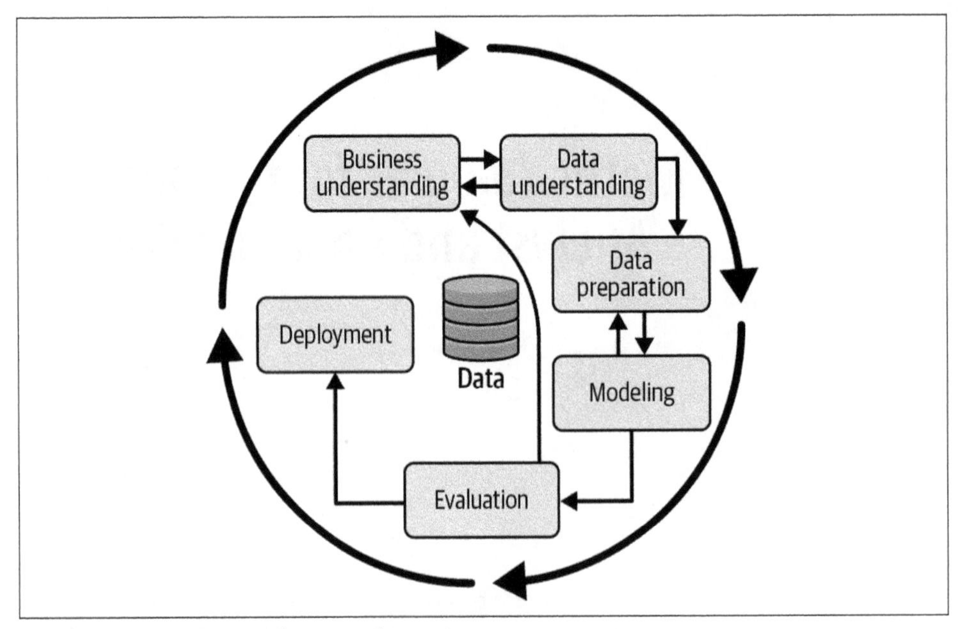

Figure 2-1. CRISP-DM phases and iterations

Business Understanding

The start of the analytics project is the most important as this is where the problem to be addressed is determined. Business analysts are involved during this process to investigate issues, understand impacts, and formulate a problem definition. The problem definition includes context, a statement, assumptions, timing, constraints, and the expected outcomes of solving the problem. It is important for business analysts to be involved to ensure the right problem is being addressed. Too often the wrong problem is focused on, or a solution is proposed before the real problem is understood. The primary outcomes for this phase include determining the business objectives, assessing the situation (context), identifying the goals (outcomes), and confirming the approach and plan of the project.

Determine Business Objectives

Good problems are often clearly understood and solved by defining what the desired outcomes are. The common notion "Begin with the end in mind" is a great strategy for framing the problem. Clearly defining business objectives brings the stakeholders together on what the expectations are for the project and what success means. Business objectives must be clear, specific, and measurable; for example:

> We want to lower customer churn by 10% within three months by predicting the likelihood a customer might churn.

There are some characteristics about this business objective that will be particularly valuable in planning the analytics project. First, there is a measurable goal within a period of time. Second, there is a focus on what is to be predicted. This provides detail that can be leveraged to determine the analytical approach, data required, and specifics around success. While not all analytical projects will be focused on a predictive model, creating business objectives at this level makes the problem statement very clear.

Assess Situation

Once the business problem is fully defined, the business analysts help define the requirements of the analytics project. Requirements are not about what functionality is needed in software applications but rather what we need the data product to do. Does the data product need to predict something or segment customers? Requirements for an analytics project focus on the information outcomes and how these information outcomes will be used.

The next step is to understand what resources are available to support the project. Resources include technology tools, computing platforms, team members available with the required skills, and budget. Analytics projects often leverage existing technology tools and computing platforms, but what if the existing infrastructure is not enough to support the project? Additionally, does the organization have the skill sets needed to support the project? Resource gaps will need to be closed before the project can commence.

It is not always clear how the outcomes of an analytics project will impact an organization. Sometimes there are unintended consequences from analytics projects. For example, Target reportedly predicted a teenager's pregnancy (*https://oreil.ly/w7P7r*), which was not the intended outcome of a marketing algorithm. There are risks with using any data product, and it is important to highlight these ahead of time.

Last, it is important to understand what the budget for the analytics project is. This will be influenced by how your organization allocates funds for projects, but often it is not always clear what budget is needed due to the iterative nature of the project. Often a fixed amount is allocated at the start and more will be provided after the pilot of the data product.

Determine Goals

Outlining specific goals for the project is important for success. The business objectives are used to guide and create the specific goals of the project. Just like the business objectives, the business goals should be specific, timely, and measurable. Success criteria are derived directly from the goals outlined for the project. The business analyst's feedback is critical here as the success criteria should clearly outline how the project will deliver business value. Keep in mind that data products, such as a predictive model, enable improved decision making or other capabilities. The precise outcomes may not be achieved as mostly it is not known how a data product will contribute to specific results.

Establish Approach and Plan

At this point, there is enough information gathered to determine the analytical approach and plan. Analytical approaches are what will be the primary focus of the project to create the data product. For instance, if the data product will be a predictive model, then predictive approaches will be the focus. If the goal is to do customer segmentation, it is likely that unsupervised approaches will be considered here. You should consider this step defining the scope of the project. Business analysts are consulted here since there may be constraints that need to be considered as well as timing requirements to drive the schedule.

Keep in mind that analytics projects are not sequential and don't follow a traditional project approach. The plan tends to be a schedule of expected milestones and timelines that will be the focus. Due to the iterative nature of the project, it is possible to go back to a prior stage.

Assessment of Tools and Techniques

The last step of business understanding is to determine if the required tools are available to support the project. It is possible that new technology may be required as existing technology does not support the goals of the project. A business analyst can be involved in assessing existing technology capabilities or evaluating new technology. The goal is to ensure that the required technology is available to start the project.

Data Exploration and Preparation

Data exploration and preparation can be time-consuming and take a large portion of the project schedule. The challenge with this stage is it is focused on discovery and the outcome is data that is cleaned, integrated, and formatted, with the correct data attributes to be used in building the analytical model. What data to review and explore is a challenging question, and the business analyst can assist with this. It goes back to what the primary problem is to be solved. For instance, if the goal is to create a predictive model, then the focus would be on what data sources and attributes should be considered in scope?

If the data exploration and preparation stage does not produce a valuable dataset, it is possible that the project would return to the business understanding stage. Sometimes data does not exist to solve a problem.

Assess Data Content and Quality

The review of data content can take different perspectives. One perspective is that the content of the data attributes needs to be understood. Answering basic questions such as data type, range, structure, valid values, and patterns will be necessary to understand the content and how the data could be used. Business analysts are key here to help the understanding since many organizations do not have metadata that can be used to answer this question. In addition to understanding what the data represents in a business context, quality needs to be assessed to determine if the data is usable for the analytics project.

Once the content is understood, then the quality is reviewed. Quality is reviewed for completeness and correctness. Completeness is focusing on whether or not all of the data is present, and correctness is where rules are applied to determine if the data is high quality. Correctness can include rules for accuracy, consistency, granularity, and timeliness.

Select and Clean Data

Once you've assessed content and quality, the next step is selecting and cleaning data. Some data will be eliminated as it does not meet the criteria to be used for prediction or due to data quality issues. Data that is not eliminated and is needed for the modeling step will then be cleansed and formatted. An example could include deriving new attributes from existing ones or changing a date format from month, day, year (MM-DD-YYYY) to year, month, day (YYYY-MM-DD). Selecting the cleaning data is typically done by source (system or subject area, as an example) in order to proceed to the next step of constructing and integrating the final dataset.

Construct and Integrate Data

To proceed to the modeling and evaluation stage, a final dataset needs to be constructed. Constructing the final dataset often involves taking different datasets and integrating the data. Integration usually occurs by matching common data elements across the dataset to match and merge the data. An example is taking a common grouping element like a city and merging the data or linking the data based on a shared key like a customer ID. Constructing and integrating data often uses the skill of a data engineer. They are skilled at cleaning, constructing, and integrating data. Business analysts can be called on to review the final dataset for quality and consistency.

Produce Dataset for Model Development

Once the dataset is constructed and integrated, the data is often split into subsets of data such as training, testing, and validation sets. These sets are mutually exclusive datasets from the same sample that will be used to build models. The structure of the dataset is broad and deep and can be an external file or a table in a database.

Modeling and Evaluation

The modeling and evaluation stages of the analytics life cycle focus on selecting an analytical technique to apply, producing a model from the technique, and evaluating a model in how it solves the business problem. Business analysts are consulted in the selection of the technique to ensure it aligns with the project's expected outcomes and are directly involved in assessing the ability of the model to solve the business problem. If the evaluation does not prove positive, it is possible that the project may go back to the data exploration and preparation stage to identify new data sources to work with.

Select Analytics Technique

Depending on the business problem to be solved, there are hundreds of analytical techniques that may be considered. Prediction problems select a regression or classification technique to be used. Other analytical problems may not be predictive, which means unsupervised or a combination of supervised and unsupervised techniques will be applied. Different techniques will be covered in later chapters.

Build and Assess Model

Building the model requires an induction step (using training data to create the model) and a deduction step (using testing data to assess the model). Building the model is often an iterative process, where continuous refinement is applied to achieve the best possible performance. Often multiple models are built using different techniques to create a benchmark. Business analysts are consulted in the outcomes and the evaluation of each model. A model or models may be chosen as candidates to move to the deployment stage of the project. The evaluation steps of a model will be covered in greater detail in later chapters.

Deployment

Deployment of a model is where the model is put to the test in business operations. Deployment can take different formats, but it often starts with a trial or a pilot. Models are implemented into a technology production environment where a model (the software version) scores (predicts) real data. The outcome of the model is then used for decision making in the organization, and the results of the decisions are evaluated. For instance, a model could be created to predict whether a customer has a high probability to buy a particular suggested product. If the lift—the ratio of results obtained with and without the model—on the product is high, likely the model would be evaluated positively. Business analysts are directly involved in deployment and assessing model performance because this is where business value materializes.

Assess Model Performance

Model performance is based on how accurate a model predicts against historical data before deployment. After deployment, the model is assessed on how well the business problem is addressed. Business analysts play a large role here because they understand how a model is used by the business and because they can evaluate business value. A pilot or trial is done with the model to ensure a controlled environment and to manage impacts. If the model is determined to be valuable to the business, then the model is deployed to a larger area.

Determine Assessment Intervals

Models may work well at the start, but it is important to determine how often a model would be assessed. Data changes throughout business cycles. For instance, a new marketing campaign may create new patterns in the data that impact model performance. Business analysts understand business operations and should provide insight into how often model performance should be assessed.

Model Operations

Model operations involve the ongoing support and monitoring of the model after it is deployed in production, ensuring it continues to function effectively and contribute to business operations. Not all models make it to the operational stage due to not passing the assessments. Models that do make it to this stage are considered valuable data products to a company because of their contribution to improved decision making and business value. Business analysts are involved here and often become the subject matter experts on the model and determine how it is used in the business.

Monitoring Models

As part of model deployment, models will have assessment criteria created that will become the threshold for monitoring. Accuracy rates, model drift, and feature drift are some of the metrics that are tracked to understand if a model needs to be retrained or retired. Business analysts are consulted to determine if model performance is impacting business value.

Life of a Model

Models are created on historical data, which means a model is only going to perform well when the data is consistent. Data changes as the business changes, so unless a business is standing still, data will change. Models have a finite life directly aligned to the change rate of the business. The more change an organization undergoes (new products, services, offers, mergers, or acquisitions, for example), the shorter the life of a model. Business analysts help to understand how business changes may impact the life of the model.

Retraining

Models may be updated by simply retraining them on new data. A retraining cycle can update the model, and it may have better performance and a longer life. Sometimes retraining is not possible due to the amount of change that has occurred in an organization. If a model cannot be retrained, a new version should be considered or the model could be retired.

Summary

This chapter reviewed the stages that are typical to an analytics project, and highlighted where a business analyst gets engaged. Business analysts provide value because they have subject matter expertise and knowledge of business impacts. They are critical to the business understanding of a problem, selecting data to be used in problem solving and model building, and evaluating whether a model is truly solving the business problem at hand. Chapter 3 will introduce the primary languages used in analytics projects: R and Python.

CHAPTER 3

Introduction to R and Python

As an analytics professional, you will need to work with the popular and powerful tools. This chapter introduces you to R and Python, both open source programming languages used in business analytics. Both R and Python are highly used, and they have similar functionality. As open source programming languages, R and Python functionality is provided through the creation of packages or libraries by the open source community. (In R, these are referred to as libraries, and in Python, these are referred to as packages.) Both languages are also object-oriented, which is a programming paradigm that supports reuse, modularity, and flexibility.

R was created by statisticians in 1994, and it became the primary analytics tool during the early 2000s. R was used for statistical analysis, and it is now used by analysts and researchers globally and includes many built-in capabilities that support the analytics process. R is often accessed via RStudio, an integrated development environment (IDE), and there is an R repository (*https://cran.r-project.org*). RStudio (*https://posit.co*) is publicly available. R is interpreted and noncompiled, which means the code runs directly from the script created.

Python is an interpreted (noncompiled) programming language that is used for analytics as well as software engineering. Python is similar to R, where a large repository of packages is available for analytics. Python has a wider community due to its portability and the general-purpose capabilities for software engineering. Python (*https://python.org*) is available via many IDEs, and documentation and code downloads are available. One of the most popular IDEs is Anaconda (*https://anaconda.com*).

This chapter will introduce you to object-oriented concepts. We'll walk through how to install R and Python, and we'll also cover data structures and types, programming constructs, and how to interact with relational databases using R and Python.

R and Python Installation and Setup Options

When considering whether business analysts should learn R and Python, it's essential to recognize the distinct advantages each language offers and how they complement the skill set of a business analyst. Both R and Python are powerful tools for data analysis, but they serve slightly different purposes and have different strengths. Learning both can be highly beneficial, but the approach to learning them depends on the analyst's current skill level, the demands of their role, and their specific projects.

Why Learn R and Python?

R is particularly strong in statistical analysis and visualization. It has a rich ecosystem of packages tailored for data analysis, making it a go-to choice for analysts working heavily with statistics and needing to produce detailed visualizations. Python, on the other hand, is a versatile language that excels in data manipulation, machine learning, and automation. It is also widely used in various domains beyond data analysis, such as web development and automation, making it a valuable skill for analysts looking to broaden their technical capabilities. We will show both options throughout the book but recommend focusing on Python and using R as a reference.

Learning Both at Once Versus One at a Time

Learning both R and Python simultaneously might seem appealing to maximize efficiency, but it can also be overwhelming, especially for beginners. Each language has its own syntax, libraries, and ecosystem, which can create a steep learning curve if tackled together. However, learning both at once could be beneficial if the analyst is working on projects that require the strengths of both languages, as it would allow them to immediately apply their learning in a practical context.

Focusing on one language at a time is often a more manageable approach. It allows the analyst to build a solid foundation in one language before moving on to the next. Starting with Python might be advisable for those who are interested in a broader range of applications, including data analysis, machine learning, and automation. Once comfortable with Python, transitioning to R can be easier, as many of the analytical concepts will carry over, but the analyst will then be equipped to handle more specialized statistical tasks.

Pros and Cons of Different Learning Strategies

Learning both languages at once can lead to faster acquisition of a versatile skill set, allowing analysts to switch between tools as needed. However, this approach may result in confusion due to the different syntaxes and methodologies, and it might slow down the learning process if the analyst struggles to keep both languages straight.

On the other hand, focusing on one language at a time allows for deeper understanding and mastery of that language, which can build confidence and efficiency. The downside is that it may take longer to become proficient in both languages, potentially delaying the ability to work on projects that require the other language's strengths.

Business analysts should definitely consider learning both R and Python to fully leverage their data analysis capabilities. However, the decision on whether to learn both at once or one at a time should be based on their current workload, the demands of their projects, and their capacity to manage learning new skills. Starting with one language, mastering it, and then moving on to the next is generally a more sustainable and less overwhelming approach.

R Installation

To install R, you can follow these steps:

1. Go to the R Project website (*https://cran.r-project.org*).
2. Click the Download R link for your operating system (Windows, Mac, or Linux).
3. Download and run the installation file for your system.
4. Follow the prompts to install R on your computer.
5. Once the installation is complete, you can launch R from your Start menu or application launcher.

The popular options for R are as follows:

RStudio
: A widely used IDE that offers many features, such as syntax highlighting, debugging, and package management

Jupyter notebook
: A web-based interactive environment (*https://oreil.ly/qJomj*) that allows users to create and share documents that contain live code, equations, visualizations, and narrative text

Visual Code Studio
: An open source code editor (*https://oreil.ly/5kYaY*) that offers extensions for R development

Python Installation

To install Python, you can follow these steps:

1. Go to the Python website (*https://python.org*).
2. Choose the Downloads menu and then the operating system that applies to you.
3. Download and run the installation file for your operating system (Windows, Mac, or Linux).
4. Follow the prompts to install Python on your computer.
5. Once the installation is complete, you can launch the Python interpreter from your terminal, command prompt, or Start menu.

There are many IDE options for both R and Python, depending on your preferences and needs. An IDE is a software application that provides a comprehensive set of tools for software developers to write, edit, debug, test, and deploy code. Popular IDEs include Visual Studio, IntelliJ IDEA, Eclipse, Xcode, and PyCharm, each catering to different programming languages and development needs. We will be using RStudio and Anaconda in most of the examples included in each chapter, but the code examples can be used with many IDEs and GUIs.

For Python, there are several popular IDE options:

PyCharm
: A feature-rich IDE with code highlighting, debugging, and code completion

Spyder
: An open source IDE (*https://oreil.ly/pJA96*) specifically designed for scientific computing with features such as code introspection, variable explorer, and integrated IPython console

Jupyter notebook
: A web-based interactive environment (*https://oreil.ly/qJomj*) that allows users to create and share documents that contain live code, equations, visualizations, and narrative text

Another option, which includes many of the IDEs just listed, is Anaconda (*https://anaconda.com*), an open source platform that allows you to write and execute code in Python. Anaconda is very popular because it simplifies package deployment and management and comes with a large number of packages included.

R and Python Scripting

Both R and Python are interpreted programming languages, which is where the code written by a programmer is executed line-by-line by an interpreter rather than being compiled into machine code and executed directly on the computer's hardware. This means that the code can be executed immediately without any intermediate compilation step. Interpreted languages are considered to be flexible and easier to work with than compiled languages, and they support faster prototyping and development and provide quick feedback when errors occur.

Both R and Python use variables, functions, control structures, and other constructs to create scripts and automate activities. In analytics, scripting is an essential skill and has many benefits. Some of the benefits include data analysis, machine learning, data science, automation, collaboration, visualization, and access to large ecosystems. There are additional similarities between these languages as well. Both R and Python are widely used for data analysis, and they offer a rich set of libraries and tools for manipulating, visualizing, and modeling data. Both languages have extensive libraries for machine learning, including classification, regression, clustering, and deep learning. They also are open source languages, and their use of scripts enables collaboration and sharing of code between data scientists, statisticians, and other stakeholders. Additionally, both R and Python offer powerful data visualization tools, making it easier to explore and understand large datasets. They also have extensive libraries and tools available, which can greatly extend their capabilities, and they provide new opportunities for exploration and discovery. Last, scripting in R and Python can automate repetitive tasks, freeing up time and reducing the likelihood of errors. Almost all of the analytics completed in R and Python will be done in a script. The next section reviews the basics of interpreted language scripting.

R Language Scripting

Scripting starts with using an R terminal or R GUI. For the examples in this book, we will be using RStudio. RStudio has four main windows, and in the bottom left you will have your console window. The console window is where you can start writing your script interactively at the command prompt. The source window is where R scripts can be written and saved for execution later. You may also start your script in the source window. Figure 3-1 depicts the command prompt and the console window. Figure 3-2 shows the menu options to start an R script, while Figure 3-3 shows the source windows that will open when the menu option to start an R script is chosen.

```
Console   Terminal ×   Jobs ×
R 4.3.1 · ~/

R version 4.3.1 (2023-06-16 ucrt) -- "Beagle Scouts"
Copyright (C) 2023 The R Foundation for Statistical Computing
Platform: x86_64-w64-mingw32/x64 (64-bit)

R is free software and comes with ABSOLUTELY NO WARRANTY.
You are welcome to redistribute it under certain conditions.
Type 'license()' or 'licence()' for distribution details.

R is a collaborative project with many contributors.
Type 'contributors()' for more information and
'citation()' on how to cite R or R packages in publications.

Type 'demo()' for some demos, 'help()' for on-line help, or
'help.start()' for an HTML browser interface to help.
Type 'q()' to quit R.

[Workspace loaded from ~/.RData]
```

Figure 3-1. Command prompt and console window

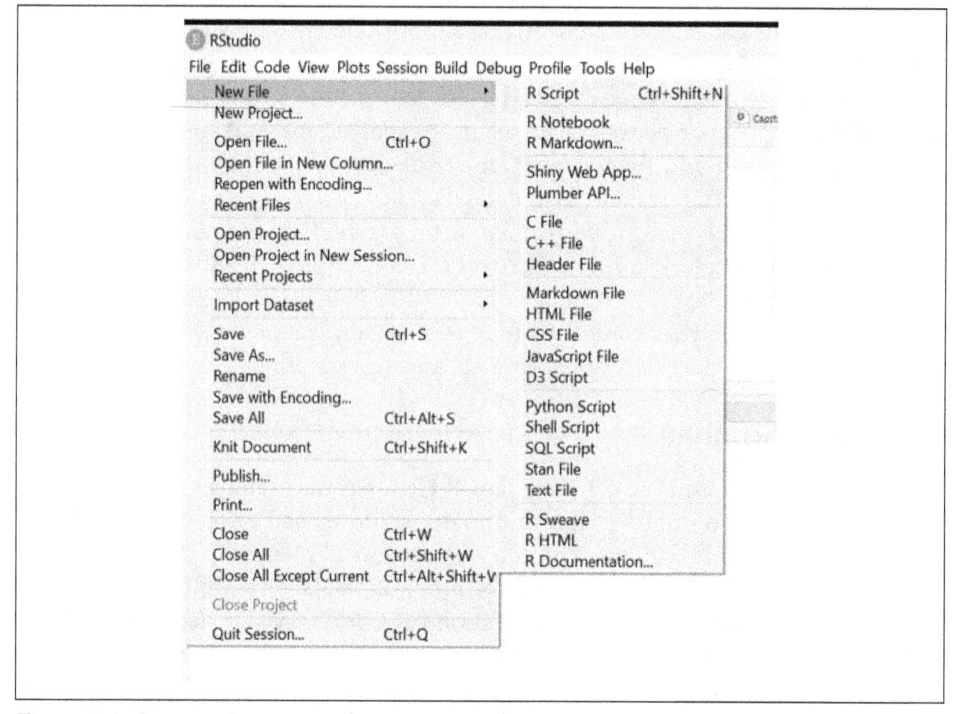

Figure 3-2. Start an R script in the source window

Figure 3-3. Source window in RStudio

To create an R script, you can start coding in either the console or source window. It is recommended you use the source window as this allows you to save your script. Starting with the console window may be a better option if you want to explore the commands first before developing the script. Writing the script includes the syntax for the functions, control structures, or other expressions for your script. Once the script is completed, you can save the script with a .R extension (e.g., *script.R*). Running the script can be executed by typing in source (e.g., "script.R") in the R terminal or clicking the Run button in the source window. We will be reviewing many different scripts in the book. Figure 3-4 shows a basic example of an R script that calculates the average from a set of numbers.

Knit on Save   Knit
Source   Visual

```
1  #Create a number of vectors
2  numbers <- c(1, 2, 3, 4, 5)
3
4  # Calculate the mean of numbers
5  mean_of_numbers <- mean(numbers)
6
7  # Print the Result
8  print(mean_of numbers)
9
```

Figure 3-4. R script using the source window

When run, this script will output the mean of the numbers in the numbers vector, which is 3. Next we will look at a script example in Python.

Python Language Scripting

Scripting in Python starts with using PyCharm, another text editor, or a GUI notebook such as Jupyter notebooks. We will use Jupyter notebooks to demonstrate a Python script. Figure 3-5 shows an example of launching a Jupyter notebook from Anaconda.

Figure 3-5. Starting Jupyter notebooks

Once you launch a Jupyter notebook, a tab will open in your default browser that looks like Figure 3-6. To open a new notebook, click the "New" button on the top right. A drop-down menu will give you an option of what type of notebook to create. Select Python3.

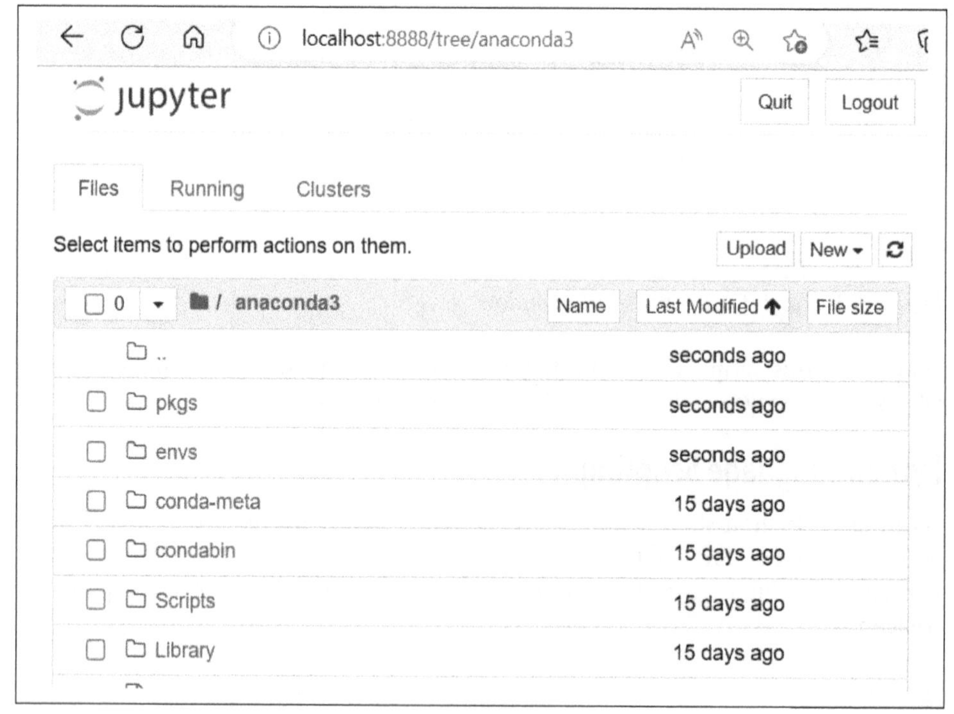

Figure 3-6. Jupyter interface

A new notebook is opened, and here is where you can start your Python script. In Figure 3-7, a simple script has been placed in a notebook cell. By clicking the arrow button, the cell executes and provides the result. The script calculates the mean of a set of numbers and will output the mean of the numbers in the numbers list, which is 3. Notice how comments are provided to highlight what the code is expected to do. Comments are a best practice in scripting to ensure others reading your code understand the expected results.

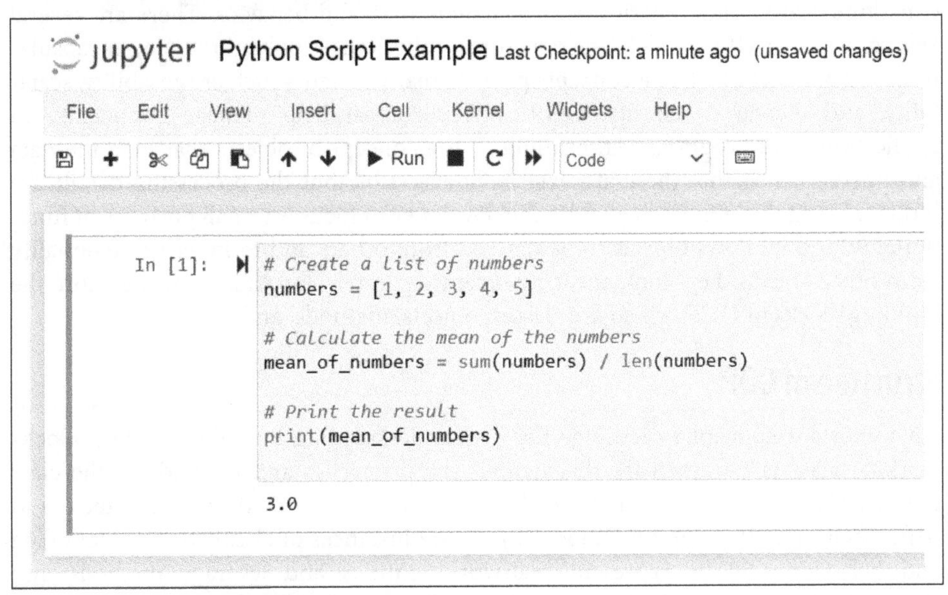

Figure 3-7. Jupyter notebook with a Python script

This section provided a brief overview of how R and Python scripts are created. Scripts can be simple, as in the examples provided, or complex, depending on the scope. Scripts can focus on data manipulation, by creating predictive models, or they can contain the commands to automate predictive modeling scripting. There will be many examples of R and Python scripts throughout the book demonstrating how to address different scenarios and functionality. The next section will introduce how to apply object-oriented concepts in R and Python.

Business analysts using R and Python can greatly benefit from understanding object-oriented programming (OOP) concepts, as these concepts enable them to write more modular, scalable, and maintainable code. OOP allows analysts to create reusable components and organize their code in a way that models real-world entities, making complex data analysis tasks more manageable. By leveraging classes, objects, inheritance, and encapsulation, analysts can build more sophisticated and efficient data processing pipelines, automate repetitive tasks, and better manage large codebases. This understanding also enhances their ability to collaborate with software developers, integrate analytical models into larger applications, and contribute to more robust, production-ready solutions.

Object-Oriented Concepts

OOP is a programming approach that leverages objects as the basis of designing and building computer programs and applications. Objects are versions of classes that constitute real-world scenarios, which include data and behavior. There are several key concepts in OOP, which are structures, inheritance, polymorphism, encapsulation, and abstraction. These concepts provide many benefits such as reusability, scalability, and increased security. OOP concepts assist help manage complexity in applications and programs. An example is the concept of objects that hide the data and operations so that these are externally accessible but the processing details are hidden. Using this approach enables the developer to focus on small portions of functionality, and it allows objects to be derived from others and to inherit functionality from other objects. Let's look at each concept separately, but first we will explore the building blocks of OOP, which are classes, objects, methods, and attributes.

Structure of OOP

An important concept to leverage OOP is understanding how the building blocks work. A class is a framework that defines the properties and methods of the class objects. Classes can create a particular data structure, provide initial values, and define behaviors through functions. Objects are instances of classes that contain data and behavior. Objects represent real-world entities and leverage methods and attributes of classes. Let's look at examples of these building blocks.

Class

First, let's examine the concept of class. A class is a template for creating different objects, which include a data structure, initial values for member attributes, or member functions or methods. Here is an example of a Python code for class:

```
class cat:
    def __init__(self, name, breed):
        self.name = name
        self.breed = breed

    def sound(self):
        print(f"{self.name} says 'Meow!'")

# Create an instance of the cat class
Cat = cat("Ginger", "Red Tabby")

# Call the sound method on the cat instance
cat.sound()
```

When run, this script will result in the following output:

```
Ginger says "Meow!"
```

The R syntax of a class is very similar to Python. Here is an example of a class in R code:

```
library(R6)

Cat <- R6Class("Cat",
  public = list(
    name = NULL,
    breed = NULL,
    initialize = function(name, breed) {
      self$name <- name
      self$breed <- breed
    },
    hello = function() {
      paste("My name is", self$name, "and I am a", self$breed, ".")
    }
  )
)

Cat1 <- Cat$new(name = "Ginger", breed = "Tabby")
Cat1$hello()
> "My name is Ginger and I am a Tabby."
```

This section demonstrated how to create and use a class and to apply it.

Object

An object is a specific instance of a class, which has its own set of variables and methods that are defined within the class. An example could be a class `Home` with variables `location` and `age`, and you could create several instances of the `Home` class to represent different Homes. Each instance would be an example of an object created from the `Home` class. Objects can be interacted with using attributes and methods to complete operations on the object's data. This would be an example of encapsulation, where the internals of the object are hidden. Encapsulation will be explored shortly. Let's look at an example of code for a Python object:

```
class Home:
    def __init__(self, location, age):
        self.location = location
        self.age = age

    def introduce(self):
   print(f"This home is located in {self.location} and is {self.age} years old.")

# Create an instance of the Home class
home = Home("Denver", 30)
```

```
# Call the introduce method on the home object
home.introduce()
```

When this code is executed, the results would be:

```
This home is located in Denver and is 30 years old.
```

In this example, we have defined a class `Home` with two variables, `location` and `age`, and a method `introduce()` that outputs a string describing the home location and home age. We then create an instance of the `Home` class, called `home`; the location is set to `"Denver"` and the age is set to `30`. Then the `introduce()` method is called on the home object, and the string of `"This home is located in Denver and is 30 years old"` is the result of the call.

Here, an example of an R object can be taken from the class example earlier:

```
Cat1 <- Cat$new(name = "Ginger", breed = "Tabby")
Cat1$hello()
> "My name is Ginger and I am a Tabby."
```

`Cat1` is an object of the `Cat` class, where the name is set to `"Ginger"` and the breed is set to `"Tabby"`. The `Cat1` object can use the methods and properties of the `Cat` class; in this case, the `hello` method was used.

Methods and attributes

By examining the examples of an object and a class, you also can see the uses of methods and attributes. Methods (sometimes referred to as functions) are defined as part of a class and outline the behavior of an object. They are reusable and are internal to the class. Attributes are defined within the class, represent the status of an object, and contain data.

In the example of the class `Cat`, both `sound()` and `hello()` are methods. An example of attributes would be `self$name` in R or `self.name` in Python. Both of these attributes hold data when the method is being called.

The building blocks provide an understanding of what is used by applying OOP. The principles of OOP provide an understanding of how OOP is applied, and we will explore these principles next.

Principles of OOP

OOP has four basic concepts: encapsulation, abstraction, inheritance, and polymorphism. Each principle provides specific outcomes that are valuable in OOP. Knowing how they work together can help you understand the basic functionality of an OOP computer program and also the benefits of using OOP. The benefits of the principles include reusability, modularity, and scalability. Reusability means classes and objects can be used in other programs, reducing code and maintenance. Modularity refers to creating smaller parts of the program increasing maintainability. Scalability is important as it enables the ability to handle more complex capability by combining objects to fit growing requirements. Let's start with encapsulation.

Encapsulation

Encapsulation focuses on keeping the internals of an object hidden and only exposing the methods that are necessary for object interaction. This principle assists in ensuring objects keep a consistent state and protecting from unintended modification. An example where encapsulation can be applied is when data within an object needs to be protected against modification or access by external code. The overall goal of encapsulation is to make code modular and easy to maintain and to increase quality.

Abstraction

Abstraction is where complexity is hidden by leveraging objects, classes, methods, and attributes to represent more complex data and code. Abstraction supports reuse of objects, modularity, and scalability.

Another way to think of it is using simple things to represent complexity. We all know how to turn the TV on, but we don't need to know how it works in order to enjoy it. In Java, abstraction means simple things like objects, classes, and variables represent more complex underlying code and data. This is important because it lets you avoid repeating the same work multiple times. In other words, OOP helps to abstract complex systems into manageable and modular objects, making the code easier to understand and maintain.

In Python, an abstract class is one that is not instantiated and is defined using the ABC module. In other words, an abstract method is declared but not implemented and provides a way to separate the class from its definition:

```
from abc import ABC, abstractmethod

class Shape(ABC):
    @abstractmethod
    def area(self):
        pass

    @abstractmethod
    def perimeter(self):
        pass
```

The `Shape` class defines one area method and one perimeter method. Any class that leverages `Shape` will implement these methods, and abstraction is applied by focusing on the features of area and perimeter. One could leverage `Shape` by creating a square or rectangle class that leverages the same methods.

In R, abstract classes are not directly supported in the same way as in Python, but you can achieve a similar effect using R's S4 object-oriented system. In S4, you can define a virtual class that cannot be instantiated and create methods that must be implemented by any subclass:

```
# Define a virtual class 'Shape' with abstract methods 'area' and 'perimeter'
setClass("Shape", contains = "VIRTUAL")

# Define a generic function for area
setGeneric("area", function(object) standardGeneric("area"))

# Define a generic function for perimeter
setGeneric("perimeter", function(object) standardGeneric("perimeter"))

# Create a subclass 'Rectangle' that inherits from 'Shape'
setClass(
  "Rectangle",
  slots = list(length = "numeric", width = "numeric"),
  contains = "Shape"
)

# Implement the 'area' method for the 'Rectangle' class
setMethod(
  "area",
  "Rectangle",
  function(object) {
    return(object@length * object@width)
  }
)

# Implement the 'perimeter' method for the 'Rectangle' class
setMethod(
  "perimeter",
  "Rectangle",
  function(object) {
    return(2 * (object@length + object@width))
  }
)

# Create an instance of 'Rectangle' and calculate area and perimeter
rect <- new("Rectangle", length = 4, width = 3)
rect_area <- area(rect)
rect_perimeter <- perimeter(rect)

rect_area      # Output: 12
rect_perimeter # Output: 14
```

The `Shape` class in R is defined as a `virtual` class using the `VIRTUAL` keyword, meaning it cannot be instantiated directly, similar to an abstract class in Python. Generic functions such as area and perimeter are created, serving as abstract methods that must be implemented by any subclass. The `Rectangle` class is a subclass of `Shape`, where these methods are specifically defined. When an instance of `Rectangle` is created, the area and perimeter can be calculated using these implemented methods. This structure ensures that any subclass of `Shape` must define the area and perimeter methods, maintaining a similar level of abstraction as in Python's abstract classes.

Inheritance

Inheritance occurs when creating a new class based on an existing class. The new class inherits all of its properties, such as methods and data, from the existing class. Using inheritance, it is possible to create a new class using an existing class and add or include overriding methods. Inheritance supports the creation of new classes from existing ones, which reduces code and time to develop new functionality.

In Python, inheritance is supported through the use of classes and the `class` keyword, allowing for the definition and organization of complex data structures. Here is an example of inheritance in Python:

```
class Shape:
    def __init__(self, length, width):
        self.length = length
        self.width = width

class Rectangle(Shape):
    def area(self):
        return self.length * self.width

r = Rectangle(7, 4)
```

In this example, the `Shape` class is defined with two attributes: `length` and `width`. The `Rectangle` class is then defined as a subclass of `Shape` using the `class` keyword, inheriting the `length` and `width` properties. The `Rectangle` class also has an additional method, `area`, which calculates the area based on the inherited length and width. A new instance of the `Rectangle` class is created with a length of 7 and a width of 4. The `Rectangle` class has access to the `length` and `width` attributes inherited from the `Shape` class, and it can also calculate the area using the `area` method. This demonstrates how inheritance allows the `Rectangle` class to build upon the foundation provided by the `Shape` class, making `Rectangle` a subclass of `Shape`.

In R, inheritance is supported through the use of S4 classes, which are defined using the `setClass()` function. S4 classes provide a way to define and organize complex data structures. Here is an example of inheritance in R using S4 classes:

```
setClass("Shape", representation(length = "numeric", width = "numeric"))
setClass("Rectangle", contains = "Shape")

r <- new("Rectangle", length = 7, width = 4)
```

The Shape class is defined using the setClass function and has two numeric properties of length and width. The Rectangle class is then defined using the setClass function, and it contains the Shape class. The Rectangle class will inherit the properties of the Shape class. A new instance of the Rectangle class is then created leveraging the function to set width to 7 and length to 4. The Rectangle class has access to width and length as these are inherited from the Shape class. The Rectangle class is a subclass of Shape.

Polymorphism

Polymorphism represents the ability of different class objects to react to methods with the same name differently. It enables you to write general code that works with objects of different classes using a common interface. Polymorphism supports different class objects to respond to the same method in different ways, which allows the code to be more flexible. Objects of different classes can be used interchangeably, as long as they have the same method names and arguments. Here is an example in Python:

```
class Shape:
    def area(self):
        pass

class Rectangle(Shape):
    def __init__(self, length, width):
        self.length = length
        self.width = width

    def area(self):
        return self.length * self.width

class Circle(Shape):
    def __init__(self, radius):
        self.radius = radius

    def area(self):
        return 3.14 * (self.radius ** 2)

# Create objects of different classes
rectangle = Rectangle(7, 4)
circle = Circle(4)

# Call the area method on each object
print(rectangle.area())  # Output: 28
print(circle.area())     # Output: 50.24
```

In this example, a Shape class is defined with an area method that is not used. Then two subclasses are defined as Rectangle and Circle. Each class inherits from the Shape class but implements unique versions of the area method. The area method is then called on each object, and the area method is used differently for each subclass, which is an example of polymorphism.

In R, polymorphism can be achieved using the S4 object-oriented system, where different classes can implement the same method in their own unique ways. This allows objects of different classes to respond to the same method name, making the code more flexible and extensible.

Here's an example of polymorphism in R using S4 classes:

```
# Define a generic function 'area'
setGeneric("area", function(object) standardGeneric("area"))

# Define the Shape class (no implementation of 'area')
setClass("Shape", contains = "VIRTUAL")

# Define the Rectangle class as a subclass of Shape
setClass("Rectangle",
         slots = list(length = "numeric", width = "numeric"),
         contains = "Shape")

# Define the area method for Rectangle
setMethod("area", "Rectangle", function(object) {
  return(object@length * object@width)
})

# Define the Circle class as a subclass of Shape
setClass("Circle",
         slots = list(radius = "numeric"),
         contains = "Shape")

# Define the area method for Circle
setMethod("area", "Circle", function(object) {
  return(3.14 * (object@radius ^ 2))
})

# Create objects of different classes
rectangle <- new("Rectangle", length = 7, width = 4)
circle <- new("Circle", radius = 4)

# Call the area method on each object
rectangle_area <- area(rectangle)  # Output: 28
circle_area <- area(circle)        # Output: 50.24

rectangle_area
circle_area
```

In this R example, the Shape class is defined as a virtual class with no implementation of the area method, functioning similarly to an abstract base class in Python. The Rectangle and Circle classes are subclasses of Shape, each providing its own implementation of the area method. Specifically, the Rectangle class calculates the area based on its length and width, while the `Circle` class calculates the area based on its radius. This setup demonstrates polymorphism, where the `area` function can be called on objects of both `Rectangle` and `Circle`, with each object using its specific implementation of the `area` method. The example illustrates how polymorphism allows the same method name, `area`, to behave differently depending on the class of the object, similar to how it works in Python.

R and Python Data Types

Both R and Python have data types and structures that are used in programming. A data type refers to the value that a variable can store. Most programming languages include different types such as integers, floats, characters, and Boolean values. Each data type has unique operations that can be performed. A data structure organizes and stores data in memory. An example could be a list or an array. Data structures can store multiple variables of the same or different data types. Since programming and analytics is about data, it's important to have a strong foundation with the various data types, including proper data handling, data analysis, debugging, and code reuse.

Understanding data types and structures enables you to process and manipulate the data being worked with. Using the proper types and structures reduces the risks of errors and unexpected results. Data structures are also the foundation for analysis. An example includes using the data structure of a dataframe, which allows for data to be stored in a virtual table of columns and rows. For example, when storing data in a virtual table of columns and rows, you'll want to leverage a dataframe for the structure of your table. This structure is used highly in analytics and enables easy organization and manipulation of data for analysis.

Using the correct data types and structures avoid defects in the code. For instance, if you expected a variable to be an array and it is a string, this would result in an error. Using the common data types in R and Python ensures code is reusable and supports collaboration on analytical projects. Knowledge of data types and structures is essential for data handling, correct analysis, code reuse, and debugging.

Let's examine the different data types and structures used in R and Python.

R Data Types

Data types and structures are used for processing and manipulating data, but each is slightly different. R's basic data types are character, numeric, integer, complex, and logical. R's basic data structures include the vector, list, matrix, dataframe, and factors. Some of these structures require that all members be of the same data type (e.g., vectors, matrices) while others permit multiple data types (e.g., lists, dataframes). See Table 3-1 for the R data types.

Table 3-1. R data types

Data type	Use	Code example
Numeric	For real numbers and integers	`x <- 4.5`
Character	For string or text data	`x <- "Hello World"`
Logical	For True/False values	`x <- TRUE`
Integer	For whole numbers	`x <- 5`
Complex	For complex numbers	`x <- 2 + 3i`
Raw	For bytes	`raw_data <- as.raw(c(0x41, 0x42, 0x43, 0x44))`
Factor	For categorical data that can be leveled	`x <- factor(c("A", "B", "C"))`

In R, you can use the class function to determine the class of an object. For example:

```
x <- 2.5
class(x)
# Output: "numeric"
```

In this example, the class function returns the class of `x`, which is `"numeric"`. The next section provides more explanation of R data structures.

R Structures

As mentioned, a data structure organizes and stores data in memory and is used to process and manipulate data that supports the main objectives of analytics: analyzing data. The list, factor, dataframe, matrix, array, and time/dates are used to store and manipulate data in R. See Table 3-2 for a list of the data structures and what the purpose of each is.

R data structures can be used leveraging the conversion functions. The code examples in Table 3-2 show the conversion functions.

Table 3-2. R data structures

Data structure	Purpose	Code example with output

List	For collection of different data type objects or other lists. E.g., you can use a list to store a set of dataframes that contain different sets of data.	`x <- list(A = c(1, 2, 3), B = c("A", "B", "C"))`
		
		`$A`
		`[1] 1 2 3`
		
		`$B`
		`[1] "A" "B" "C"`

<table>
<tr><th>Data structure</th><th>Purpose</th><th>Code example with output</th></tr>
<tr><td>Dataframe</td><td>For tabular structures with columns and rows that can be different data types.
E.g., you can use a dataframe to store a set of data that includes columns for a person's name, age, and address.</td><td><pre>x <- data.frame(Name = c("John", "Jane"), Age = c(30, 28))

  Name Age
1 John 30
2 Jane 28</pre></td></tr>
<tr><td>Matrix</td><td>For tabular structures with the same data type in columns and rows.
You can use a matrix to store a set of values from multiple columns in a dataframe.</td><td><pre>x <- matrix(1:6, nrow = 2, ncol = 3)

 [,1] [,2] [,3]
[1,]   1   3   5
[2,]   2   4   6</pre></td></tr>
<tr><td>Array</td><td>For multidimensional structures that are of the same data type.
You can use an array to store a set of values from multiple columns and multiple rows in a dataframe.</td><td><pre>x <- array(1:12, dim = c(2, 3, 2))

, , 1

     [,1] [,2] [,3]
[1,]   1   3   5
[2,]   2   4   6

, , 2

     [,1] [,2] [,3]
[1,]   7   9   11
[2,]   8   10   12</pre></td></tr>
<tr><td>Time and date</td><td>For storing data and time information.</td><td><pre>x <- as.Date("2023-01-01")

[1] "2023-01-01"</pre></td></tr>
<tr><td>Vector</td><td>For storing one-dimensional arrays of the same data type.</td><td><pre>x <- c(1, 2, 3, 4, 5)

[1] 1 2 3 4 5</pre></td></tr>
</table>

In R, you can use the `str` function to get the structure of an object. For example:

```
x <- c(1, 2, 3, 4, 5)
str(x)
# Output: num [1:5] 1 2 3 4 5
```

In this example, the `str` function returns the structure of `x`, which is a `num` (numeric) vector of length 5.

The appropriate data structure depends on the specific requirements of your problem and the data you need to store. It's important to choose the right data structure for the job in order to make your code more efficient and easier to maintain. Data types and structures are common across programming languages, but each will have its own specific data types and structures. Next we review Python data types and structures.

Python Data Types

Just like R, Python has several built-in data types. Again, data types are used for storing data based on the data definition, but there are some differences between Python and R. See Table 3-3 for the different data type categories.

Table 3-3. Python data categories

Data type	Use	Code example with output

Numeric	This data category includes integers, floating-point numbers, and complex numbers (int, float, complex).	Integers are used to store whole numbers. For example:
		`x=4`
		`#Output`
		`x`
		`4`
		Floats are used to store real numbers. Here is an example:
		`x = 3.5`
		`#Output`
		`x`
		`3.5`

Sequence	This category represents an ordered and indexed collection of objects including list, tuple, and range (list, tuple, range).	A list is an order collection of data elements, which can include different data types. An example:
		`x = [2, 4, 6, 8]`
		`#Output`
		`x`
		`[2, 4, 6, 8]`
		A tuple is an ordered, unchangeable collection of data elements, which can include other data types. Here is an example:
		`x= (3, 5, 7, 9)`
		`#Output`
		`x`
		`(3, 5, 7, 9)`

String	This category represents sequences of characters (str).	`x = 'Here I am!'`
		`#Output`
		`x`
		`Here I am!`

Data type	Use	Code example with output

Mapping	This category represents unordered, key-value pairs (dict).	The first position is the key, and the second is the value, otherwise known as key-value pairs. In the following example, the A, B, and C are the keys, and the 1, 2, and 3 are the values.
		`x = {"A": 1, "B": 2, "C": 3}`
		`#Output`
		`x`
		`{'A': 1, 'B': 2, 'C': 3}`

Set	The category represents collections of unique elements (set, frozenset).	`my_set = {1, 2, 3, 4, 5}`
		`#Output`
		`my_set`
		`{1, 2, 3, 4, 5}`

Boolean	The category represents logical values true and false (bool).	`x = True`
		`#Output`
		`True`

As outlined in R, Python has a function to determine the type of an object. This is important for verification. The `type()` function can be used to determine the data type. Here is an example:

```
x = 4

print(type(x))

# Output: <class 'int'>
```

The `type` function returns the data type of `int` or integer for the variable of `x`.

Table 3-4 outlines some scenarios where Python data types might be used. In each scenario, the appropriate data structure depends on the specific requirements of your problem and the data you need to store. It's important to choose the right data structure for the job in order to make your code more efficient and easier to maintain.

Table 3-4. Data type use cases

Data type	Purpose	Code example with output

Lists	Stores a collection of items that need to be ordered and accessible by index. E.g., you can use a list to store products for a shopping cart application.	`my_list = [1, 2, 3, 4, 5]`
		`print("List:", my_list)`
		`List: [1, 2, 3, 4, 5]`

Data type	Purpose	Code example with output

Tuples	Stores a collection of items that need to be ordered, but should not be changed. E.g., you can use a tuple to store a date, where the day, month, and year should not be changed once set.	`my_tuple = (10, 20, 30, 40, 50)`
		`print("Tuple:", my_tuple)`
		`Tuple: (10, 20, 30, 40, 50)`

Dictionaries	Stores a collection of key-value pairs, where each key is used to access its corresponding value. E.g., you can use a dictionary to store a user's profile information, where the keys are the field names and the values are the field values.	`my_dict = {`
		`    "name": "John",`
		`    "age": 30,`
		`    "city": "New York"`
		`}`
		`print("Dictionary:", my_dict)`
		`Dictionary: {'name': 'John', 'age': 30, 'city': 'New York'}`

Sets	Stores a collection of unique items. E.g., you can use a set to store a list of unique words in a document.	`my_set = {100, 200, 300, 400, 500}`
		`print("Set:", my_set)`
		`Set: {400, 100, 500, 200, 300}`

The next section addresses Python advanced data structures, which can include different data types.

Python Data Structures

Python has several advanced data structures to store and manipulate data. These data structures are called advanced because they can be combined and used in different ways to solve analytical problems. They can support multiple dimensions, sequencing, status checking, hierarchical data, and tabular data. Table 3-5 describes each.

Table 3-5. Python advanced data structures

<table>
<tr><th>Advanced data structure</th><th>Description</th><th>Code example</th></tr>
<tr><td>Array</td><td>A multidimensional data structure that stores a collection of elements, all of the same type.</td><td><pre>import array as arr

my_array = arr.array('i', [1, 2, 3, 4, 5])
print("Array:", my_array)

Array: array('i', [1, 2, 3, 4, 5])</pre></td></tr>
<tr><td>Linked list</td><td>A data structure consisting of a sequence of nodes, where each node contains data and a reference to the next node in the sequence.</td><td><pre>class Node:
    def __init__(self, data):
        self.data = data
        self.next = None

class LinkedList:
    def __init__(self):
        self.head = None

    def append(self, data):
        new_node = Node(data)
        if self.head is None:
            self.head = new_node
            return
        last = self.head
        while last.next:
            last = last.next
        last.next = new_node

    def print_list(self):
        current = self.head
        while current:
            print(current.data, end=" -> ")
            current = current.next
        print("None")

# Create a linked list and append elements
my_linked_list = LinkedList()
my_linked_list.append(10)
my_linked_list.append(20)
my_linked_list.append(30)
print("Linked List:", end=" ")
my_linked_list.print_list()

Linked List: 10 -> 20 -> 30 -> None</pre></td></tr>
</table>

<table>
<tr><th>Advanced data structure</th><th>Description</th><th>Code example</th></tr>
<tr><td>Stacks and Queues</td><td>A last in, first out (LIFO) data structure, which means that the last element added to the stack is the first one to be removed. A queue is a first in, first out (FIFO) data structure, which means that the first element added to the queue is the first one to be removed.</td><td>

```
# Stack: LIFO (Last In, First Out)
stack = []

# Push elements onto the stack
stack.append(1)
stack.append(2)
stack.append(3)
print("Stack after pushing:", stack)

# Pop elements from the stack
stack.pop()
print("Stack after popping:", stack)

Stack after pushing: [1, 2, 3]
Stack after popping: [1, 2]

from collections import deque

# Queue: FIFO (First In, First Out)
queue = deque()

# Enqueue elements
queue.append(1)
queue.append(2)
queue.append(3)
print("Queue after enqueuing:", queue)

# Dequeue elements
queue.popleft()
print("Queue after dequeuing:", queue)

Queue after enqueuing: deque([1, 2, 3])
Queue after dequeuing: deque([2, 3])
```

</td></tr>
</table>

<table>
<tr><th>Advanced data structure</th><th>Description</th><th>Code example</th></tr>
<tr><td>Trees</td><td>A hierarchical data structure consisting of nodes connected by edges. Each node in the tree can have zero or more child nodes, and each child node can have its own child nodes, forming a branching structure.</td><td>

```
class TreeNode:
    def __init__(self, value):
        self.value = value
        self.left = None
        self.right = None

def inorder_traversal(root):
    if root:
        inorder_traversal(root.left)
        print(root.value, end=" ")
        inorder_traversal(root.right)

# Create the tree
root = TreeNode(10)
root.left = TreeNode(5)
root.right = TreeNode(20)
root.left.left = TreeNode(3)
root.left.right = TreeNode(7)

print("Binary Tree (In-order Traversal):", end=" ")
inorder_traversal(root)
print()

Binary Tree (In-order Traversal): 3 5 7 10 20
```

</td></tr>
<tr><td>Heaps</td><td>A a specialized tree-based data structure that satisfies the heap property. The heap property requires that the value of each node in the tree is greater than or equal to (in a max heap) or less than or equal to (in a min heap) the values of its child nodes.
Heaps are often used to implement priority queues, where the highest-priority element is always at the front of the queue.</td><td>

```
import heapq

# Min-Heap
heap = []

# Push elements onto the heap
heapq.heappush(heap, 20)
heapq.heappush(heap, 10)
heapq.heappush(heap, 30)
print("Heap after pushing:", heap)

# Pop the smallest element
heapq.heappop(heap)
print("Heap after popping:", heap)

Heap after pushing: [10, 20, 30]
Heap after popping: [20, 30]
```

</td></tr>
</table>

Advanced data structure	Description	Code example

Dataframe	A two-dimensional table-like data structure. It is a way to represent and manipulate data in a tabular format, similar to a spreadsheet or an SQL table.	`import pandas as pd`
		
		`# Data Frame: A tabular data structure with rows`
		`# and columns`
		`data = {`
		`    'Name': ['Alice', 'Bob', 'Charlie'],`
		`    'Age': [25, 30, 35],`
		`    'City': ['New York', 'Los Angeles', 'Chicago']`
		`}`
		`df = pd.DataFrame(data)`
		`print("Data Frame:\n", df)`
		
		`Data Frame:`
		`       Name  Age         City`
		`0    Alice   25     New York`
		`1      Bob   30  Los Angeles`
		`2  Charlie   35      Chicago`

The Python advanced data structures are more complex than the R data structures but best illustrated by examples. Let's first look at an array. Arrays can be created by importing the array module or the NumPy package:

```
import numpy as np

# Creating a numpy array
arr = np.array([1, 2, 3, 4, 5])

# Printing the array
print(arr)

[1 2 3 4 5]
```

In this example, we first import the NumPy library using the `import numpy as np` statement. Then, we create a numpy array using the `np.array()` method, passing a Python list `[1, 2, 3, 4, 5]` as input. Finally, we print the array using the `print()` function.

Linked list is a data structure consisting of a sequence of nodes, where each node contains data and a reference to the next node in the sequence. Here's an example:

```
class Node:
    def __init__(self, data):
        self.data = data
        self.next = None

class LinkedList:
    def __init__(self):
```

```
        self.head = None

    def append(self, data):
        new_node = Node(data)
        if self.head is None:
            self.head = new_node
            return
        last_node = self.head
        while last_node.next:
            last_node = last_node.next
        last_node.next = new_node

    def prepend(self, data):
        new_node = Node(data)
        new_node.next = self.head
        self.head = new_node

    def print_list(self):
        current_node = self.head
        while current_node:
            print(current_node.data)
            current_node = current_node.next
```

This code defines two classes: `Node` and `LinkedList`. Each `Node` object contains a `data` attribute and a `next` attribute that points to the next node in the list. The `LinkedList` class contains a head attribute that points to the first node in the list. The `LinkedList` class has three methods:

append
: Adds a new node to the end of the list

prepend
: Adds a new node to the beginning of the list

print_list
: Prints the data of each node in the list

You would use a linked list in this manner.

The initial linked list shows no entries: None. When 10 is appended it is now before None: 10 -> None. When 20 is appended, then the list looks like this: 10 -> 20 -> None. When 30 is appended, the list would look like this: 10 -> 20 -> 30 -> None. Here is an example on how the linked list can be used:

```
my_list = LinkedList()
my_list.append(1)
my_list.append(2)
my_list.append(3)
my_list.prepend(0)
my_list.print_list()
```

```
0
1
2
3
```

This code is a simple implementation of a singly linked list. There are many other operations you can perform on a linked list, such as inserting a node at a specific position, removing a node, and so on.

Stacks are LIFO data structures, which means that the last element added to the stack is the first one to be removed. A stack follows the LIFO principle, meaning the last element added is the first one to be removed.

The initial stack would be empty. After the first push (Push 1) only the value 1 is in the stack. After the second push (Push 2), both 1 and 2 are in the stack. The last push (Push 3) adds 3 to the stack resulting in 1, 2, and 3. The last action of Pop removes the last entry which was 3 resulting in the stack having only 1 and 2

Here is an example of a stack using a list:

```
class Stack:
    def __init__(self):
        self.items = []

    def push(self, item):
        self.items.append(item)

    def pop(self):
        return self.items.pop()

    def peek(self):
        return self.items[-1]

    def is_empty(self):
        return len(self.items) == 0

    def size(self):
        return len(self.items)
```

In the preceding code, we define a `Stack` class that has various methods to interact with the stack. The `__init__` method initializes an empty list that will be used to store the stack's elements.

The `push` method takes an item and adds it to the top of the stack by appending it to the end of the list. The `pop` method removes and returns the top element from the stack. The `peek` method returns the top element without removing it. The `is_empty` method checks if the stack is empty and returns a Boolean value. The `size` method returns the number of elements in the stack.

Here's an example of how to use this stack:

```
stack = Stack()

stack.push(1)
stack.push(2)
stack.push(3)

print(stack.peek())  # output: 3

stack.pop()

print(stack.peek())  # output: 2

print(stack.is_empty())  # output: False

print(stack.size())  # output: 2
```

In this example, we create a `Stack` object and push three items onto the stack. We then use the `peek` method to see the top item without removing it, and the `pop` method to remove and return the top item from the stack. Finally, we use the `size` method to see how many items are left on the stack.

Queues are similar to stacks but are FIFO data structures, which means that the first element added to the queue is the first one to be removed. An inverse of the previous example could demonstrate a queue. Trees and heaps will be explored further as these are more complex examples that have specific use cases.

Dataframes are used frequently in analytics. Here's an example of a Pandas DataFrame in Python:

```
import pandas as pd

data = {'name': ['Alice', 'Bob', 'Charlie', 'David'],
        'age': [25, 30, 35, 40],
        'city': ['New York', 'Los Angeles', 'Chicago', 'Houston']}

df = pd.DataFrame(data)
print(df)

      name  age         city
0    Alice   25     New York
1      Bob   30  Los Angeles
2  Charlie   35      Chicago
3    David   40      Houston
```

In this code, we first import the Pandas library using the import statement. We then define a Python dictionary data that contains the data we want to store in the dataframe. The keys of the dictionary are the column names, and the values are lists of data for each column.

We then create a Pandas `dataframe df` using the `pd.DataFrame()` constructor, and pass in the data dictionary as an argument. The resulting dataframe has four rows and three columns, with the column names `'name'`, `'age'`, and `'city'`, and the corresponding data from the data dictionary. The dataframe is then printed out.

You're welcome to refer back to this section for additional details about which options to select for a given scenario in the future. However, if there's one key takeaway, remember: data types and data structures are used in analytics for data processing and manipulation. Data processing and manipulation are managed through scripting. The next section reviews R and Python scripting.

Interaction with Relational Databases

To begin understanding databases, it's essential to recognize the different types available, each suited for various needs and use cases. Databases can be broadly categorized into types such as relational, hierarchical, NoSQL, and others. Hierarchical databases organize data in a tree-like structure, ideal for applications with a clear parent-child relationship. NoSQL databases are designed for unstructured or semi-structured data, offering flexibility and scalability, particularly in handling large-scale, distributed data environments.

Amidst these options, relational databases stand out for several reasons. They are particularly effective for managing structured data, where relationships between data points are crucial. Relational databases emphasize data integrity, scalability, and security, ensuring that data remains accurate and accessible even as the system grows. Moreover, the use of SQL, a standard language for querying and managing data, makes interacting with relational databases straightforward and consistent, providing a reliable foundation for data analytics and large-scale data management. These features collectively make relational databases a preferred choice in many analytical and business environments.

Why Relational Databases?

Interpreted languages, such as Python and R, offer several benefits when interacting with relational databases. Interpreted languages are often easier to learn and use, even for nontechnical users, compared to compiled languages. This makes it easier for users to interact with relational databases and extract meaningful insights from the data. Interpreted languages offer a high level of flexibility, allowing users to write scripts and programs that can easily be adapted to changing requirements. This makes it easier to integrate relational databases into existing workflows and processes. Last, interpreted languages are well-suited for data analysis and manipulation tasks, as they provide a wide range of libraries and tools for data analysis, visualization, and modeling. This makes it easier to perform data analysis on data stored in relational databases. There are many different relational databases, but we will primarily be using SQLite.

R Connection to Relational Databases

R can connect to several different types of relational databases. R can connect to several relational databases in order to manipulate data. This is not an exhaustive list but reflects many of the popular relational databases:

SQLite
: A lightweight, file-based relational database that is well-suited for small and medium-sized datasets. R can connect to SQLite databases using the RSQLite library.

MySQL
: An open source relational database that is widely used for web applications and other large-scale applications. R can connect to MySQL databases using the RMySQL library.

PostgreSQL
: An open source relational database that is widely used for web applications and other large-scale applications. R can connect to PostgreSQL databases using the RPostgreSQL library.

Microsoft SQL Server
: A commercial relational database that is widely used for enterprise applications and business intelligence. R can connect to Microsoft SQL Server databases using the RODBC library or the Rmssql library.

Oracle
: A commercial relational database that is widely used for enterprise applications and business intelligence. R can connect to Oracle databases using the ROracle library.

R can also connect to other relational databases, including IBM DB2, Teradata, Amazon Redshift, and Snowflake, using specialized libraries and packages. Additionally, R can also connect to nonrelational databases, such as NoSQL databases, using specialized libraries and packages.

Examples of R and Relational Databases

Connecting to relational databases in R requires using specific libraries. Common libraries include RODBC, RMySQL, and RPostgreSQL, among others that are tailored to the database being used. These libraries are available through the CRAN repository (*https://cran.r-project.org*).

Each database connection is similar, but let's look at a code example to connect to an SQL Server database. The RODBC library is used:

```
library(RODBC)

# Connect to the database
conn <- odbcConnect("sqlserver", uid="user", pwd="password")

# Run a query to retrieve data from the database
result <- sqlQuery(conn, "SELECT * FROM my_table")

# Close the connection to the database
odbcClose(conn)
```

The precise syntax will differ depending on the database and library, but the general approach is to load the library, connect to the database, run a query, store the results, complete the needed data manipulation, and then close the connection. The libraries can be leveraged to complete many database operations, including executing stored procedures or data definition language. To see another example of the syntax for connecting to an SQLite database, see the following example.

SQLite

SQLite is a relational database management system that is open source and used in applications and embedded systems due it being lightweight. SQLite does not require a server and runs as an embedded library, which means it does not have the same overhead as other relational databases. SQLite is a common choice for mobile applications and supports transaction management, query optimization, and data integrity. SQLite is less complex and uses fewer compute resources than other database systems and is good for applications that need high performance and scalability.

The next section outlines how to connect to different relational databases using R and Python. While R and Python have data types and structures for data manipulation, most of the manipulation happens in memory, which means there are limited resources for large datasets. This is where the use of relational databases occurs.

To connect to SQLite in R, the RSQLite library is used:

```
library(RSQLite)

# Connect to an existing SQLite database
conn <- dbConnect(SQLite(), dbname = "database.sqlite")

# Perform operations on the database
result <- dbGetQuery(conn, "SELECT * FROM table_name")

# Close the connection
dbDisconnect(conn)
```

In this example, you first load the RSQLite library and then use the `dbConnect` function to connect to the SQLite database located at *database.sqlite*. You can then perform operations on the database using the `dbGetQuery` function, which allows you to execute SQL queries against the database. Finally, you close the connection using the `dbDisconnect` function. Next, we will look at how Python connects to relational databases.

Python Connection to Relational Databases

Python can connect to several different types of relational databases, just like R. "R Connection to Relational Databases" on page 54 lists the several common relational databases that R can connect to, and these also apply to Python. Python can also connect to other relational databases, including IBM DB2, Teradata, Amazon Redshift, and Snowflake, using specialized libraries and packages. Python can also connect to nonrelational databases, such as NoSQL databases, using specialized libraries and packages.

Examples of Python and Relational Databases

To connect to a relational database in Python, you can use a library such as sqlite3, mysql-connector-python, or psycopg2, or a library specialized for the specific relational database you want to connect to (e.g., cx_Oracle for Oracle databases). Here's an example using the `sqlite3` module to connect to an SQLite database:

```
# Connect to the database
conn = sqlite3.connect("my_database.db")

# Run a query to retrieve data from the database
cursor = conn.cursor()
cursor.execute("SELECT * FROM my_table")
rows = cursor.fetchall()

# Close the connection to the database
conn.close()
```

The exact syntax will vary depending on the library and the relational database you are connecting to, but the general process is to import the library, connect to the database, create a cursor, run a query to retrieve data, fetch the results, and then close the connection. You can also use the cursor to run other types of database operations, such as inserting, updating, or deleting data, as well as executing stored procedures and transactions.

Summary

This chapter introduced you to R and Python, two open source interpreted languages used in business analytics. We covered OOP concepts such as structures, classes, methods, objects, attributes, and principles (encapsulation, abstraction, inheritance, polymorphism). R and Python data types and structures were addressed with coding examples, and scripting was reviewed. Different tools can be used to leverage R and Python such as RStudio and Jupyter notebooks, which were also introduced. Last, we covered relational databases and how R and Python can connect as relational databases are highly used in business analytics. Now that we have a foundational knowledge in R and Python, Chapter 4 will focus on getting started with statistical analysis.

CHAPTER 4

Statistical Analysis with R and Python

Statistics play an important role in analytics because they are used to analyze data, identify patterns, and determine relationships between variables. The results of statistical analysis can be used to make business decisions and predictions or to classify new data, and this wide variety of use cases make it a science worth learning. Additionally, business analysts leverage algorithms to create analytical models. An algorithm is a set of steps, instructions, and procedures that are used to solve a problem or perform a task. In analytics, they are used to perform specific computations, manipulate data, and create models. Descriptive, diagnostic, predictive, or prescriptive analytics are then applied to these models. Together, statistics and algorithms are used to preprocess data, select data elements to be used in problem solving, and compare models to select the best one to solve a problem. This chapter explores the role of statistics, different approaches used, and data visualization in R and Python.

Example Analytical Projects

Before we explore specifics, it is important to understand the context of the problem to be solved, since this determines which statistical approaches will be used. Not every problem will be solved with analytics, but the problems you'll be tackling as an analyst will be complex, requiring critical thinking to break down, examine, and interpret components. Once the components are identified, the next step is to understand the problem at a level where you can explain and solve it. To solve an analytical problem, analytical methods are applied to come to a conclusion or solution. These methods may include data analysis or hypothesis testing, for example. Problems can't be solved in isolation and often require evaluating relationships, patterns, and trends, as well as the application of domain-specific knowledge.

To best understand analytical problems, let's look at some examples that are common across industries. The analytics life cycle discussed in Chapter 3 is used to outline how the analytics problem is solved. As a quick refresher, the life cycle steps include problem identification, data understanding and preparation, model building, model evaluation, and implementation.

Telecom Churn

Consider the following: a major telecom company has contracted with your organization, asking your team to help it understand why customers are canceling their service and switching to other providers. Analytics can assist in managing customer churn by identifying the factors that cause customers to switch service providers. Analytics can also predict if a customer is likely to churn.

You'll see the techniques mentioned in this example in action over the course of the next few chapters, but we want to provide a high-level overview here so you can get a sense of the end-to-end process:

Step 1: Define the problem
: A major telecom company wants to understand why it's losing customers. Your team is tasked with understanding what is causing this churn, or turnover of customers. Churn is a large problem for telecom companies as it means a loss of revenue as well as market share.

Step 2: Data understanding
: In the data understanding phase, the focus is on collecting and exploring the telecom churn data to grasp its structure and potential insights. This begins with gathering data from sources like customer demographics, usage patterns, and billing information. Exploratory data analysis (EDA) is then performed to identify key trends and relationships, such as how contract length or monthly charges might influence churn rates. Additionally, data quality is assessed to spot and address issues like missing values or inconsistencies, which are crucial for ensuring reliable analysis later on. We will cover EDA in Chapter 5.

Step 3: Data preparation
: The data preparation phase involves refining the data to make it suitable for modeling. This includes cleaning the data by handling missing values and correcting inconsistencies. Feature engineering is performed to create new variables that could improve predictive power, such as segmenting customers based on usage. The data might also be transformed, normalized, or integrated from multiple sources to create a cohesive dataset. Finally, unnecessary or redundant features may be reduced to simplify the model while retaining essential information. The goal is to produce a clean, well-structured dataset ready for the modeling phase, where predictions about customer churn can be made. Data preparation techniques are covered in Chapter 5.

Step 4: Building a model

In the modeling phase, the prepared data is used to build predictive models that help identify patterns and predict outcomes—in this case, customer churn. Various algorithms, such as logistic regression, decision trees, or random forests, are applied to the data to create models that can accurately forecast which customers are likely to churn.

The process includes selecting the appropriate modeling techniques, training the models on the dataset, and fine-tuning parameters to optimize performance. The models are then evaluated using metrics like accuracy, precision, recall, and area under the curve (AUC) to ensure they effectively distinguish between customers who will churn and those who will stay. The goal is to develop a robust model that provides actionable insights, allowing the business to take proactive steps to reduce churn. Algorithms are covered in Chapter 6.

Step 5: Implementing a predictive model

The final step is implementing the predictive model for use and decision making. This is where the value occurs. Potential outcomes may involve offering targeted promotions or incentives to customers who are at high risk of churn.

In this example, analytics were used to help telecom companies identify the drivers of customer churn and predict which customers are most likely to leave. By using this information, companies can develop effective retention strategies and improve customer satisfaction, thereby reducing the impact of customer churn on their business. Let's explore another type of analytical project that a business analyst could be involved in: A/B testing.

A/B Testing

A/B testing is a statistical method used to compare two versions of a variable (such as a web page, product feature, or marketing campaign) to determine which one performs better in achieving a specific goal, like higher conversion rates or customer engagement. It is an analytical problem because it involves hypothesis testing, data collection, and analysis to draw meaningful conclusions about which version is more effective. For business analysts, understanding A/B testing is crucial as it informs decision making by providing evidence-based insights into what works best for the business. This knowledge is closely related to building predictive models because both processes rely on data-driven approaches to optimize outcomes, validate hypotheses, and make informed predictions about future behavior based on past data.

As part of A/B testing, the focus is on developing a hypothesis on what is expected to happen with each new version of a marketing campaign or product page. The testing of the hypothesis is accomplished using different metrics such as conversion rates, click-through rates, or other performance measures depending on the scenario. A sample is then determined by randomly selecting users or customers that represent the target audience to participate in the test.

In this example, a company has created different versions of its product and marketing campaigns, which are then presented to a sample group of customers. Data is then collected on the performance of each version, and analysts assess the data using statistical methods to determine which variation performs better. The company uses techniques including hypothesis testing, confidence intervals, and Bayesian analysis. Then, the company interprets the results to determine which variation performed better and whether the hypothesis was supported or rejected. Typically, the version that performed the best will then be implemented.

A/B testing, paired with analytics, is used by marketing and product teams to make data-driven decisions and determine which version of a given campaign or product performs the best. This approach enables organizations to improve products and marketing campaigns.

Marketing Campaigns

An analytical project for marketing campaigns involves systematically evaluating the effectiveness of different marketing strategies to optimize outcomes such as customer acquisition, engagement, and conversion rates. This type of project typically includes segmenting the target audience, analyzing past campaign performance, and identifying key factors that drive success. By leveraging data analytics, business analysts can assess the impact of various marketing tactics, refine targeting approaches, and develop more effective, personalized campaigns. Understanding this process is essential for business analysts, as it enables them to provide actionable insights that enhance the efficiency and ROI of marketing efforts, similar to the data-driven methodologies used in building predictive models. Marketing campaigns are used by organizations to promote products and services as well as engage customers. Marketing campaigns are most effective when using the right channels, sending the right messages, and targeting the right audiences. This can be challenging to do well, and analytics play a role in determining the right channels, messages, and audiences.

To optimize a marketing campaign, data is collected on customer behavior and marketing campaign performance. Additional data may be collected for prior or similar marketing campaigns, including website traffic, social media engagement, email open rates, click-through rates, and conversion rates. Then, data can be preprocessed and analyzed to determine patterns, trends, and data relationships.

What does this look like in practice? Using the insights gained from data analysis, campaigns may be optimized by targeting specific customer segments, optimizing messaging and creative content, and adjusting the marketing mix to improve campaign performance. Additionally, A/B testing can be leveraged in marketing campaign optimization. This can involve testing different messaging, creative materials, or call-to-action (CTA) variations to see which performs better. Another analytical technique that may be useful here is attribution modeling. This includes techniques used to determine which marketing channels and actions are having the largest return. Once a campaign is being used, campaign performance is monitored to identify areas that need adjustment or improvement. Analytics is used to ensure marketing campaigns are effective and, ultimately, provide more revenue for the organization.

Financial Forecasting

Financial forecasting is an important analytical project within the finance industry, involving the use of historical data and advanced analytics to predict future organizational performance. Business analysts play a pivotal role in this process by gathering, analyzing, and interpreting data to generate forecasts that inform strategic decisions. Whether it's assessing potential risks, identifying growth opportunities, or optimizing resource allocation, financial forecasting enables organizations to plan for the future with greater accuracy and confidence. Business analysts collaborate with financial professionals to develop models, conduct scenario and sensitivity analyses, and ensure that forecasts are grounded in robust data insights, making them essential contributors to the financial forecasting process.

Financial forecasting is a crucial process in finance that involves using historical data to predict future organizational performance. This process is essential across various sectors of finance, each with its specific applications and challenges:

Investment banking
: In investment banking, financial forecasting is vital for mergers and acquisitions (M&A). Analysts use scenario analysis to forecast the financial outcomes of potential mergers, evaluating different deal structures, market conditions, and synergies. For example, they might analyze how a merger between two companies could impact future cash flows and valuations under different economic scenarios.

Trading
: In the trading sector, time series forecasting is frequently used to predict stock prices, interest rates, and market trends. Traders rely on these forecasts to make informed buy or sell decisions. For instance, trend analysis might help a trader predict a stock's future performance based on past price movements and market conditions, allowing them to optimize their trading strategies.

Retail banking
: Retail banks use financial forecasting to predict customer demand for loans, deposits, and other banking products. Sensitivity analysis can help these institutions understand how changes in interest rates or economic conditions could affect loan default rates or deposit inflows. For example, a bank might use these

analyses to adjust its interest rate offerings or to plan for potential increases in loan defaults during economic downturns.

Commercial B2B banking

In commercial banking, forecasting is crucial for managing credit risk and liquidity. Banks use scenario analysis to assess how different economic conditions could impact the creditworthiness of their business clients. For example, they might forecast the impact of a recession on the repayment abilities of companies in their loan portfolio, allowing them to adjust their lending strategies accordingly.

Across all these sectors, responsible financial forecasting is about making informed predictions based on rigorous data analysis. Scenario analysis and sensitivity analysis are key tools that help identify potential risks and opportunities, enabling financial institutions to make strategic decisions, allocate resources effectively, and manage risk. Whether it's forecasting market trends for trading, predicting customer behavior in retail banking, or evaluating M&A scenarios in investment banking, accurate financial forecasting is essential for success in the finance industry. Business analysts are integral to this process, ensuring that forecasts are data-driven and aligned with organizational goals.

Healthcare Diagnosis

Healthcare analytics uses medical data and clinical expertise to identify the underlying causes of a patient's symptoms or health conditions. Analytics is used to identify patterns and relationships that can be used to make accurate diagnoses. Medical data such as electronic health records, medical images, laboratory results, and other relevant information are all used in this analysis.

There are many practical applications for analytics in healthcare. For instance, predictive models can be used to identify patients who are at risk of developing certain health conditions or complications. This can help clinicians to provide early interventions or preventive care to improve patient outcomes. Ultimately, analytics provides clinicians with decision support tools that can assist with diagnosis and treatment planning. This may involve using predictive models or decision support algorithms to suggest diagnostic tests, treatment options, or medication recommendations.

Another application is population health management. This is leveraging analytics to manage the health of populations by identifying health trends, risk factors, and areas for improvement. This can help healthcare organizations to implement targeted interventions that improve health outcomes and reduce healthcare costs. By using different analytical techniques such as classification, the healthcare industry can inform diagnosis and treatment decisions, improve the quality of care, and achieve better health outcomes for patients.

Starting with the Problem Statement

The problem statement is one of the most important parts of an analytical project as it defines the scope and objective of the project. Without a clear problem statement, the project may lack direction and focus, making it difficult to achieve the desired outcomes. The benefits of a clear problem statement include helping to determine the specific problem to be solved and what kind of analytical problem is at hand. Also, a problem statement helps stakeholders to understand what the project aims to achieve, what its limitations are, and what to expect and when.

Once a problem statement is defined, it guides decision making and assists stakeholders to make informed decisions about project design, data collection, analysis techniques, and other aspects of the project. Additionally, it enables the measurement of project success by defining clear objectives and outcomes that can be tracked and evaluated.

Let's look at an example problem statement regarding telecom churn:

> The problem we plan to address is the high rate of customer churn in our telecommunications company. We need to develop a predictive model that can accurately predict which customers are most likely to churn in the next 30 days, so that we can implement targeted retention strategies to retain those customers.

This problem statement clearly defines the problem we aim to solve (high rate of customer churn) and the objective of the project (to develop a predictive model to identify customers most likely to churn). It also provides a time frame for the prediction (next 30 days) and the reason for the prediction (to implement targeted retention strategies). With this problem statement, we can design a predictive modeling project that focuses on developing a model that accurately predicts which customers are at risk of churning and enables the implementation of targeted retention strategies to reduce churn rates.

Let's consider another example problem statement about low conversion rates for a checkout page:

> Our ecommerce website is experiencing a low conversion rate for our checkout page. We want to increase the conversion rate by optimizing the checkout page design. We will conduct an A/B test by creating two versions of the checkout page and randomly assigning visitors to each version. The problem we aim to solve is to identify which version of the checkout page leads to a higher conversion rate.

The problem is clearly identified (low conversion rate), as is the objective (to use A/B testing to determine the best version of the checkout page) that will result in a higher conversion rate (outcome).

The problem statements in both examples outline the approach that will be used in analytics. The telecom churn problem will be addressed with a predictive model, and

A/B testing will use experimental design and inferential statistics to better understand the low conversion rates. From these problem statements, it is also likely to identify the data sources and elements that would be collected to start the analytical project (problem statement examples will be discussed in the next section).

Getting to the Analytical Problem

A primary challenge is understanding what kinds of analytical problems are at hand and which analytical approach to take. In the example of telecom churn or A/B testing, these are well-known analytical problems, and the approach to take is clear. However, not all problems are straightforward, and they may require multiple analytical techniques to solve. Here are the most common types of analytical problems:

Prediction
: Problems that involve predicting future outcomes or events, such as predicting customer churn, stock prices, or demand for a product. Prediction problems are divided between classification and regression.

Classification (prediction)
: Use cases that involve categorizing data into different classes or categories, such as classifying emails as spam or not spam, or classifying images as dogs or cats.

Regression (prediction)
: Issues that involve predicting a continuous numerical value, such as predicting the price of a house based on its features.

Optimization
: Situations that involve finding the best solution or combination of solutions to achieve a specific objective, such as optimizing a supply chain to minimize costs or maximizing revenue.

Clustering
: Scenarios that involve grouping data into similar clusters or segments based on their characteristics, such as clustering customers based on their purchasing behavior or clustering images based on their content.

Association
: Use cases that involve finding associations or relationships between different variables, such as identifying which products customers often purchase together. Association identifies relationships between variables in a dataset, while clustering groups similar data points together based on shared characteristics.

Anomaly detection
: Problems that involve identifying unusual or anomalous behavior or data points, such as detecting fraud or faulty equipment in a manufacturing process.

These different types of analytical problems require different methods and techniques to solve them effectively, and it is important to choose the right approach for each problem based on its specific requirements and goals. In Chapter 1 we highlighted different analytical techniques: descriptive, diagnostic, discovery, predictive, and prescriptive. Each of the techniques or combinations may be adopted to solve a problem. In addition, each technique builds on the other. As an example, descriptive analytics needs to be completed before predictive analytical approaches can be applied.

Let's break this down a bit further. Descriptive analytics projects focus on describing past events or trends, such as analyzing sales data to understand which products were popular during a specific period. Often the focus of descriptive projects is to answer the question, What has happened?

Discovery analytics projects focus on discovering new patterns or insights in data, such as analyzing social media data to understand customer sentiment about a product or brand. Discover analytics projects focus on answering these questions: What else do I need to know? Is there something I have overlooked?

Diagnostic analytics projects focus on identifying the root cause of a problem or issue, such as analyzing customer complaints to understand the reasons for a decline in customer satisfaction. Diagnostic analytics projects focus on answering the question, What causes the outcome? In addition, it addresses the question, Why is there a problem? The answer to this question allows us to answer the what, why, and how.

Predictive analytics projects focus on predicting future events or trends, such as developing a statistical learning model to predict which customers are most likely to churn. The focus here is the question, What is expected to happen?

Prescriptive analytics projects focus on recommending the best course of action to achieve a specific objective, such as identifying the optimal pricing strategy to maximize revenue. Prescriptive analytics focuses on answering these questions: What is the right action for this problem? How to make something happen?

Going back to the problem statement helps us understand how to design the project. Let's consider a healthcare organization that wants to improve diagnosis accuracy. Here is the problem statement:

> Our healthcare organization needs to improve the accuracy of its breast cancer diagnosis. We aim to develop a classification model that can accurately classify mammogram images as benign or malignant. The problem we aim to solve is to improve the accuracy of breast cancer diagnosis by developing a statistical learning model that can accurately classify mammogram images and reduce the rate of misdiagnosis. We can design a classification project that focuses on developing a statistical learning model that accurately classifies mammogram images and enables accurate breast cancer diagnosis. The outcome will be given to a medical doctor, who will use it to improve their diagnosis.

This problem statement identifies that a classification project will be needed, which means descriptive, diagnostic, discovery and predictive techniques will be used. While the last example is a strong problem statement, it is important to be able to identify weak ones. Let's look at a version of the previous statement:

> Our healthcare organization wants to use statistical learning to do something with mammogram images. We hope it will help in diagnosing breast cancer better. The aim is to create a statistical learning model that does something with the data and will hopefully be useful to doctors.

This statement has a number of issues:

Lack of specificity
: The statement is vague and does not clearly define the problem or the objective. It mentions "doing something" with the mammogram images without specifying what that something is (e.g., classifying images, improving diagnosis accuracy).

No clear goal
: It fails to establish a clear goal or outcome, making it difficult to understand what success looks like. The phrase "hopefully, it will be useful to doctors" shows uncertainty and a lack of direction.

Ambiguous language
: The use of words like "do something" and "hopefully" reflects a lack of focus and commitment to a defined solution, which weakens the purpose and direction of the project.

Lack of impact
: The statement doesn't clearly articulate the impact or importance of the project, unlike the strong example, which emphasizes the need to reduce misdiagnosis and improve diagnostic accuracy.

In contrast, the strong problem statement is clear, is focused, and provides a concrete goal (improving the accuracy of breast cancer diagnosis through a specific statistical learning model), which makes it a solid foundation for a successful project.

Let's consider another problem statement:

> Our manufacturing company wants to improve the efficiency of its production process by reducing energy consumption. We aim to develop a regression model that can accurately predict energy consumption based on various production factors such as temperature, humidity, and production volume. The specific problem we aim to solve is to reduce energy consumption and costs by developing a statistical learning model that can accurately predict energy consumption.

This problem statement identifies that a regression project will be needed, which means descriptive, diagnostic, discovery, and predictive techniques will be used.

A good portion of analytical projects lead to prediction techniques, and this requires building up to creating a predictive model. Now that we've walked through a few example problem statements, you have a better understanding of what information they should convey and how this impacts the analytical approaches you use. If you receive a project with a weak problem statement, don't be afraid to ask questions of the stakeholders to better understand the parameters before you begin work. This will save time for everyone involved and will lead to greater quality in your work.

Many analytical problems will focus on prediction. As outlined in this section, both classification and regression are called out as prediction approaches.

Classification

Classification is a type of supervised learning problem that is used to predict a categorical or discrete target variable based on a set of input features or variables. All forms of classification models focus on determining the relationship between the input features (predictors) and the target variable (prediction), leveraging this relationship for predictive capabilities. For example, a classification model could predict whether an email is spam or not based on the email's content. The predicted value or target would be a flag of Spam or Not Spam. While classification can result in a number of categories, when the output is limited to two results, this is referred to as a binary classification.

Regression

Regression is also a type of supervised learning problem that involves predicting a continuous numerical target variable based on a set of input features. A regression model could predict the price of a house based on its location, size, number of bedrooms, and other relevant features. Regression is widely used in various domains such as finance, healthcare, and engineering to model the relationships between variables and make predictions about future values of the target variable.

What Do We Want to Measure?

Determining what needs to be measured is a critical step in formulating a plan for an analytics project because it helps to define the project's scope and goals. It provides clarity on the specific business problem that needs to be addressed, which in turn helps to determine the appropriate analytical approach, methods, and data required for the project. Without a clear understanding of what needs to be predicted, it is easy to get lost in the sea of available data and analytical methods, leading to misguided analyses.

It is important to determine the specific measurement that will be the outcome of a model. For example, if the goal of the project is to predict customer churn in a telecommunications company, the focus will be on analyzing customer behavior patterns and factors that drive customer retention. Churn is actually a churn rate, which is a continuous number, which also identifies the analytical problem further: this would be a regression problem. We described classification and regression earlier in the chapter, but knowing what we want to predict also tells us more about our analytical problem. If we were predicting a label, this would be a classification problem.

Analysis Approaches

Regression and classification problems have different goals and require different techniques for EDA in statistical learning. How you approach EDA will be influenced by the type of prediction problem you have. For predictive analytic problems, the primary goal is to identify the relationships between the input variables (known as predictors or features) and the target (known as the predicted value). Depending on the type of problem—either regression or classification—different statistical analysis will be completed.

For regression problems, the focus is on analyzing the distribution and correlation of the input variables, identifying outliers, and detecting nonlinear relationships. EDA visualization techniques such as scatter plots, histograms, and correlation matrices are commonly used in regression analysis. For classification problems, analyzing the distribution and correlation of the input variables for each class, identifying class imbalances, and detecting nonlinear relationships are common activities. EDA visualization techniques such as bar charts, box plots, and heatmaps are commonly used in classification analysis. Examples for both cases are covered in a later section.

In both cases, EDA plays a critical role in identifying data quality issues, understanding the relationships between variables, and selecting appropriate feature engineering and preprocessing techniques. Effective EDA can help ensure that the model is able to learn the underlying patterns in the data and make accurate predictions. Let's explore this further.

EDA

EDA is the process of analyzing and visualizing data to understand its characteristics, patterns, and relationships. EDA is an important step in the statistical learning process as it helps to identify important features, detect anomalies and outliers, and understand the underlying structure of the data. The main goals of EDA in statistical learning are as follows:

- To gain a deeper understanding of the data and its characteristics, such as its distribution, range, and correlation between variables
- To identify any data quality issues, such as missing or incorrect values, outliers, or duplicates

- To explore potential relationships between the input variables and the target variable
- To determine appropriate feature engineering and data preprocessing techniques that can improve the performance of the model

Expected outcomes of EDA in statistical learning may include the identification of important variables, the detection of data quality issues, the identification of relationships between variables, and the selection of appropriate feature engineering and data preprocessing techniques. Statistical analysis techniques such as correlation analysis and hypothesis testing are also used. Effective EDA can help ensure that the statistical learning model is trained on high-quality data and is able to make accurate predictions.

Unsupervised Learning

Unsupervised learning is a type of statistical learning where the goal is to find patterns or structure in data without any labeled output or target variable. In other words, the algorithm is given a set of input features and is expected to identify meaningful patterns or relationships in the data on its own, without any guidance. During EDA, unsupervised learning can be explored to gain a deeper understanding of the data and to identify hidden patterns or relationships that may not be immediately obvious. Unsupervised learning can be used to identify clusters or groups of similar data points, detect anomalies or outliers, and reduce the dimensionality of the data.

Common unsupervised learning techniques include clustering, dimensionality reduction, and anomaly detection. Clustering involves grouping similar data points together based on their similarity, while dimensionality reduction techniques aim to reduce the number of features or variables in the data while preserving as much of the original information as possible. Anomaly detection aims to identify data points that are significantly different from the rest of the data and may indicate errors or outliers. For instance, the discovery of clusters in a dataset can result in different predictive models. When we apply clustering to a dataset, it can help to identify patterns or groups of data points that have similar characteristics or behavior.

If we use these clusters to create a predictive model, the resulting model may be different from one that is built without clustering. This is because the clusters may highlight specific relationships between the input features and the target variable that are not apparent in the original dataset.

For example, imagine that we have a dataset of customer transactions from an online store, and we want to build a predictive model to forecast customer purchases. If we use clustering to group customers based on their transaction history or demographics, we may identify distinct customer segments that have different purchase patterns. We can then use these customer segments as input features in our predictive model, which may result in a more accurate and targeted model. However, it's important to note that the resulting predictive model may not necessarily be better than a model built without clustering, and the choice of which approach to use ultimately depends on the specific problem and the available data. Code examples of clustering and other unsupervised learning methods will be reviewed in Chapter 5.

Statistical Analysis for Regression

The analysis plan that will be used in EDA is determined by the prediction approach (regression or classification). All datasets will be analyzed for content quality, patterns, relationship, and other issues that could impact the accuracy of the potential model. Linear regression is the primary regression approach used in analytics, and we will explore it in upcoming chapters. Here, I'll focus on how statistical analysis is used in preparing for a linear regression model. This allows us to understand the underlying structure, patterns, and characteristics of the data. As a result of EDA, an analyst will:

- Identify influential variables that can assist in predicting an outcome
- Discover and detect outliers and anomalies
- Test assumptions about the data to determine if they exist
- Explore relationships between variables to help in feature selection and engineering
- Visualize the data to make analysis easier
- Clean and process data

Most of this list applies to any predictive model, but with linear regression, statistical analysis relies on current properties being present in the data. For instance, understanding the distribution of the variables is important in analyzing data for regression models. In many regression models, especially linear regression, we assume that the residuals or errors are normally distributed. Checking the distribution of the dependent variable (and possibly the independent variables) can help determine if a normality assumption is reasonable or if data transformations are required to meet this assumption. Distribution can be used to identify outliers, which can impact the accuracy of the regression model.

Correlation analysis assists in identifying relationships between the predictors and the predicted value. A strong correlation between the predictor and predicted outcome means that it might be a good predictor for the regression model, which is part of determining the correct predictors. Correlation can also be used to identify multicollinearity, which occurs when two or more independent variables are highly correlated with each other. Last, correlation is used to determine independence between the variables, as required, to use regression models.

In addition to analyzing distribution and correlation, EDA focuses on determining the content quality of the data. One focus is identifying and addressing missing data. Techniques like mean imputation, median imputation, or interpolation can be used to address missing data. Another focus is to determine what transformation is needed for the data. Categorical data such as a user's favorite color cannot be used as input into a regression model as the regression model needs numerical input.

One common technique to handle categorical variables in classification problems is one-hot encoding, also known as dummy variable encoding. One-hot encoding converts each categorical variable into a binary vector with a length equal to the number of categories in the variable. Each element in the vector corresponds to a category, and its value is 1 if the data point belongs to that category, and 0 otherwise.

For example, suppose we have a categorical variable `color` with three categories: red, green, and blue. One-hot encoding this variable would result in three binary variables: `color_red`, `color_green`, and `color_blue`. If a data point has a `red` color, the `color_red` variable would be 1, and the `color_green` and `color_blue` variables would be 0. One-hot encoding is a feature engineering method and used to prepare the data for the modeling phase.

When dealing with a categorical variable like a PIN code in EDA, the challenge arises when the number of unique categories is very high, such as thousands of different PIN codes. This can make it difficult to extract meaningful patterns or use the variable effectively in a model. One approach to handle this is to aggregate the PIN codes into broader geographic regions, such as grouping them by city, state, or region. Alternatively, if geographic specificity is important, dimensionality reduction techniques like PCA or clustering can be applied to create meaningful groups from the PIN codes. Another option is to assess the impact of each PIN code category on the target variable and retain only the most significant ones, while the rest can be grouped into an "Other" category to reduce dimensionality.

The end goal of EDA is to identify the most important variables to be used as predictors for the regression model and prepare the data to build the model. Distribution, correlation, outliers, and converting data to ensure numerical inputs are the focus for statistical analysis for regression models.

Analysis for Classification

Classification, in contrast to regression, has similar but different analysis goals for EDA. The same goals exist: testing assumptions about the data, exploring data relationships, cleaning and processing the data, and identifying influential variables to predict the outcome. However, because classification does not rely on statistical relationships, the focus of analysis will slightly differ from regression. For instance, analyzing the distribution and correlation of variables depends on the classification algorithm used, but reviewing these areas can provide important insights that can improve the accuracy and interpretability of the model.

Analyzing the distribution of variables can help to identify potential issues like skewness or outliers that may affect the performance of the classification model. For example, distribution can identify if the target variable is highly imbalanced (where one class is more populous than the other). A significant imbalance can lead to poor model performance, as the model might be biased toward the majority class. If class imbalance is detected, techniques such as resampling, synthetic data generation, or adjusting the classifier's decision threshold can be applied as part of the EDA process.

Skewness refers to the asymmetry in the distribution of a variable's values, where data points are more concentrated on one side of the distribution than the other. In classification models, skewed variables can impact the model's performance by leading to biased predictions, especially if the skewed variable is influential in determining the outcome. For instance, if a feature with significant skewness has outliers or an imbalanced distribution, the model might overly rely on these extreme values, potentially misclassifying cases that fall into the tails of the distribution. Additionally, skewed features can distort the decision boundaries of algorithms like logistic regression or support vector machines. To mitigate these effects, it's common practice to transform skewed variables using techniques such as logarithmic or Box-Cox transformations, which can help create a more balanced distribution and improve the robustness of the classification model.

Correlation is used in evaluating data, and it could impact the choice of algorithm. For instance, if logistic regression is used, multicollinearity is a concern. Multicollinearity can cause unstable models or models that cannot easily be interpreted. Regression-based algorithms such as logistic regression or others, including linear discriminant analysis (LDA), least absolute shrinkage and selection operator (LASSO) regression, or ridge regression, can be impacted where highly correlated variables can influence the coefficient estimates. Support vector machine (SVM) and k-nearest neighbors (KNN) can also be impacted negatively by multicollinearity. On the other hand, some classification algorithms are more robust to correlated variables, such as tree-based algorithms and neural networks. These algorithms will be explored more in upcoming chapters.

Strong correlations between predictors and the target variable help identify the most important predictors, making correlation an important factor in selecting predictors. Analyzing correlations between features can also help identify opportunities for feature engineering. Combining or transforming correlated features can create new, more informative features that may improve the performance of your classification model.

Role of Hypothesis Testing

Hypothesis testing is a statistical tool used to determine whether a particular hypothesis is supported by the data or not. In regression or classification problems, hypothesis testing can be used to evaluate the significance of individual predictor variables or to test the overall significance of the model.

As part of EDA, hypothesis testing can be used to determine if individual predictor variables in regression models are significant in explaining the variation in the target variable. The null hypothesis could be tested to see if the coefficient of a predictor variable is equal to zero, and if rejected, the predictor is determined to be significantly related to the target variable.

With classification models, hypothesis testing is used to determine whether the model performs better than a baseline model. The null hypothesis in this context typically states that there is no difference in accuracy between the model and the baseline, which could be a model predicting the most frequent class or simply predicting the average outcome. For example, the null hypothesis might state, "The accuracy of the classification model is equal to the accuracy of a baseline model that predicts the most frequent class." Hypothesis testing can also be applied to compare the performance of different models, such as a simple model versus a more complex one. If the null hypothesis is rejected, it indicates that the classification model provides a statistically significant improvement over the baseline model or the simpler model. This process ensures that any observed improvement in model performance is not due to random chance but rather reflects a meaningful enhancement.

Overall, hypothesis testing can play a critical role in assessing the significance of predictor variables and the model as a whole in regression and classification problems. It helps to identify the most important predictors and evaluate the overall performance of the model, which can guide further model improvement and selection.

Visualization in Analytics

Visualizing data in EDA allows analysts and data scientists to understand the underlying patterns, trends, and relationships within the data. By creating visualizations of the data, analysts can identify outliers, potential errors, or missing values, and better understand the distribution of the data. In addition, visualizations assist in identifying potential relationships between variables that are not easily identified using numerical analysis alone. Visualizations can also help in identifying any class imbalances, biases, or inconsistencies in the data that can affect the performance of any models built using the data.

There are many tools that can be used to support visualization in analytics, but our focus will be on how to leverage R and Python to complete visualizations.

Visualization in R and Python to Support EDA

Both R and Python are powerful analytical languages that contain several popular visualization libraries. Each of the libraries has similar functionality but with a different focus. The choice of library will depend on the use of the visualization. Table 4-1 outlines the different visualization libraries for Python and when each might be used, and Table 4-2 outlines the libraries for R. This is not an exhaustive list, but I've included the most popular libraries.

Table 4-1. Libraries used for Python visualization

Library	Description	Use case
Matplotlib	A versatile and widely used library that provides a comprehensive range of plotting functions.	• When creating static, 2D, and simple 3D visualizations. • When a library with similar syntax and functionality to MATLAB is preferred. • For extensive customization options for plots. • For integration with other libraries like pandas, NumPy, or Seaborn, as each integrates well with Matplotlib.
Seaborn	A statistical data visualization library built on top of Matplotlib. It provides high-level and aesthetically pleasing plots with minimal coding.	• When creating visually appealing and informative statistical plots. • For a library that integrates well with pandas and Matplotlib. • When a library with a high-level and easy-to-use interface for complex visualizations is required. • When using data in a tidy format (data in a tidy format works seamlessly with Seaborn). Tidy data is organized so that each variable is a column, each observation is a row, and each type of observational unit forms a table.
Plotly	An interactive and web-based visualization library that supports a wide range of chart types.	• When creating interactive and web-based visualizations. • For a variety of chart types, including 3D and geographic plots. • When you have a requirement to create visualizations that can be easily embedded in web applications or dashboards. • When you're already working with Plotly tools, such as Dash for building web applications.
Bokeh	Another library for creating interactive and web-based visualizations, with a focus on providing more control over the look and feel of the plots.	• When creating interactive and web-based visualizations with extensive customization options. • When you have a requirement to build interactive visualizations without requiring JavaScript knowledge. • When you need integration with pandas and other data manipulation libraries. • When you need to use Bokeh's server capabilities for deploying interactive visualizations.
ggplot2	A port of the popular ggplot2 library in R, which allows for creating visually appealing graphics. See the use case in Table 4-2.	

R also has many visualization libraries. ggplot2 is an example of a powerful visualization library and has been ported to Python as referenced in Table 4-1.

Table 4-2. Libraries used for R visualization

Library	Description	Use case
Base R	This graphics package comes with R and provides simple, quick, and easy-to-use plotting functions.	• When you are just getting started with R and want to learn basic plotting functions. • When you require simple and quick visualizations without customizations. • When you do not want to install additional packages.
ggplot2	A powerful and widely used library based on the Grammar of Graphics. It allows for more sophisticated and customizable visualizations.	• When you need to create complex, aesthetically pleasing, and publication-quality plots. • When you need to customize the appearance of your plots using a consistent and modular syntax. • When working with tidy data (data in a tidy format works seamlessly with ggplot2). • When you will use other packages from the "tidyverse" ecosystem, as they integrate well with ggplot2.
lattice	Another popular library for creating complex and customizable visualizations, particularly for multivariate data.	• When you require advanced visualizations for multivariate data. • When you need a grid-based approach to create multiple plots with shared axes. • When you want to use a library based on the Trellis graphics framework. • For working with data in a non-tidy format.
Plotly	An interactive and web-based visualization library, which can create interactive plots using both ggplot2 and base R graphics.	• When you need to create web-based visualizations. • When you need to add interactivity to existing ggplot2 or base R plots. • When you need easy integration with web applications, such as Shiny or R Markdown.
Shiny	Not a visualization library but rather an R package for building interactive web applications. However, it can be used in conjunction with other visualization libraries to create interactive visualizations.	• When you have a requirement for interactive web applications or dashboards using R. • When you require dynamic visualizations that respond to user input. • When you require integration with other R packages and libraries, such as ggplot2 or Plotly.

Many other visualization libraries are available in both Python and R, each with its own strengths and weaknesses, so the choice of library ultimately depends on the specific requirements and goals of the visualization project.

Regression Visualization

Exploring data through visualizations is an important step in understanding relationships and patterns within the data when building regression models. Some visualization techniques that can be used to explore data for regression problems include histograms and density plots. For example, these can be used to check for normality and to identify potential outliers. In this section, I'll walk you through examples that show the visualization libraries being applied for various scenarios in analyzing datasets for regression.

For the examples in this section, we will use the mtcars, a well-known dataset in the R programming environment, often used for teaching and demonstration purposes in statistical analysis and data visualization. It contains data on 32 different car models from the 1974 *Motor Trend* US magazine. The dataset includes 11 variables related to various aspects of automobile performance and design, such as miles per gallon (`mpg`), number of cylinders (`cyl`), displacement (`disp`), horsepower (`hp`), rear axle ratio (`drat`), weight (`wt`), quarter mile time (`qsec`), engine type (`vs`), transmission type (`am`), number of forward gears (`gear`), and number of carburetors (`carb`). These variables allow for extensive analysis of relationships between car characteristics, such as the correlation between engine size and fuel efficiency or the impact of weight on performance metrics. The mtcars dataset is particularly useful for exploring regression models, creating visualizations, and practicing data manipulation techniques.

Scatter plots

Scatter plots show the relationship between two continuous variables. They can be used to identify patterns and relationships between the independent and dependent variables. Figure 4-1 shows an example scatter plot in R:

```
library(ggplot2)
ggplot(data = mtcars, aes(x = mpg, y = disp)) +
  geom_point()
```

Figure 4-2 shows a scatter plot in Python:

```
import statsmodels.api as sm
mtcars = sm.datasets.get_rdataset('mtcars')
mtcars = mtcars.data
import matplotlib.pyplot as plt
import seaborn as sns
sns.scatterplot(x = 'mpg', y = 'disp', data = mtcars)
plt.show()
```

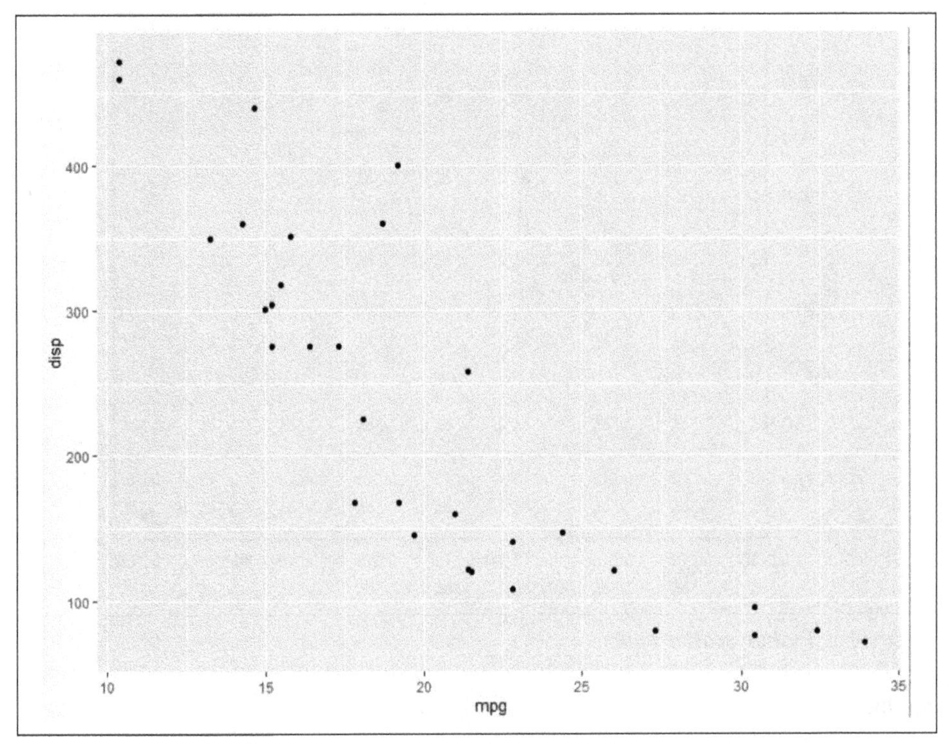

Figure 4-1. R scatter plot

Both figures feature box plots that show the distribution of a continuous variable and can be used to identify potential outliers. Figures 4-1 and 4-2 show a scatter plot that displays the relationship between miles per gallon (`mpg`) and displacement (`disp`) for the cars in the mtcars dataset. The plot typically shows a negative correlation between `mpg` and `disp`, meaning that as the displacement of a car's engine increases, the miles per gallon tend to decrease. This indicates that cars with larger engines (higher displacement) are generally less fuel-efficient. This visualization helps in understanding how engine size (displacement) impacts fuel efficiency (`mpg`) in the dataset.

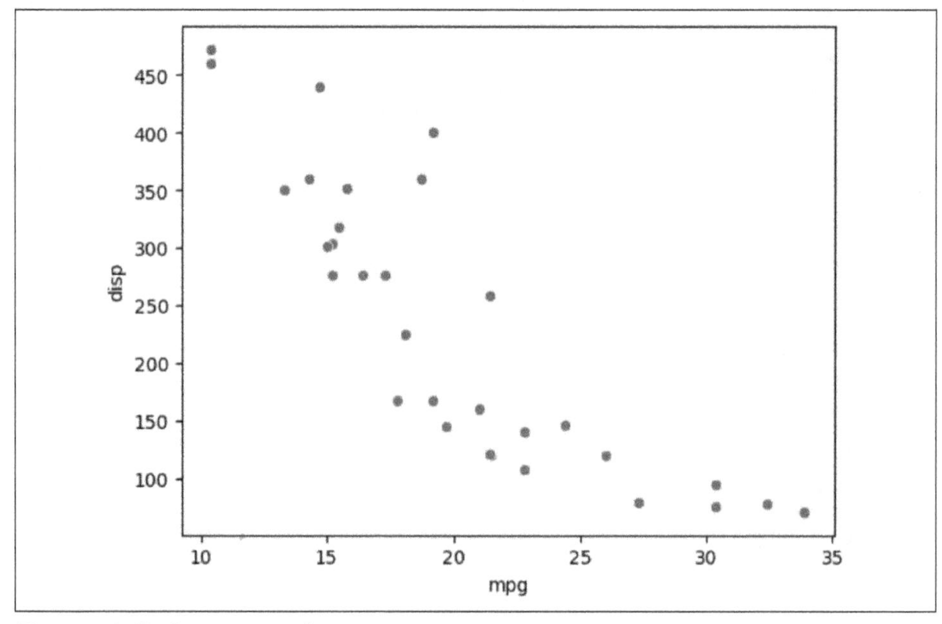

Figure 4-2. Python scatter plot

Box plots

Figure 4-3 shows an example box plot in R:

```
ggplot(data = mtcars, aes(x = factor(cyl), y = disp)) +
  geom_boxplot()
```

Figure 4-4 shows an example box plot in Python:

```
sns.boxplot(x = 'cyl', y = 'disp', data = mtcars)
plt.show()
```

A box plot is a standardized way of displaying the distribution of data based on a five-number summary: minimum, first quartile (Q1), median, third quartile (Q3), and maximum. Its purpose is to provide a visual summary of key aspects of the data distribution.

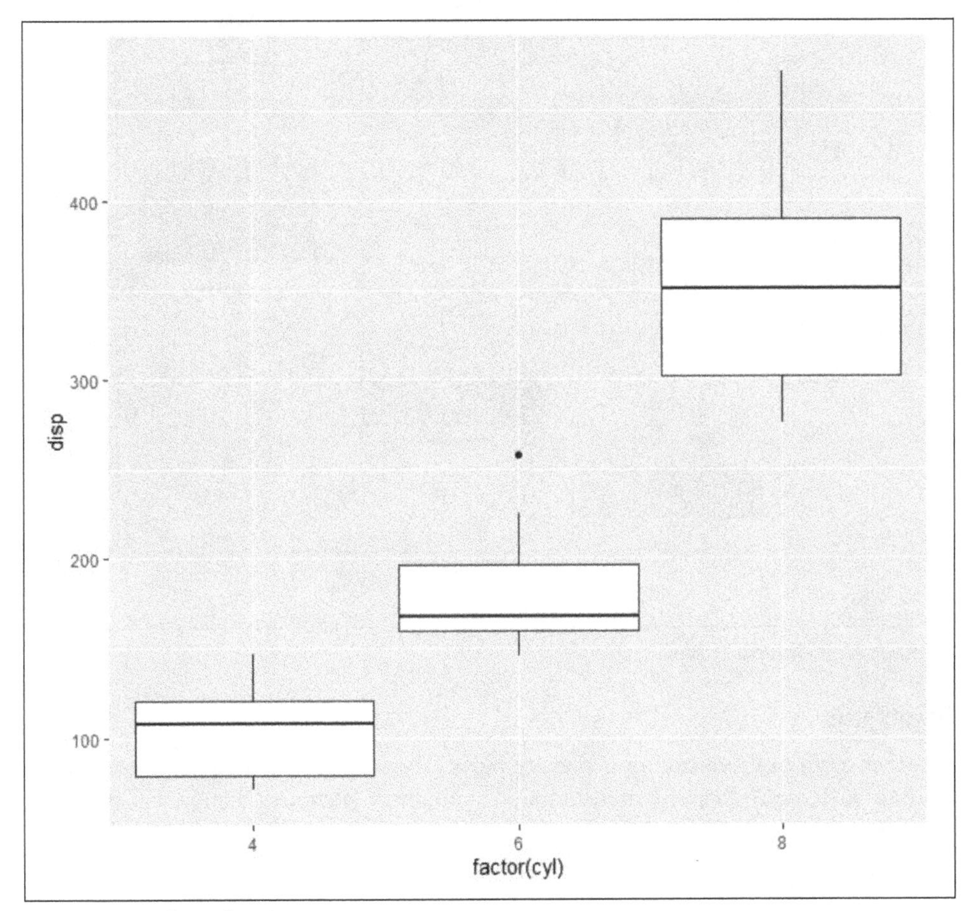

Figure 4-3. R box plot

Both figures show a boxplot that displays the distribution of engine displacement (`disp`) across different categories of cylinders (`cyl`) in the mtcars dataset. The boxplot shows how engine displacement varies for cars with different numbers of cylinders (4, 6, and 8 cylinders). Each box represents the interquartile range (IQR) of displacement for that cylinder group, with the line inside the box indicating the median displacement. The plot typically shows that cars with more cylinders (e.g., 8 cylinders) tend to have higher engine displacements compared to cars with fewer cylinders (e.g., 4 cylinders). This is visible through the shift in the median and the overall higher range of values for higher cylinder counts.

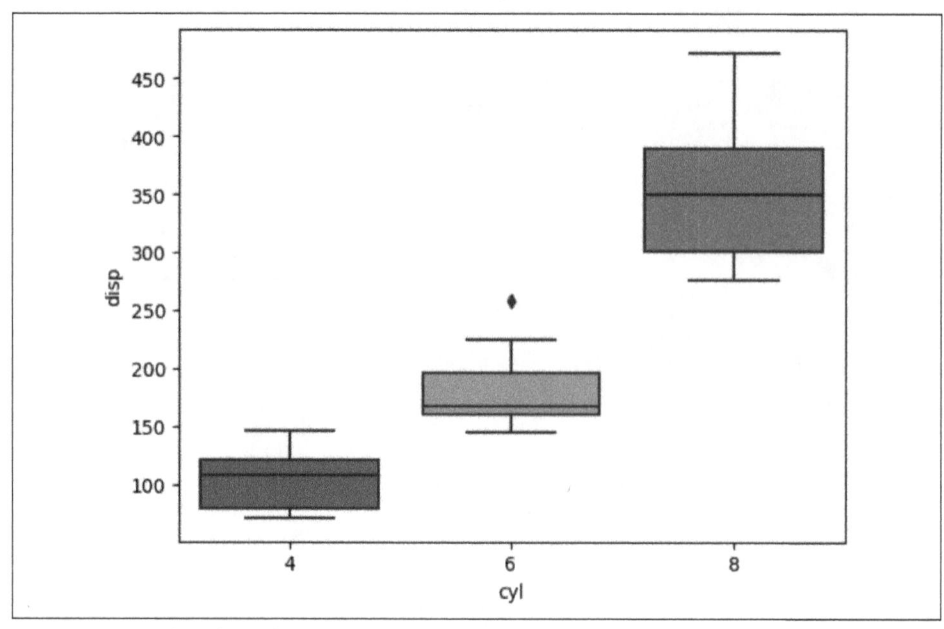

Figure 4-4. Python box plot

Density plots

Another form of visualization is density plots. These show the distribution of a continuous variable and can be used to identify potential skewness. Figure 4-5 shows an example of density plots in R:

```
ggplot(data = mtcars, aes(x = mpg)) +
  geom_density()
```

Figure 4-6 shows an example density plot in Python:

```
sns.kdeplot(x = 'mpg', data = mtcars)
plt.show()
```

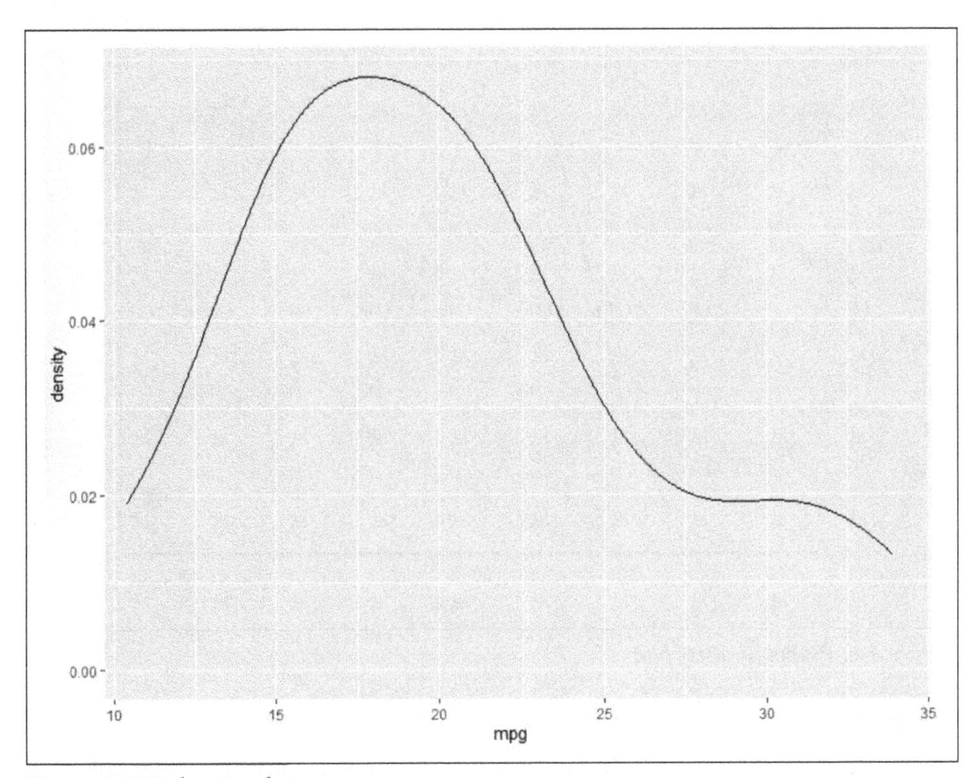

Figure 4-5. R density plot

The density plots shown in Figures 4-5 and 4-6 visualize the distribution of miles per gallon (`mpg`) values in the mtcars dataset. It highlights the most common mpg ranges, with the peak of the curve indicating where the majority of the cars' fuel efficiencies lie. The plot also shows the spread of mpg values and any skewness, helping to understand the variability in fuel efficiency across different car models. This visualization is significant for quickly assessing the overall distribution of fuel efficiency in the dataset.

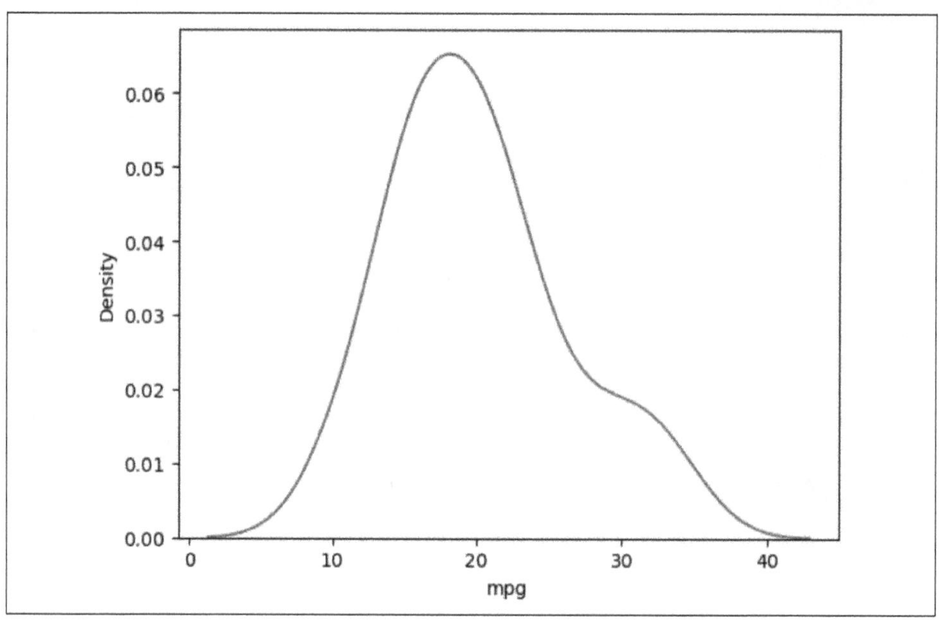

Figure 4-6. Python density plot

Heatmaps

Heatmaps show the relationship between two continuous variables using colors to represent the magnitude of the relationship. Figure 4-7 shows an example of heatmaps in R:

```
ggplot(data = mtcars, aes(x = hp, y = disp)) +
  geom_bin2d() +
  scale_fill_gradient(low = "white", high = "blue")
```

Figure 4-8 shows an example heatmap in Python:

```
sns.histplot(x = 'hp', y = 'disp', data = mtcars, cmap = 'Blues')
plt.show()
```

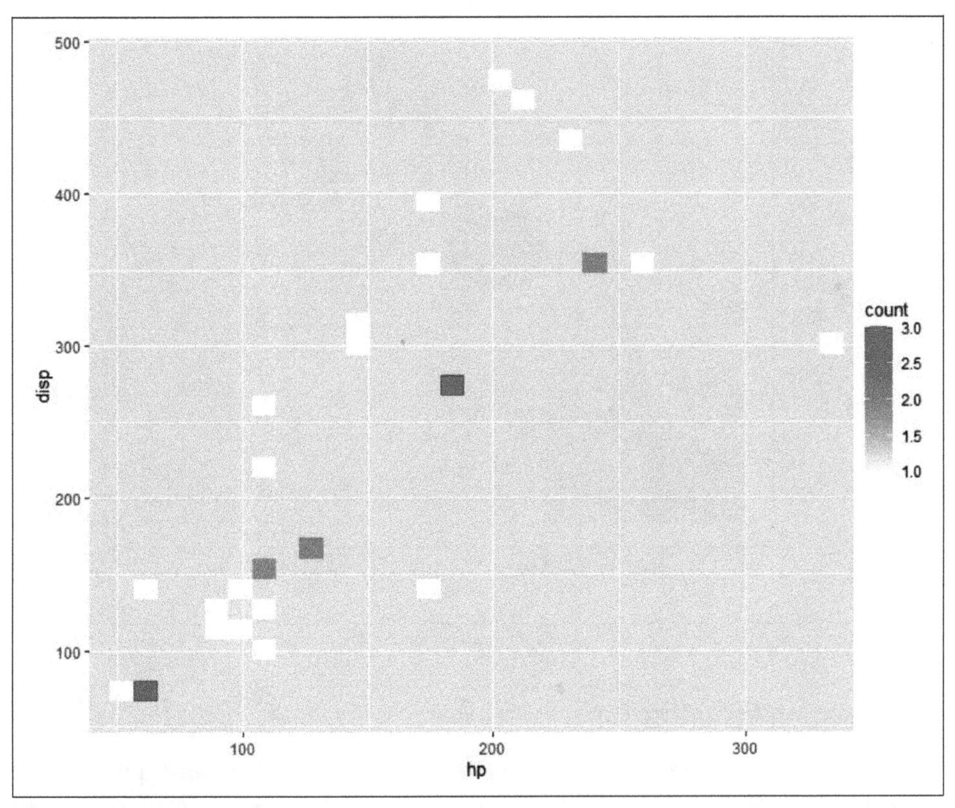

Figure 4-7. R heat map

These visualization techniques can help identify patterns, trends, and outliers in the data, which can guide feature selection and engineering efforts to prepare the data for regression modeling. Both figures show the relationship between horsepower (hp) and engine displacement (disp) in the mtcars dataset. The data points are grouped into bins, with the color intensity (from white to blue) indicating the density of points within each bin. Darker blue areas represent higher concentrations of cars with similar horsepower and displacement values. This visualization is useful for identifying patterns and clusters in the data, such as regions where cars tend to have similar performance characteristics.

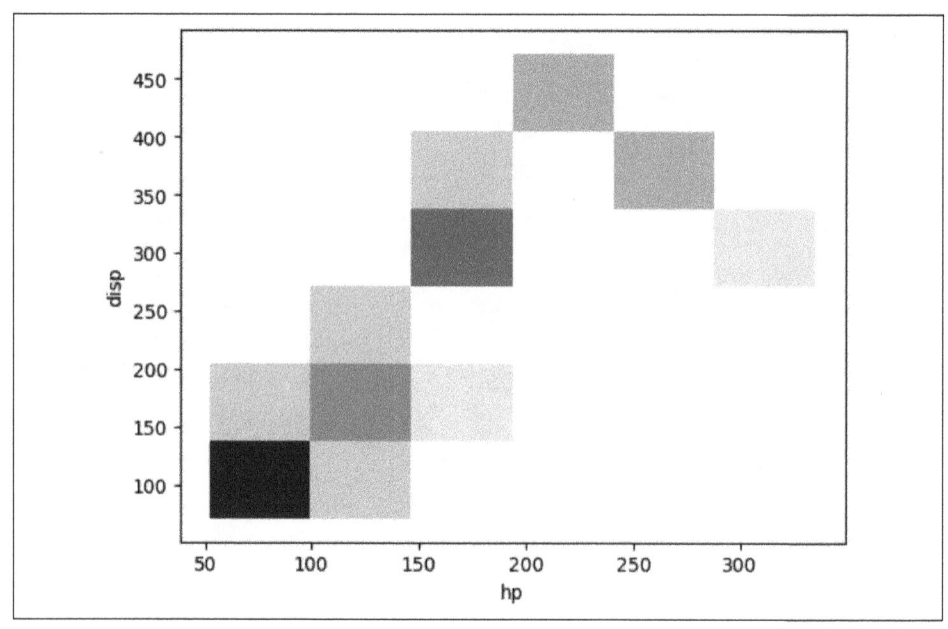

Figure 4-8. Python heat map

Classification Visualization

Visualization for classification problems is completed for the same purposes as regression in EDA. One visualization technique that can be used to explore data for classification problems is examining the distribution of categorical variables. Let's consider the visualization techniques that can be applied here.

For these examples, we will use the Iris dataset. The Iris dataset is a classic dataset in machine learning and statistics, often used for benchmarking algorithms and exploring data analysis techniques. It contains 150 observations of iris flowers, with each observation belonging to one of three species: Iris setosa, Iris versicolor, and Iris virginica. The dataset includes four numerical features: sepal length, sepal width, petal length, and petal width, all measured in centimeters. These features describe the physical dimensions of the iris flowers, and the dataset is commonly used for tasks such as classification, clustering, and visualization to differentiate between the three species based on these measurements.

Bar plots

Bar plots show the distribution of categorical variables and can be used to identify potential class imbalances. Figure 4-9 shows an example of bar plots in R:

```
library(ggplot2)
library(dplyr)
```

```
# Create a new categorical variable based on Petal.Length
iris <- iris %>%
  mutate(PetalLengthCategory = cut(Petal.Length,
                                  breaks = c(0, 2, 4, 6, Inf),
                                  labels = c("Short", "Medium", "Long", "Very
                                                                   Long")))

# Create the bar chart
ggplot(data = iris, aes(x = Species, fill = PetalLengthCategory)) +
  geom_bar(position = "dodge") +
  labs(title = "Distribution of Petal Length Categories by Species",
       x = "Species",
       y = "Count",
       fill = "Petal Length") +
  theme_minimal()
```

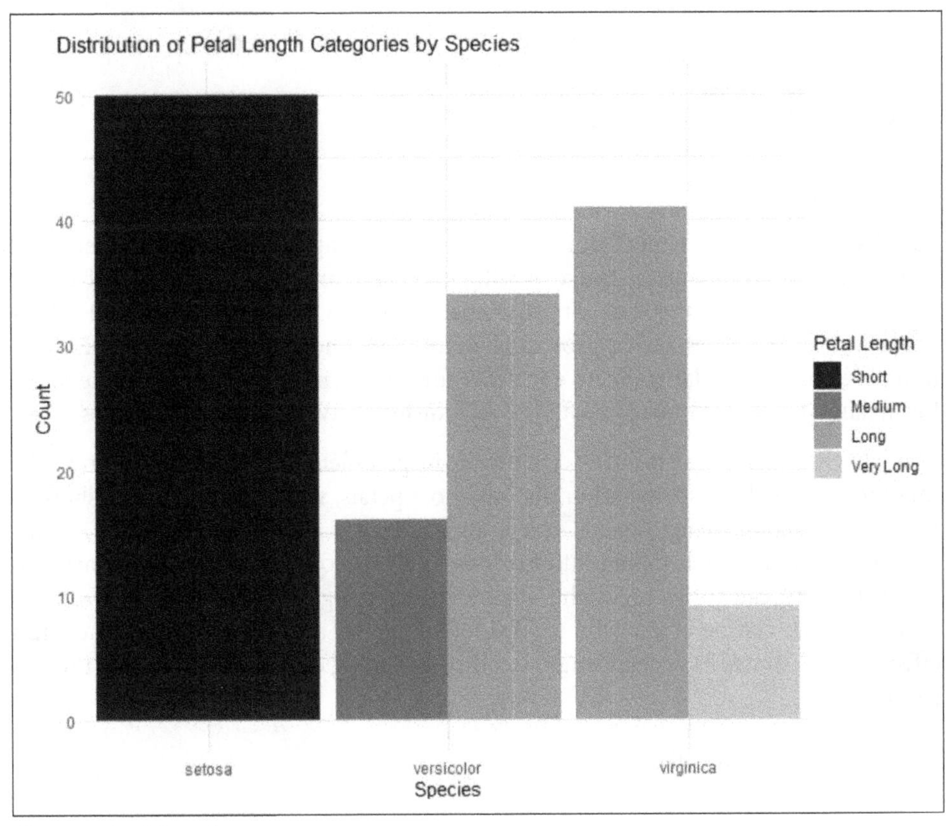

Figure 4-9. R bar chart

Figure 4-10 shows a bar chart in Python:

```
import pandas as pd
import seaborn as sns
import matplotlib.pyplot as plt
```

```
# Load the iris dataset
iris = sns.load_dataset('iris')

# Create a new categorical variable based on Petal.Length
bins = [0, 2, 4, 6, float('inf')]
labels = ['Short', 'Medium', 'Long', 'Very Long']
iris['PetalLengthCategory'] = pd.cut(iris['petal_length'], bins=bins,
labels=labels)

# Create the bar chart
plt.figure(figsize=(10, 6))
sns.countplot(data=iris, x='species', hue='PetalLengthCategory',
palette='viridis')

# Add labels and title
plt.title('Distribution of Petal Length Categories by Species')
plt.xlabel('Species')
plt.ylabel('Count')
plt.legend(title='Petal Length')

# Show the plot
plt.show()
```

Bar charts are a fundamental tool in EDA for visualizing and comparing categorical data. Each bar represents a category with its height or length proportional to the value or frequency of that category. Bar charts excel in revealing trends and differences among categories and in providing an intuitive understanding of data distribution. They are particularly effective in EDA for identifying patterns and outliers and for making initial assessments about the relationship between discrete variables.

The bar chart reveals significant differences in petal length distribution among the three iris species. Iris setosa primarily has short petals, while Iris versicolor shows a majority of medium-length petals. Iris virginica displays a wider range, with a significant portion of its petals falling into the long and very long categories. This variation in petal length categories highlights distinct morphological differences between the species, which can be crucial for species identification and classification. The chart effectively illustrates how petal length is a distinguishing characteristic among the iris species.

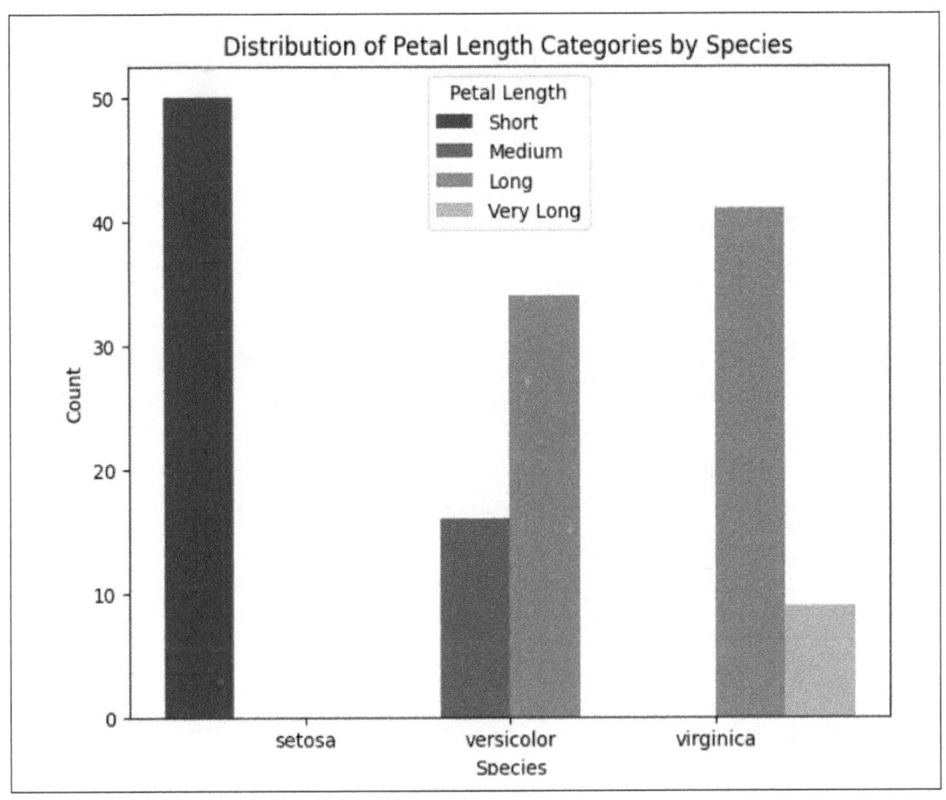

Figure 4-10. Python bar chart

Parallel coordinates plot

A parallel coordinate plot is used to visualize multivariate data by plotting each feature on a separate axis and connecting the data points across axes. This chart helps to identify how different features contribute to the classification of each instance. Figure 4-11 shows an example of a parallel coordinate plot in R:

```
install.packages("GGally")
# Load required libraries
library(ggplot2)
library(GGally)

# Load the iris dataset
data(iris)

# Create a parallel coordinates plot
ggparcoord(data = iris, columns = 1:4, groupColumn = 5, scale = "uniminmax") +
  theme_minimal() +
  labs(title = "Parallel Coordinates Plot for Iris Dataset",
       x = "Features",
       y = "Scaled Values")
```

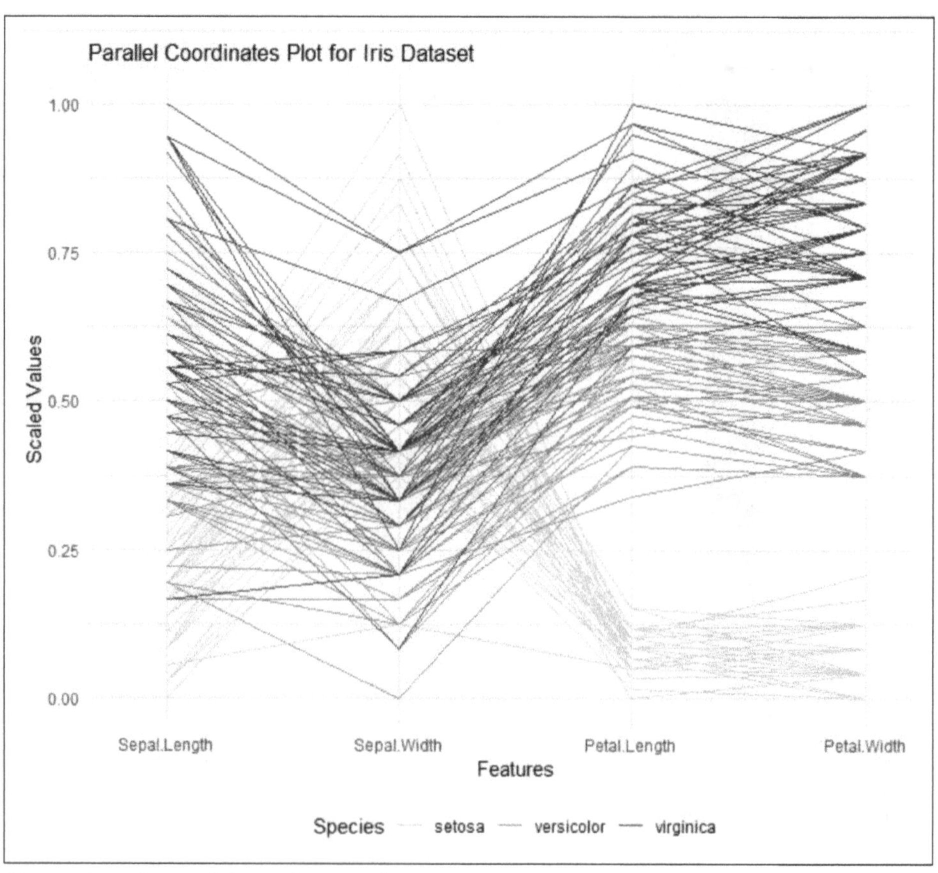

Figure 4-11. R parallel coordinates plot

Figure 4-12 shows a parallel coordinate plot in Python:

```
import pandas as pd
import seaborn as sns
import matplotlib.pyplot as plt
from pandas.plotting import parallel_coordinates

# Load the iris dataset
iris = sns.load_dataset("iris")

# Create a parallel coordinates plot
plt.figure(figsize=(10, 5))
parallel_coordinates(iris, 'species', colormap=plt.get_cmap("Set1"))

# Set the title and labels
plt.title("Parallel Coordinates Plot for Iris Dataset")
plt.xlabel("Features")
plt.ylabel("Values")
```

```
# Show the plot
plt.show()
```

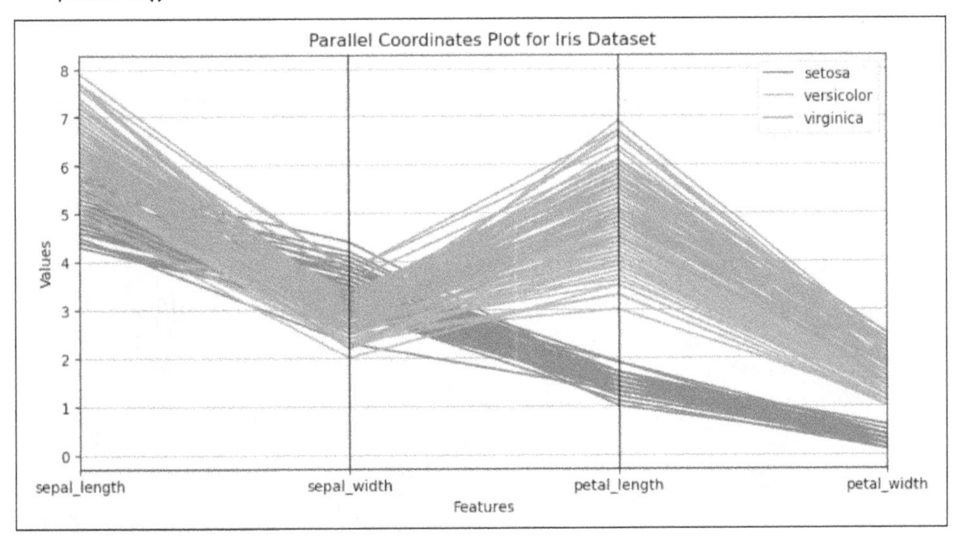

Figure 4-12. Python parallel coordinates plot

The parallel coordinates plot provides a comprehensive view of how the different features (sepal length, sepal width, petal length, and petal width) vary across the three iris species. Each line represents an individual observation, and the plot visually highlights the distinct patterns among species. For example, Iris setosa exhibits clear separation from the other two species, particularly in petal length and petal width, indicating these features are key differentiators. Iris versicolor and Iris virginica show more overlap, but differences in petal measurements still allow for distinction. This plot effectively demonstrates the multidimensional relationships between features and their role in distinguishing iris species.

Violin plots

Violin plots display the distribution of a feature across different target classes. Violin plots combine aspects of box plots and kernel density plots, providing additional insights into the data distribution. Figure 4-13 shows an example violin plot in R:

```
# Load required libraries
library(ggplot2)

# Load the iris dataset
data(iris)

# Create a violin plot
p <- ggplot(iris, aes(x = Species, y = Sepal.Length)) +
  geom_violin(fill = "lightblue", draw_quantiles = c(0.25, 0.5, 0.75)) +
  labs(title = "Violin Plot for Sepal Length by Iris Species",
```

```
       x = "Species",
       y = "Sepal Length")

# Show the plot
print(p)
```

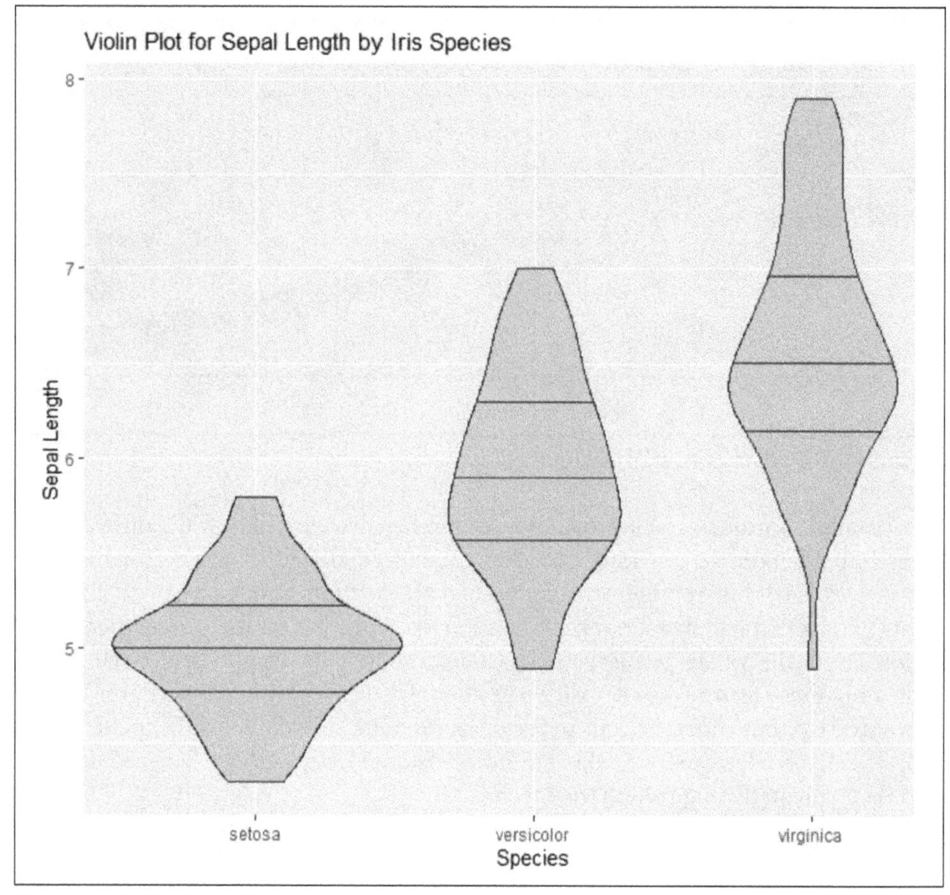

Figure 4-13. R violin plot

Figure 4-14 shows an example violin plot in Python:

```
import seaborn as sns
import matplotlib.pyplot as plt

# Load the iris dataset
iris = sns.load_dataset("iris")

# Create a violin plot
sns.violinplot(data=iris, x='species', y='sepal_length')

# Set the title and labels
```

```
plt.title("Violin Plot for Sepal Length by Iris Species")
plt.xlabel("Species")
plt.ylabel("Sepal Length")

# Show the plot
plt.show()
```

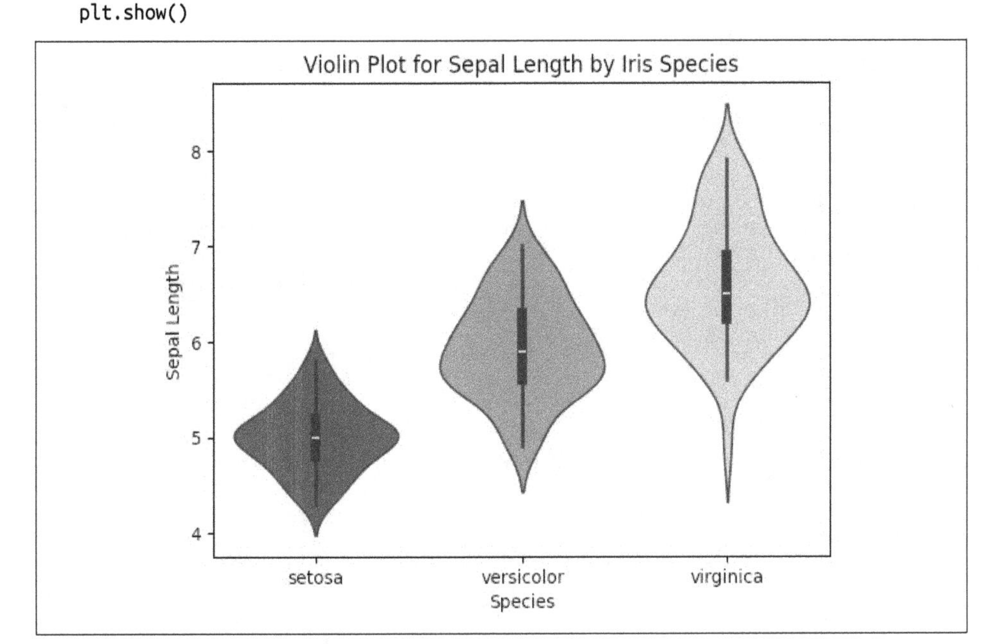

Figure 4-14. Python violin plot

Violin plots are valuable in EDA for visualizing the distribution of numeric data across different categories. They combine elements of box plots and density plots, showing the median, interquartile range, and density of the data, thereby offering a deeper insight into the data distribution than traditional box plots. Violin plots are particularly useful in comparing multiple distributions and highlighting differences in both the central tendency and variability of data across categories. Violin plots are also adept at revealing multimodal distributions.

The violin plot provides a detailed visualization of the distribution of sepal length across the three iris species. It reveals not only the central tendency but also the variability and distribution shape of sepal length within each species. Iris setosa shows a relatively tight distribution with a higher concentration of lower sepal lengths, while Iris virginica has a broader distribution with higher sepal lengths. Iris versicolor's distribution lies in between, with a moderate spread. This plot highlights the differences in sepal length among the species, making it clear that sepal length is a distinguishing feature, particularly between Iris setosa and the other two species.

Contour plots

Contour plots (or 2D density plots) help visualize the joint distribution of two continuous features. This can help you identify clusters, trends, or patterns in the data. Figure 4-15 shows an example contour plot in R:

```
# Load required libraries
library(ggplot2)

# Load the iris dataset
data(iris)

# Create a contour plot
p <- ggplot(iris, aes(x = Sepal.Length, y = Sepal.Width, color = Species)) +
  geom_density_2d() +
  labs(title = "Contour Plot for Sepal Length vs. Sepal Width by Iris Species",
       x = "Sepal Length",
       y = "Sepal Width")

# Show the plot
print(p)
```

Figure 4-16 shows an example contour plot in Python:

```
import seaborn as sns
import matplotlib.pyplot as plt

# Load the iris dataset
iris = sns.load_dataset("iris")

# Create a contour plot
sns.kdeplot(data=iris, x="sepal_length", y="sepal_width", hue="species",
cmap="viridis")

# Set the title and labels
plt.title("Contour Plot for Sepal Length vs. Sepal Width by Iris Species")
plt.xlabel("Sepal Length")
plt.ylabel("Sepal Width")

# Show the plot
plt.show()
```

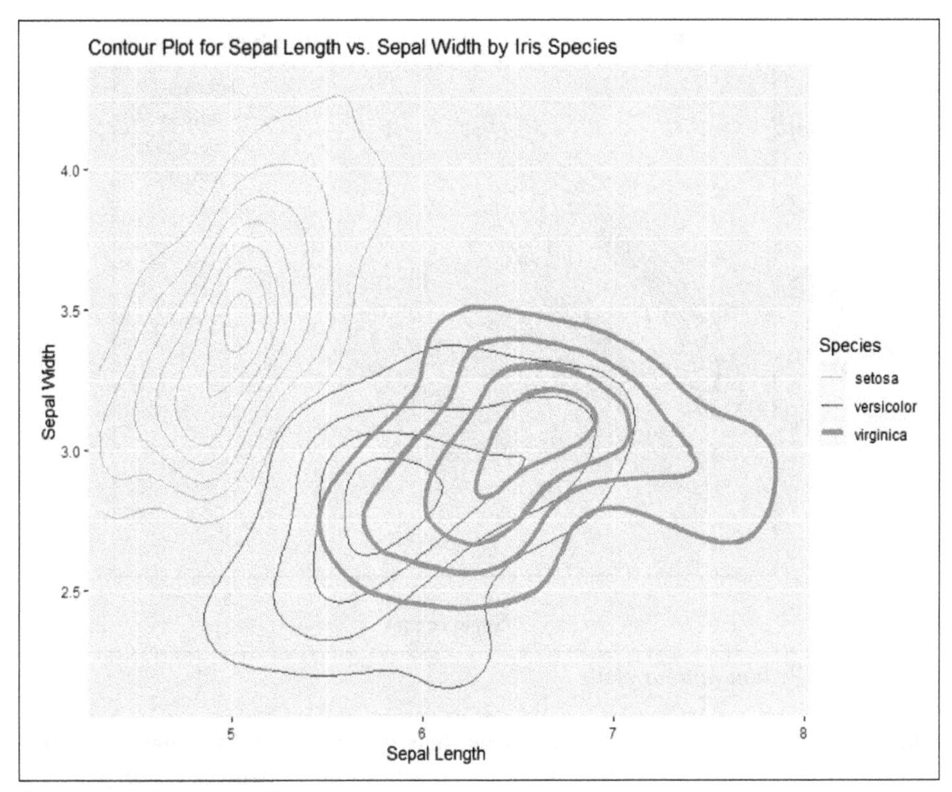

Figure 4-15. R contour plot

The contour plot visualizes the relationship between sepal length and sepal width across the three iris species, highlighting regions of high data density. The contours indicate where the data points for each species are most concentrated, with distinct patterns emerging for each species. Iris setosa is well-separated from the other two species, clustering in a region with shorter sepal lengths and higher sepal widths. Iris versicolor and Iris virginica show some overlap, but Iris virginica generally occupies areas with longer sepal lengths. This plot effectively demonstrates the variation and clustering of sepal dimensions across species, aiding in understanding how these features differentiate the iris species.

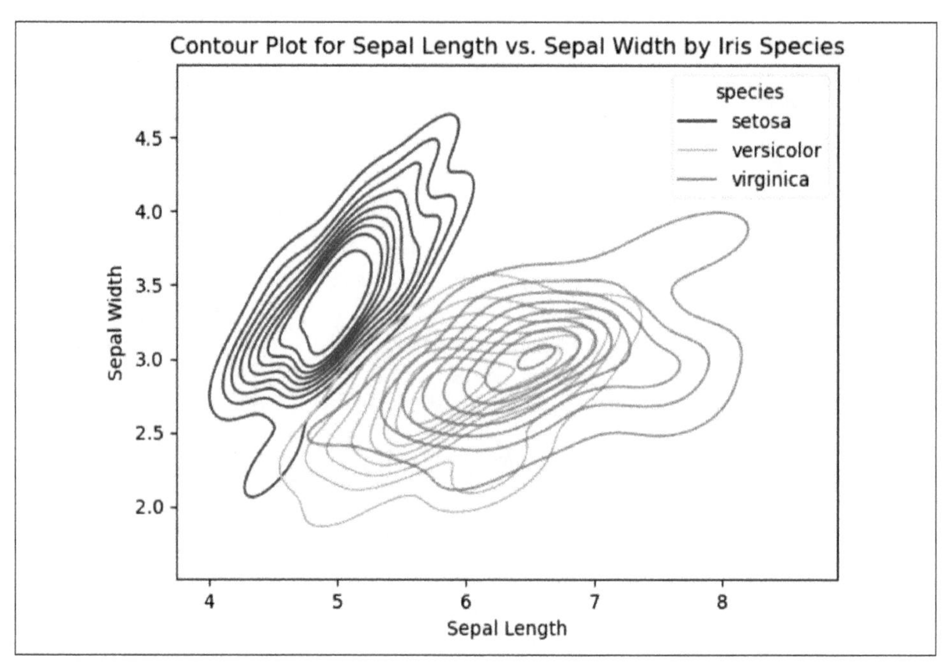

Figure 4-16. Python contour plot

These different visualization techniques can help analyze feature imbalance and feature contribution to create more accurate classification models. Contour plots are essential in EDA for visualizing three-dimensional data in two dimensions. They represent data points on a grid, using contour lines (or level curves) to show areas of similar values. These plots are particularly useful for identifying patterns, trends, and gradients in the data, as well as potential outliers. Contour plots excel in displaying the topography of a surface, making them invaluable for analyzing geographical data, heatmaps, and various scientific and engineering applications.

Summary

This chapter introduced you to different types of analytical projects. Understanding the problem to be solved is the start to understanding how an analytical project will be approached. For example, this impacts the approach to analysis and determines whether the project is a regression or classification project. You should now have an understanding of different considerations for how EDA should be approached and reviewed and be able to tackle visualization in R and Python to support EDA. In the next chapter, we dive deeper into EDA.

CHAPTER 5

Exploratory Data Analysis with R and Python

Exploratory data analysis is an important preparation step that influences all subsequent steps in the business analytics cycle and ensures that models are built on a solid foundation of well-understood, appropriately processed data. EDA is used to understand the characteristics of data, identify errors and inconsistencies, uncover relationships between features, validate assumptions, and make informed decisions about which models may perform the best. This chapter will explore each of the major steps involved in EDA.

John Tukey, a pioneering statistician, played a crucial role in the development of EDA, emphasizing the importance of using visual methods to understand data. Tukey advocated for EDA as a way to uncover underlying patterns, spot anomalies, and test assumptions before applying formal statistical models. His approach encourages analysts to interact with data through visualization and summary statistics to gain insights and intuition about its structure and relationships. Tukey's work laid the foundation for modern data analysis, highlighting the value of EDA in the initial stages of data understanding. The first step in the EDA process is exploring the quality of the data to be used in the analytics project, so we'll start the chapter with this topic.

Data Quality

Data quality refers to the condition of a dataset. It's particularly important in EDA because the value of the different data points depends on the quality of the data content.

Data Quality Characteristics

Assessing data quality is an important aspect of EDA for a number of reasons. First, it influences the accuracy of results. Low-quality data can lead to incorrect or misleading results, even with sophisticated models and algorithms. Next, it is important to maximize the use of resources because cleaning and managing poor-quality data can be time-consuming and costly. Most importantly, the performance of machine learning models heavily depends on the quality of the training data. Inaccurate or inconsistent data can lead to poor model performance.

High-quality data typically has the following characteristics: accuracy, completeness, consistency, reliability, and relevance. It is free of duplicate entries, irrelevant information, inaccuracies, and missing values. Timeliness, which means the data is up-to-date and therefore still useful, is also an important variable.

The following list defines each of these characteristics:

Completeness
: This checks if there are any missing values in the dataset. Missing data can lead to inaccurate machine learning models.

Consistency
: Consistency refers to the uniformity of data across the dataset. Data inconsistencies, such as variations in the format of data or redundant entries, can distort results.

Accuracy
: Accuracy checks ensure the data correctly represents the real-world construct it is supposed to model. These checks involve identifying and correcting errors in the data.

Relevance
: This involves assessing if the data is suitable and enough to address the problem statement or if irrelevant data columns need to be dropped.

Uniqueness
: Checking for uniqueness involves ensuring that there are no duplicate records in the dataset, as these can affect the performance of some machine learning algorithms.

Validity
: This checks whether the data follows specified formats and conventions. For instance, negative values for an age column would be considered invalid.

Timeliness
: In some cases, it's important to consider whether the data is still relevant or up-to-date enough for the problem at hand.

Balanced data

Particularly for classification problems, it's crucial to check whether the classes are balanced or imbalanced in the dataset. Imbalanced classes can lead to biased machine learning models. Balanced data in classification refers to a dataset where the classes (or categories) have roughly equal numbers of observations. This is important because, in a balanced dataset, the classification model is less likely to be biased toward the majority class, leading to more accurate and fair predictions across all classes.

By analyzing these characteristics, it is possible to improve the quality of your data and, in turn, the efficacy of machine learning models. In order to assess data quality, analysts often perform data profiling, which we'll explore next.

Data Profiling

Data profiling in the context of EDA is the process of examining, describing, and summarizing a dataset to better understand its structure, content, and quality. The process of data profiling typically involves a combination of various methods. The first step focuses on variable identification. Descriptive statistics is often the next step, especially if the variable is numeric. This includes calculating measures like mean, median, mode, range, variance, and standard deviation. This provides a summary of continuous variables and helps to understand the distribution of data. Other analysis that can be completed with profiling includes:

Variable type identification

This involves identifying the types of variables in a dataset (e.g., numerical, categorical, ordinal).

Checking for missing values

Data profiling helps to uncover fields that may have missing or null values, which are important to address before moving on to data analysis.

Cardinality check

This involves understanding the number of unique values each variable has. High cardinality in certain variables might affect the performance of some models.

Value frequency

This involves examining the most frequent and least frequent values in the dataset.

If you aren't performing these tasks in R or Python, you are likely using tools like Excel or BI platforms such as Tableau or Power BI. While these tools are user-friendly and familiar, it's important to highlight that many tasks can actually be easier and more efficient in R or Python, despite the common perception that coding is more difficult. For example, R and Python offer powerful libraries for data manipulation, visualization, and automation, allowing for more complex analyses with less repetitive effort. Additionally, these programming languages enable reproducibility and scalability, which can be challenging to achieve in Excel or BI tools. By leveraging their existing knowledge of Excel or BI platforms, readers can smoothly transition to R or Python, building on their understanding of data analysis concepts while gaining the benefits of more advanced and versatile tools.

In both R and Python, business analysts can conduct data profiling using additional libraries like DataExplorer in R and pandas_profiling in Python, which generate comprehensive reports with a single line of code. When these libraries are used for data profiling, outputs from these libraries include displaying the first few rows, checking the data structure or information, and producing basic summary statistics.

Here is an example in R using the DataExplorer library:

```
# Install and load the DataExplorer package
install.packages("DataExplorer")
library(DataExplorer)

# Load built-in mtcars dataset
data(mtcars)

# Perform detailed data profiling
create_report(mtcars)
```

Once run, this code provides a detailed HTML report that contains a table of contents and all of the output contents of the R data profiling report as outlined in the following list. The output is comprehensive and includes summary statistics, data structure analysis, and missing data as examples. Figures 5-1 through 5-8 show the data profiling results for the mtcars dataset.

The report generated by the DataExplorer package on the built-in mtcars dataset provides a comprehensive overview of the dataset's structure, distributions, and relationships among variables. The dataset consists of 32 observations on 11 variables, including information on miles per gallon (mpg), cylinder count (cyl), horsepower (hp), and weight (wt), among others.

Significant findings from the report include the identification of key relationships between variables. For example, there is a noticeable negative correlation between mpg and hp, suggesting that cars with higher horsepower tend to have lower fuel efficiency. Similarly, a strong positive correlation exists between wt and hp, indicating that heavier cars generally have more horsepower. The report also highlights the distribution of each variable, with mpg and hp showing substantial variability across the dataset, while variables like am (transmission type) are categorical with clear separation between automatic and manual transmissions. Overall, the report offers valuable insights into the data, aiding in further analysis and decision making.

Data Profiling Report

- Basic Statistics
  - Raw Counts
  - Percentages
- Data Structure
- Missing Data Profile
- Univariate Distribution
  - Histogram
  - QQ Plot
- Correlation Analysis
- Principal Component Analysis

Basic Statistics

Raw Counts

Name	Value
Rows	32
Columns	11
Discrete columns	0
Continuous columns	11
All missing columns	0
Missing observations	0
Complete Rows	32
Total observations	352
Memory allocation	5.8 Kb

Figure 5-1. Basic statistics from profiling

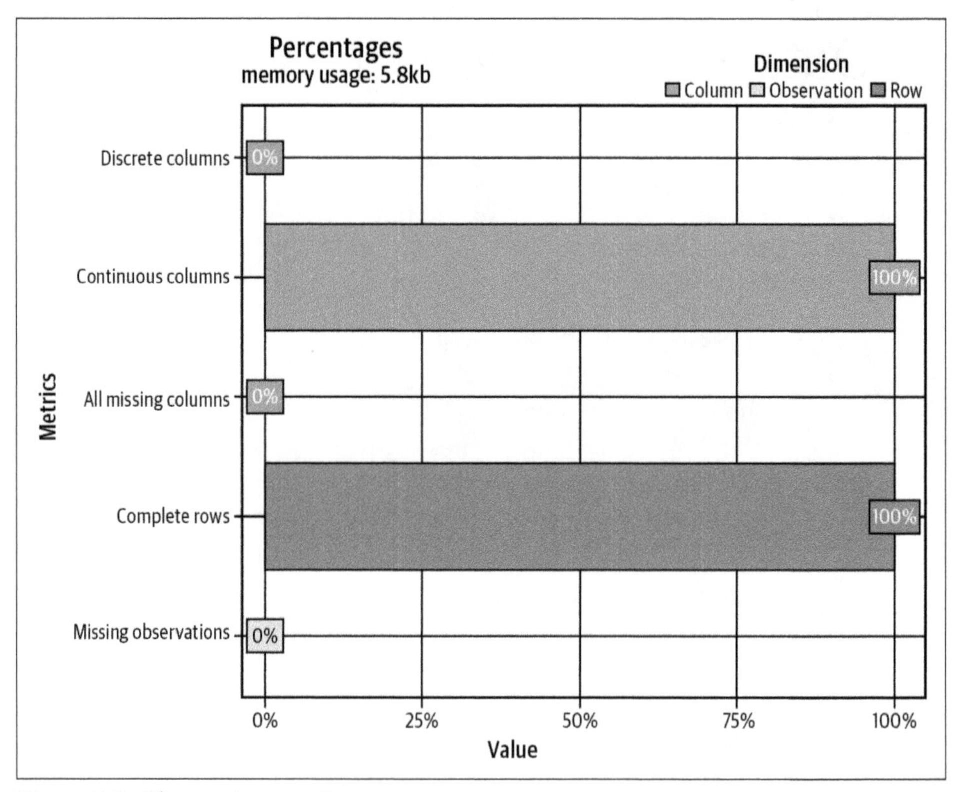

Figure 5-2. Observation metrics

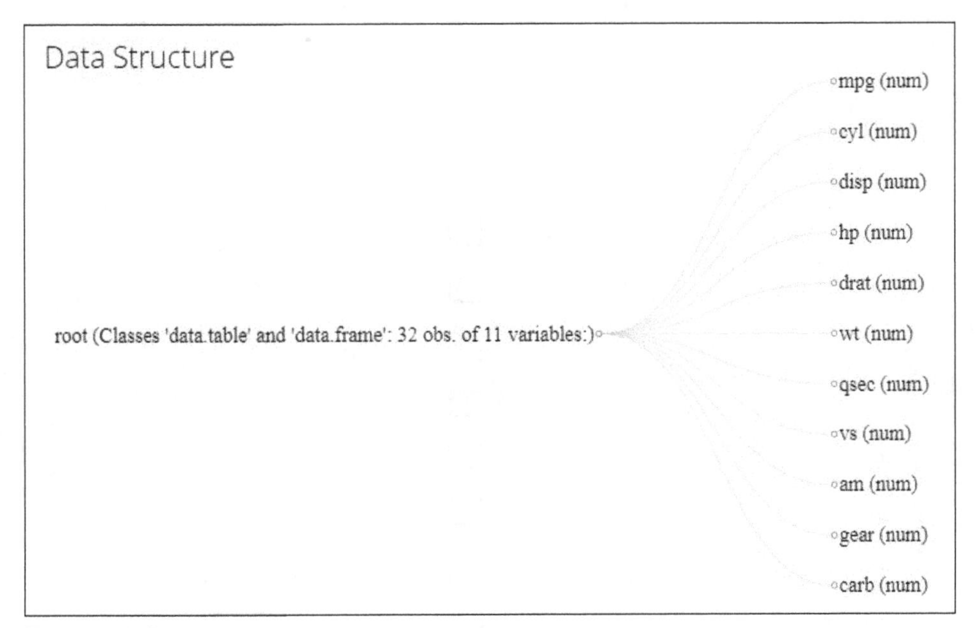

Figure 5-3. Data types

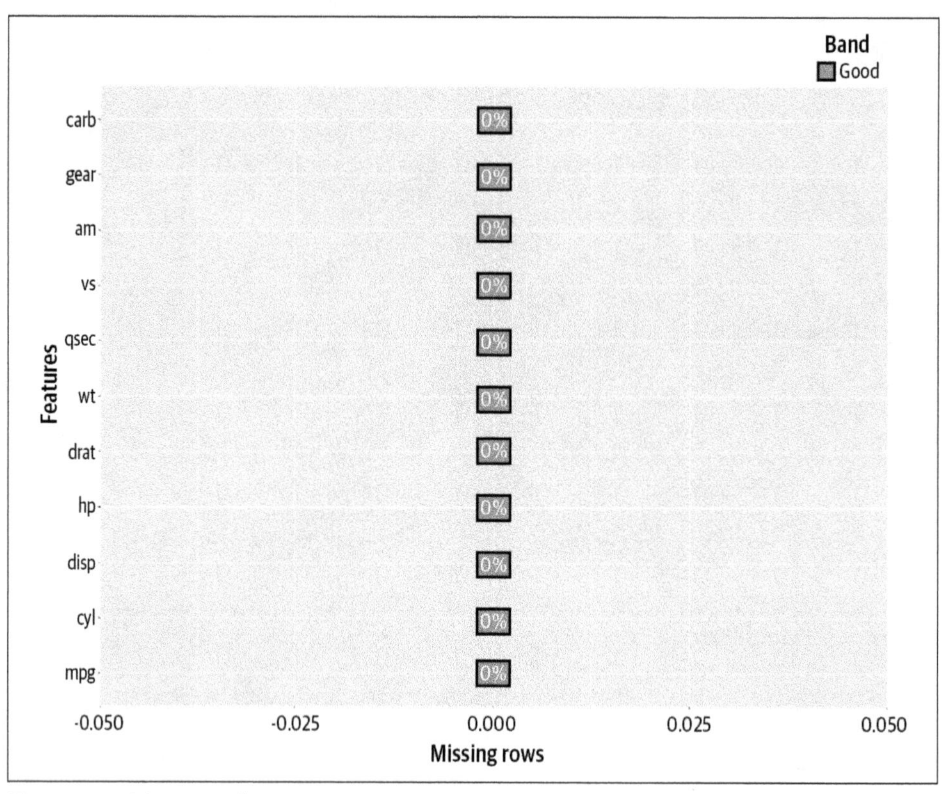

Figure 5-4. Missing values

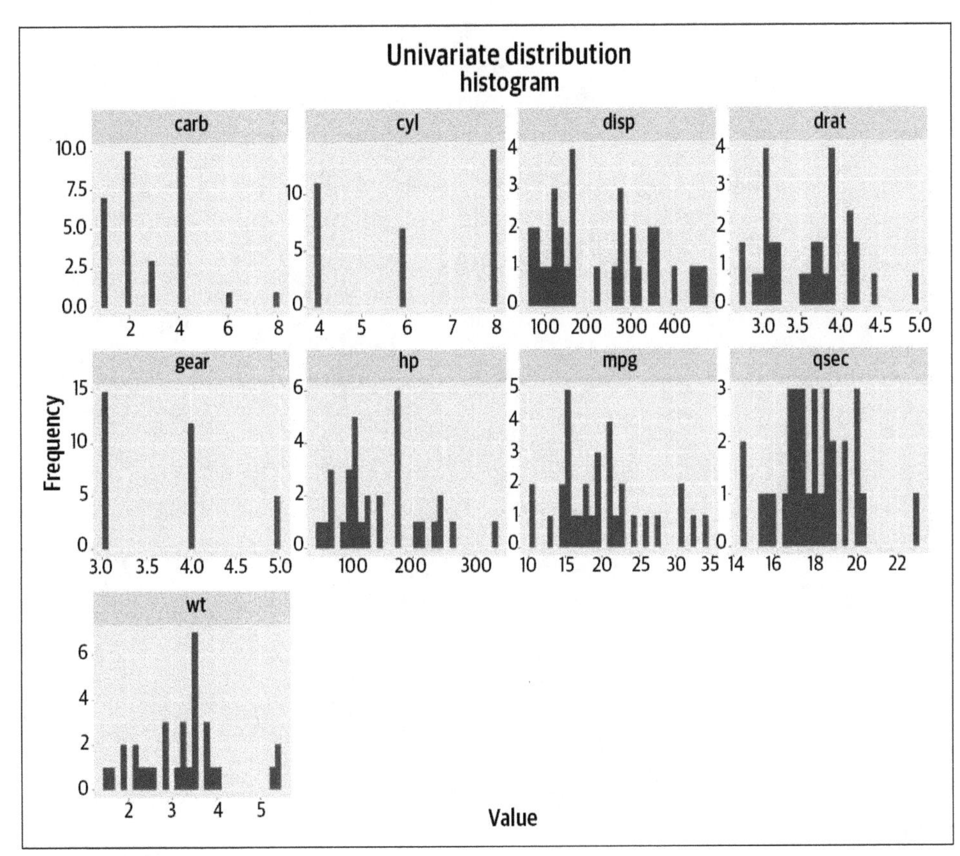

Figure 5-5. Variable distribution

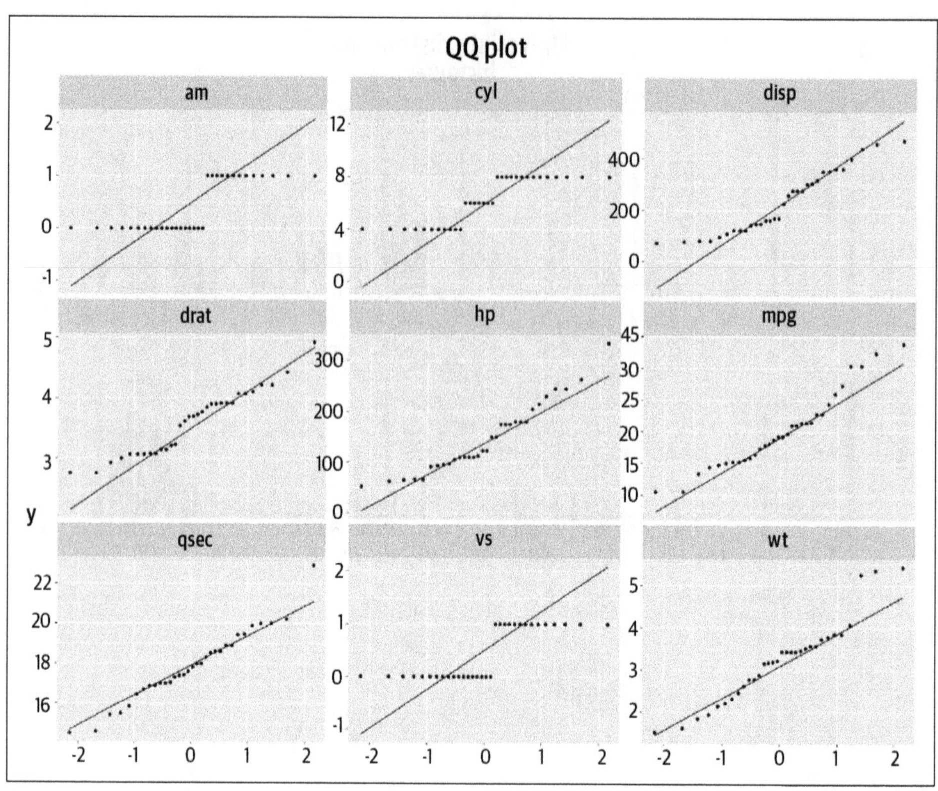

Figure 5-6. QQ plots

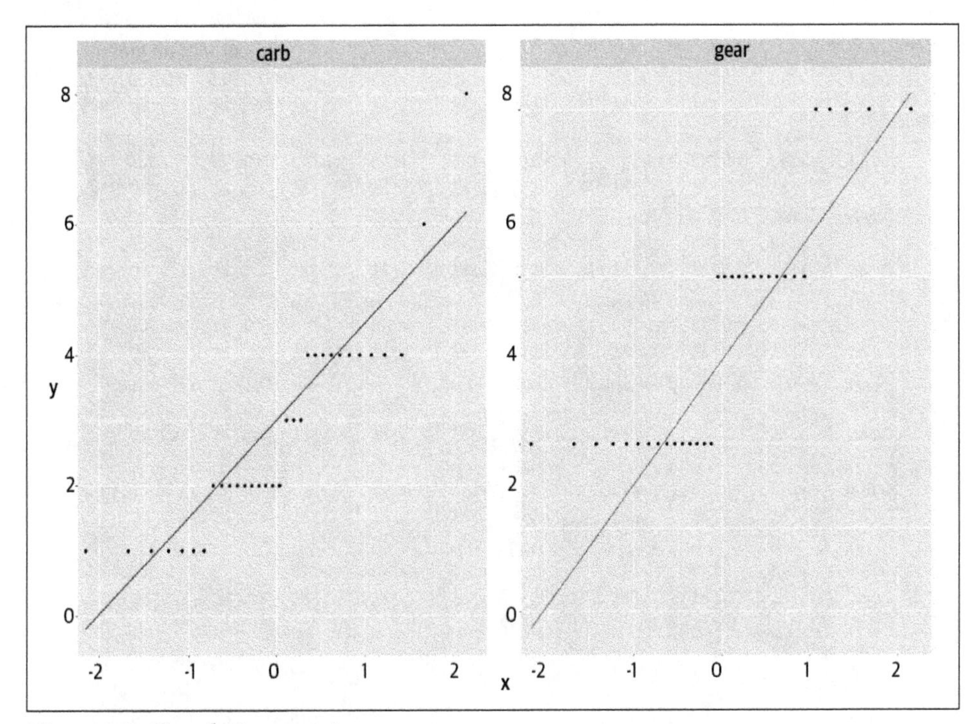

Figure 5-7. Correlation matrix

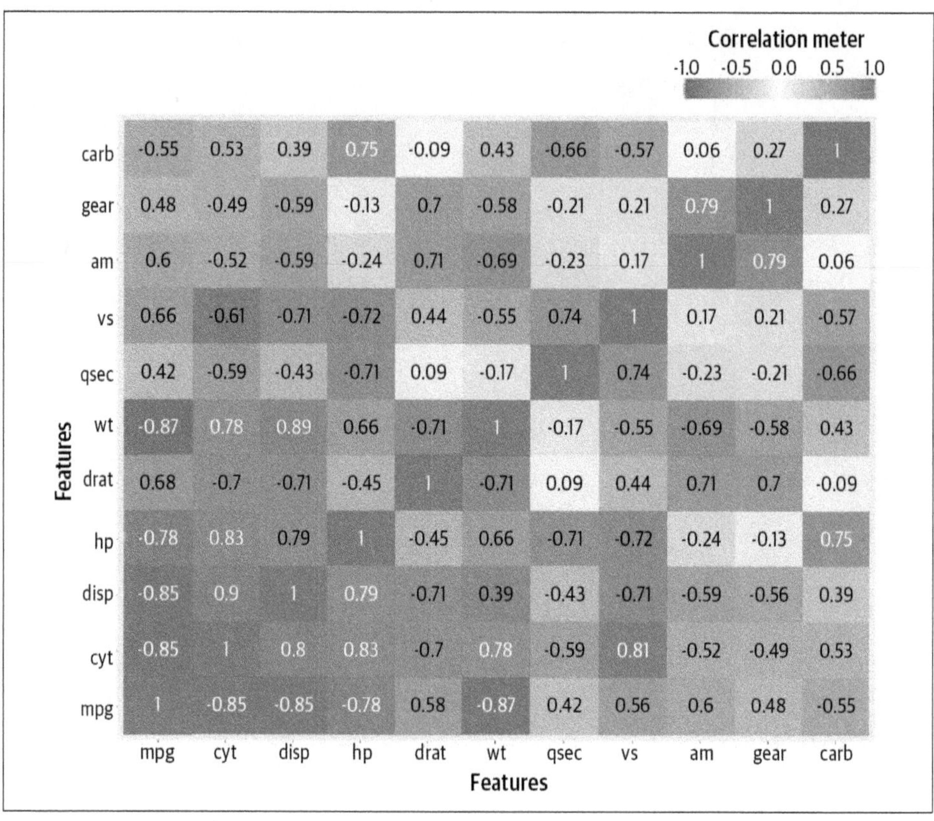

Figure 5-8. Principal component analysis (PCA) and variance charts

In Python, the sweetviz library is used:

```
pip install sweetviz
import sweetviz as sv
import seaborn as sns

# Load the mtcars dataset
mtcars = sns.load_dataset('mpg').dropna()  # mtcars is not in seaborn,
                                           # but mpg is similar

# Profile the dataset using sweetviz
report = sv.analyze(mtcars)

# Display the report in a web browser
report.show_html("mtcars_report.html")
```

sweetviz also generates a comprehensive report. Figure 5-9 shows a view of the information provided on the dataset. This information includes correlation, interaction, missing values, uniqueness, and data type analysis as examples.

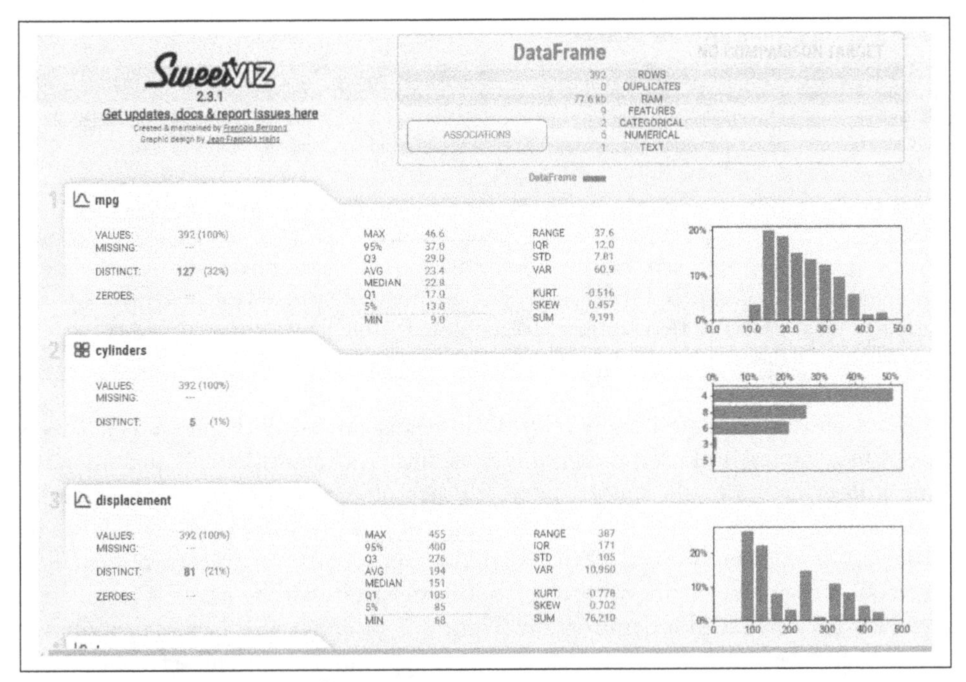

Figure 5-9. Python data profiling report output

In addition to examining data quality, data profiling can support activities that will be completed in EDA. When completing EDA for regression models, you can leverage the profiling output for analyzing correlation and distribution as well as variable interactions. The recommended best practice for starting EDA is profiling and using the output reports to support the multiple analysis activities outlined in this chapter. We will provide more detail on this as we go along.

Once data quality is examined as part of EDA, the next step is to examine the data for clustering. It's important to complete clustering because subsets of data may need to be treated differently as part of a predictive analytics project. Let's explore this in the next section.

Clustering and Unsupervised Learning

Clustering is a type of unsupervised machine learning method, used to group similar instances on the basis of features. Unsupervised learning is a type of machine learning where the algorithm identifies patterns and relationships in data without using labeled outputs or predefined categories. Table 5-1 summarizes the three types of clustering methods: distance-based (k-means), hierarchical, and density-based (DBSCAN), highlighting their main features and characteristics.

Table 5-1. Different clustering methods

Clustering type	Description	Key characteristics
Distance-based clustering (k-means)	Partitions n observations into k clusters, where each observation belongs to the cluster with the nearest mean.	Simple and commonly used; clusters are defined by the nearest mean.
Hierarchical clustering	Creates a hierarchy of clusters, which can be agglomerative (merging individual data points into clusters) or divisive (splitting one large cluster into smaller clusters).	Results in a tree-like diagram called a dendrogram; useful for understanding the structure of data.
Density-based clustering (DBSCAN)	Group points based on the density of data points, clustering closely packed points and marking isolated points as outliers.	Discovers clusters of arbitrary shape; effective for identifying outliers.

Distance-based clustering usually refers to k-means clustering (Figure 5-11). This is one of the simplest and most commonly used clustering algorithms. It aims to partition n observations into k clusters in which each observation belongs to the cluster with the nearest mean. Hierarchical clustering creates a hierarchy of clusters. It can be either agglomerative (starting with individual data points and merging them into clusters) or divisive (starting with one large cluster and dividing it). This results in a tree-like diagram called a dendrogram (Figures 5-14 and 5-15). Last, density-based clustering or density-based spatial clustering of applications with noise (DBSCAN) is based on the density of data points. It groups together points that are packed closely together (points with many nearby neighbors), marks points that lie alone in low-density regions as outliers, and discovers clusters of arbitrary shape in the dataset.

Purpose of Unsupervised Learning

Clustering is completed as part of EDA for several reasons. These include uncovering structure and patterns in the data, reducing datasets based on identified clusters, creating new features, labeling data that is unlabeled, and understanding variable interactions.

Clustering can help reveal the natural structure and patterns in the data that may not be immediately apparent. For example, customers in a dataset might naturally group into different segments based on their purchasing behaviors, or documents might group into different topics based on their content. Clustering can serve as a form of data reduction, especially when dealing with large datasets. By identifying clusters, we can summarize the data using a smaller number of representative examples or centroids.

The results of clustering can be used to create new features that can be useful for machine learning models. For example, the cluster assignments themselves can be used as new categorical features. In addition, clustering can be used to label unlabeled data. Once labeled, supervised learning techniques can be applied. For example, the clusters identified in a customer dataset can be used as labels for a customer segmentation model. Last, in multidimensional data, clustering can help in understanding how different variables interact with each other and the role they play in the formation of these clusters.

Clustering is an exploratory technique, and the clusters identified are dependent on the method and parameters chosen. Therefore, different clustering methods or parameters may yield different results. Always consider this when interpreting the results of a clustering analysis. It is possible to complete clustering analysis and have results that are not meaningful to solving the business problem or creating a predictive model. Despite completing clustering analysis, the results may lack relevance for solving business problems or creating predictive models. This could occur due to inappropriate choice of clustering algorithm, irrelevant features, or insufficient understanding of the underlying data structure, leading to clusters that fail to capture actionable insights or predictive patterns. Or clusters may not be present in the data.

Example of Clustering Impacting Supervised Learning

Now that we understand that clustering is a helpful technique that can provide insights for supervised learning, let's walk through a common scenario to demonstrate this.

Imagine a telecom company wants to reduce customer churn. As part of its EDA, the company decides to use clustering to better understand its customers. The company has access to a variety of data about each customer, such as age, monthly bill amount, types of services used, length of customer relationship, and frequency of customer service contact. To infer clustering results and make business sense out of them, you must first analyze the characteristics of each cluster to understand the underlying patterns that group similar data points together. For instance, in customer segmentation, clusters might reveal groups of customers with similar purchasing behaviors, demographic profiles, or product preferences. By interpreting these clusters, you can identify actionable insights, such as tailoring marketing strategies to target high-value customer segments (short-term), developing personalized product offerings (medium-term), or even shifting business strategy to focus on emerging customer trends (long-term). The key is to align the findings with specific business objectives, ensuring that the insights drive meaningful decisions that contribute to the organization's overall goals.

The telecom company's customers naturally group into three clusters (Figure 5-10):

- Long-term, high-value customers who rarely contact customer service
- Short-term, low-value customers who frequently contact customer service
- Medium-term, medium-value customers who occasionally contact customer service

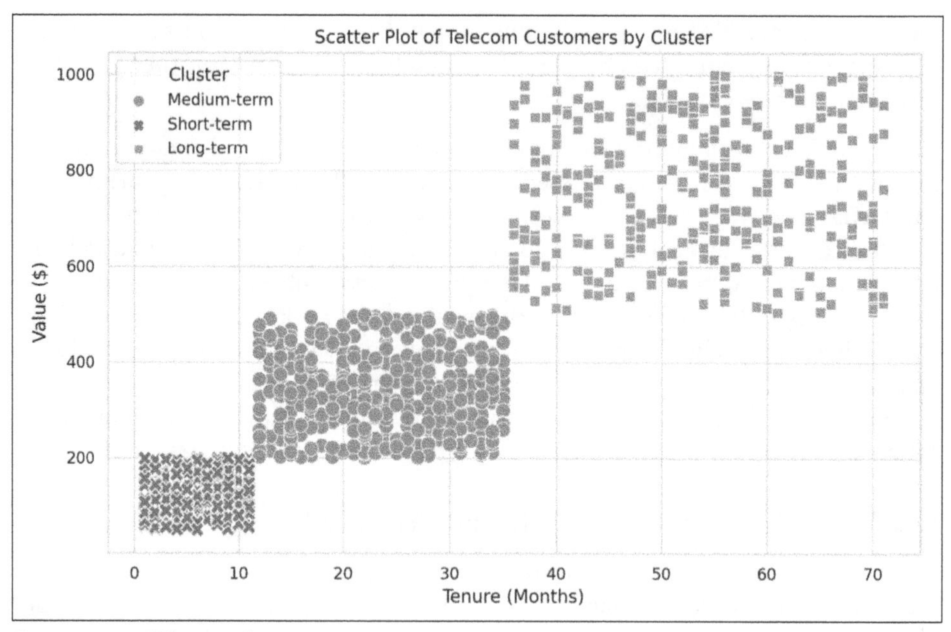

Figure 5-10. Telecom clusters

After identifying these clusters, the company decides to label each customer according to their cluster and then uses these labels as a new feature in a supervised learning model to predict churn. By doing this, the telecom company incorporates the information learned during the clustering process into its predictive model. For example, if the model learns that customers from the second cluster (short-term, low-value, frequent customer service contact) are more likely to churn, it can use this information to more accurately predict churn for new customers based on their predicted cluster. After the churn model is applied, the telecom company finds that this new feature improves the accuracy of its churn prediction model, and the company is able to improve its targets for the company's customer retention efforts.

Segmenting the data by clusters and building separate models for each cluster can enhance the overall accuracy of predictive analytics. Since customers within a specific cluster exhibit similar behaviors—such as purchase patterns, service usage, or churn tendencies—tailoring a model to each cluster allows for more precise predictions that align with the unique characteristics of that group. By developing three distinct models, one for each cluster, the analysis can capture nuances that a single, generalized model might overlook, thereby improving the accuracy and effectiveness of predictions. This approach leverages the homogeneity within clusters to refine predictions and better address the specific needs or risks associated with each customer segment.

K-Means Clustering

As outlined, the k-means algorithm iteratively assigns each data point to one of the K groups based on the features provided. Data points are clustered based on feature similarity, which can be measured using different distance measures like Euclidean (*https://oreil.ly/a5u95*) or Manhattan distance (*https://oreil.ly/g8tGr*).

Here are the steps in the k-means algorithm:

1. Initialize K centroids randomly.
2. Assign each data point to the nearest centroid, forming k clusters.
3. Recalculate the centroid (mean) of each cluster based on the members of that cluster.
4. Repeat steps 2 and 3 until the centroids do not change significantly or a predetermined number of iterations is reached.

In the R example, one set of data points is used with a normal distribution. Then, k-means clustering is applied to classify these data points into two clusters. In the Python example, the Iris dataset is used to create the clusters. Now, let's look at a simple code example for k-means clustering in both R and Python.

Setting a seed in unsupervised learning, especially for clustering algorithms like k-means, ensures reproducibility. Since many clustering methods involve random initialization (e.g., selecting random cluster centroids), setting a seed guarantees the same starting conditions across different runs. This consistency helps avoid variability in the results and allows for more reliable comparisons or evaluations of the clustering performance.

Here is the R example (Figure 5-11 demonstrates this output):

```
# Load required library
library(cluster)

# Create some data
set.seed(20)
data <- rbind(matrix(rnorm(100), nc=2),
              matrix(rnorm(100, mean=3), nc=2))

# Perform kmeans clustering
clust <- kmeans(data, centers=2)

# Print the clustering results
print(clust)

# Plot the clusters and centroids
plot(data, col=clust$cluster)
points(clust$centers, col=1:2, pch=8, cex=2)
```

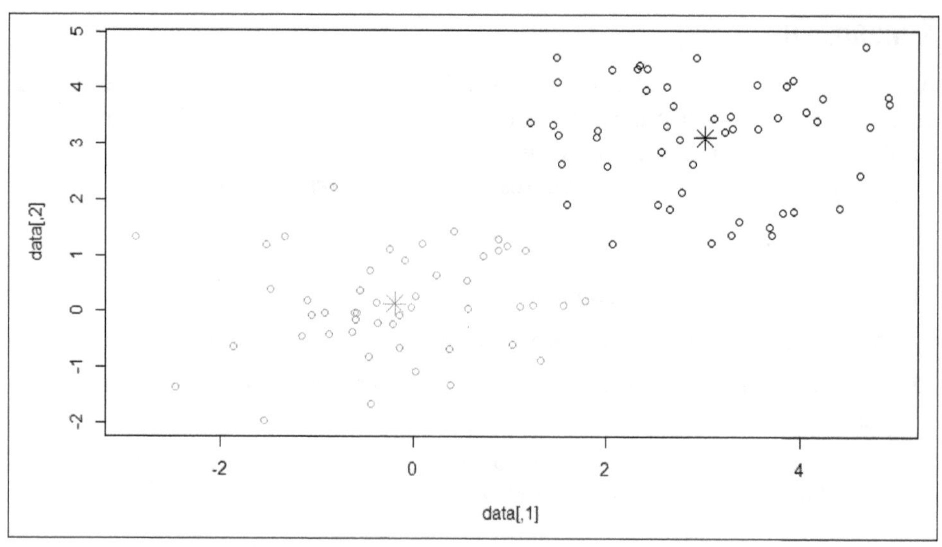

Figure 5-11. K-means clustering in R

Figure 5-12 shows that the k-means algorithm effectively separates the data into two clusters, corresponding to the two underlying distributions. The distinct separation of points and the compactness of clusters around their centroids suggest the presence of two well-defined groups in the data. This validates the use of k-means in this case for clustering.

Here is the Python example (Figure 5-12 shows this output):

```
# Import necessary libraries
import seaborn as sns
from sklearn.cluster import KMeans

# Load the Iris dataset
iris = sns.load_dataset('iris')

# Define the features and target
X = iris[['sepal_length', 'sepal_width', 'petal_length', 'petal_width']]

# Perform K-means clustering
kmeans = KMeans(n_clusters=3, random_state=0).fit(X)

# Add the cluster labels for each sample in the dataset
iris['cluster'] = kmeans.labels_

# Display clustering result using seaborn
sns.pairplot(iris, hue='cluster', palette=sns.color_palette('hls', 3));
```

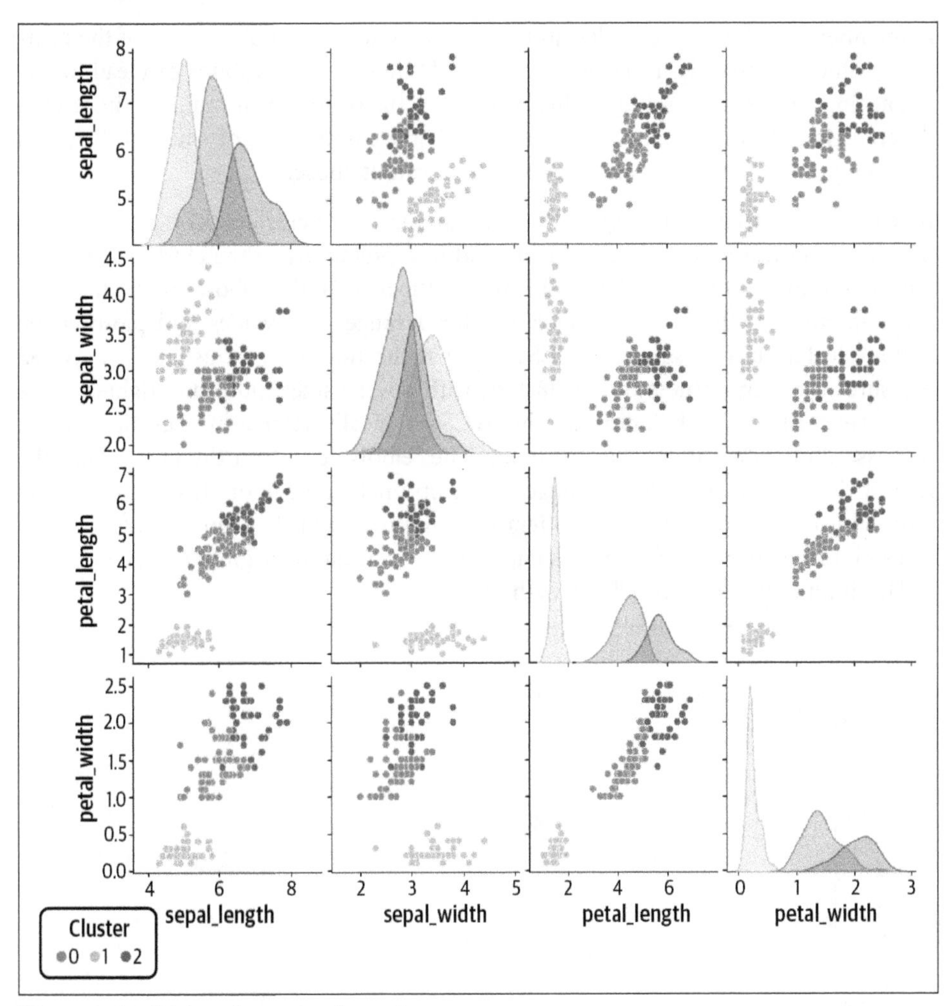

Figure 5-12. Python k-means clustering output

Figure 5-13 suggests that the algorithm is reasonably successful in identifying the three clusters in the Iris dataset. The pairplot visualization highlights that petal length and petal width are the most important features for distinguishing clusters, while sepal length and sepal width provide less separation. Cluster 0 is likely well separated and corresponds to setosa, while clusters 1 and 2 overlap more and correspond to versicolor and virginica. The plot helps you visually inspect the clustering performance and determine how well k-means can group the data based on the available features.

As you saw, Python and R each provide visual outputs, but they look quite different. However, the end outcome is k-means clustering will work the same way on different datasets.

Remember that the k-means algorithm is sensitive to the initial location of the centroids, which are usually initialized randomly. Different initializations can lead to different final clusters. To mitigate this issue, k-means is often run multiple times with different initializations, and the solution with the lowest sum of squared distances between points and their assigned cluster's centroid is chosen.

In k-means clustering, identifying the optimal number of clusters (k) is a crucial step that can significantly impact the quality and interpretability of the clustering results. One common method to determine the optimal k is the elbow method, which involves running the k-means algorithm for a range of k values and plotting the within-cluster sum of squares (WCSS) against the number of clusters. The WCSS measures the compactness of the clusters, with lower values indicating tighter, more cohesive clusters. As k increases, the WCSS typically decreases, but the rate of decrease diminishes after a certain point. The "elbow" point on the plot, where the rate of decrease sharply slows, suggests the optimal number of clusters, as adding more clusters beyond this point offers little improvement in compactness. See the scree plot in Figure 5-13 for the Iris data and how the elbow begins at PCA 2 or PCA 3. This is an example of the elbow method.

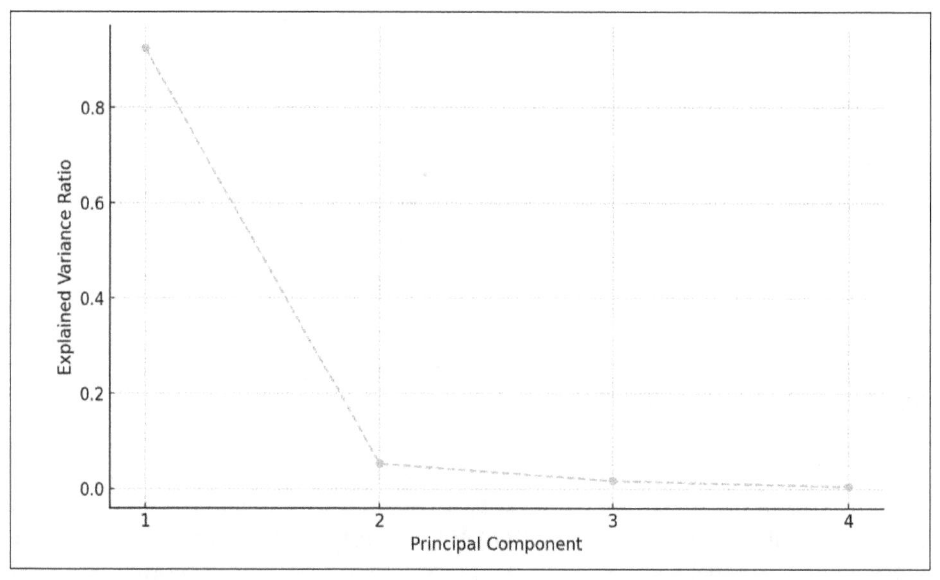

Figure 5-13. Scree plot for Iris data

Another approach is the silhouette score, which evaluates how well each data point fits within its assigned cluster compared to other clusters. The score ranges from –1 to 1, with higher values indicating that data points are well clustered. By calculating the silhouette score for different values of k, the optimal number of clusters can be identified as the k value that maximizes the silhouette score. This method provides an additional layer of validation by considering both intracluster cohesion and intercluster separation.

Hierarchical Clustering

Hierarchical clustering is an unsupervised machine learning algorithm that builds a hierarchy of clusters by merging or splitting them successively. This algorithm starts by treating each object as a singleton cluster. Then, pairs of clusters are successively merged based on the closeness of these pairs until all clusters are merged into one top-level cluster. This hierarchy of clusters is represented as a tree (or dendrogram). The root of the tree is the unique cluster that gathers all the samples, and leaves are clusters with only one sample.

Two main types of hierarchical clustering methods exist: a bottom-up approach and a top-down approach. The bottom-up approach is called agglomerative hierarchical clustering, where each data point starts in its own cluster, and pairs of clusters are merged together as one moves up the hierarchy. The top-down approach is called divisive hierarchical clustering, where data points start in one cluster and splits are performed recursively as one moves down the hierarchy.

Let's have a look at simple code examples of hierarchical clustering in R and Python.

Here is the R example (Figure 5-14 demonstrates this output):

```
# Load required library
library(cluster)

# Create some data
set.seed(10)
data <- rbind(matrix(rnorm(100), nc=2),
              matrix(rnorm(100, mean=3), nc=2))

# Compute the distance matrix
dist_matrix <- dist(data)

# Perform hierarchical clustering
clust <- hclust(dist_matrix)

# Plot the dendrogram
plot(clust)

# Cut the dendrogram into 2 clusters
groups <- cutree(clust, k=2)

# Print the clustering results
print(groups)
```

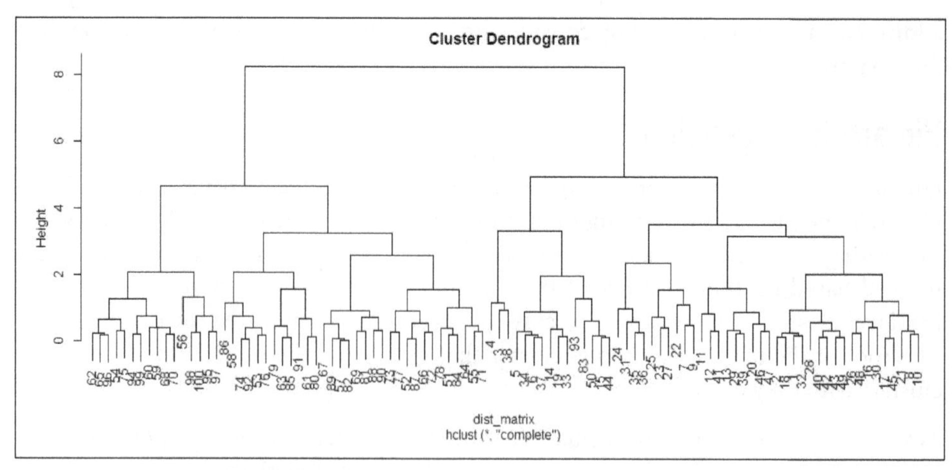

Figure 5-14. Hierarchical clustering in R

Here is the Python example (Figure 5-15 demonstrates this output):

```
# Import required libraries
from scipy.cluster.hierarchy import dendrogram, linkage
from matplotlib import pyplot as plt
import numpy as np

# Create some data
np.random.seed(10)
data = np.vstack((np.random.normal(size=(50, 2)),
                  np.random.normal(loc=3, size=(50, 2))))

# Perform hierarchical clustering
Z = linkage(data, 'ward')

# Plot the dendrogram
dendrogram(Z)

plt.show()

# Generate cluster labels
from scipy.cluster.hierarchy import fcluster
max_d = 10
clusters = fcluster(Z, max_d, criterion='distance')

# Print the clustering results
print(clusters)
```

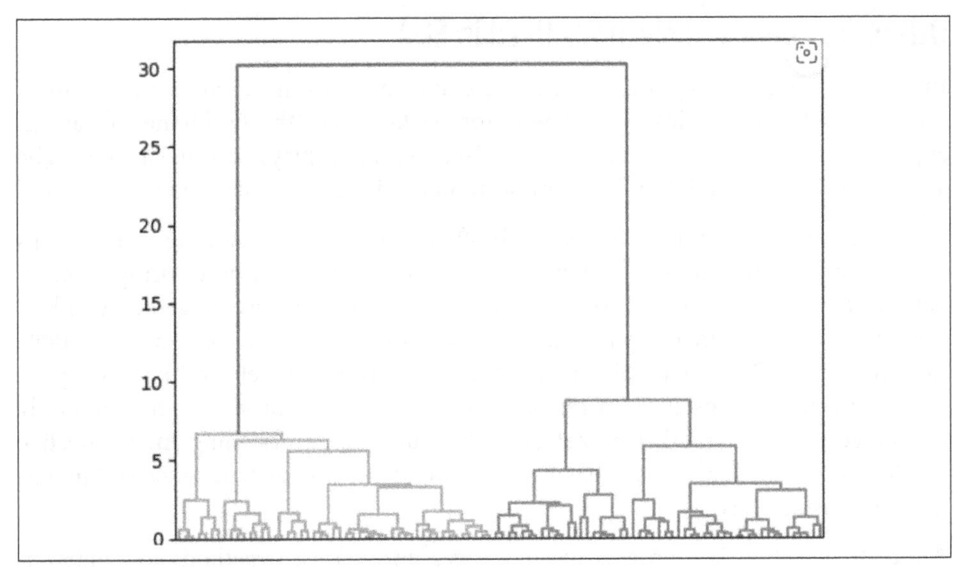

Figure 5-15. Hierarchical clustering in Python

In hierarchical clustering, determining the optimal number of clusters can be more challenging due to the nature of the algorithm, which creates a nested hierarchy of clusters. One common method for identifying the optimal number of clusters is to analyze the dendrogram, a tree-like diagram that illustrates the merging (in agglomerative clustering) or splitting (in divisive clustering) of clusters at each step. By examining the dendrogram, analysts look for the largest vertical distance between successive horizontal cuts, which represents the merging of dissimilar clusters. The height at which these merges occur can indicate a natural division in the data, suggesting the optimal number of clusters. A horizontal cut at this point can be used to define the clusters.

Another approach is the silhouette analysis applied to hierarchical clustering, similar to k-means. After the clustering process is complete, the silhouette score can be calculated for different numbers of clusters obtained by cutting the dendrogram at various levels. The optimal number of clusters is where the silhouette score is maximized, indicating well-separated and cohesive clusters.

In both the R and Python examples, two sets of data points are generated from two different normal distributions. Agglomerative hierarchical clustering is applied to classify these data points into clusters, and the dendrogram is displayed and shows the number of clusters.

Other Unsupervised Methods Used in EDA

In addition to clustering, other techniques and methods are used in the realm of unsupervised machine learning. These approaches form the backbone of various applications and fields where unsupervised learning plays a crucial role. The approaches are association rules learning, anomaly detection, and autoencoders.

Association rules learning is an approach often employed for unveiling relationships among large datasets. Its application in market basket analysis is noteworthy, where it helps discover associations between products frequently purchased together. By identifying sets of items that often occur together, you can extract meaningful insights that might be useful for prediction in machine learning models and enhance your sales strategies. For instance, in retail, understanding what products are commonly purchased together (market basket analysis) can help in designing more effective product recommendations, ultimately improving the customer experience and potentially increasing sales.

Anomaly detection is a process that leverages unsupervised methods to single out outliers or unusual instances within a dataset. This method is particularly valuable in scenarios that demand the identification of instances that deviate from the norm, such as fraud detection and network intrusion identification. Detecting these outliers allows you to make decisions about how to handle them (e.g., removing them, treating them separately, or adjusting the model to be less sensitive to outliers). You'll want to make these decisions before building your models. Anomaly detection is also essential in fields like fraud detection, where the goal is to identify unusual patterns of behavior.

Autoencoders is a specific type of neural network designed for learning efficient encodings of input data. With diverse applications, they can be used for data denoising, dimensionality reduction, and learning representations of the data as examples. Dimensionality reduction is often a crucial step in EDA, as it can help to simplify the dataset and speed up the learning process without losing too much information. In the context of EDA, autoencoders can help visualize high-dimensional data and reveal the structure that might not be apparent in the original space. They can also be used to denoise data, making the dataset cleaner for machine learning models.

Each of these unsupervised techniques provides a unique way of understanding, cleaning, and preparing data, making them valuable tools in the EDA process for machine learning.

Identifying Outliers

An outlier is a data point that differs significantly from other observations. An outlier can be a result of variability in the data or an experimental error; the latter are sometimes excluded from the dataset. Outliers can also be due to a genuine deviation in the system you're studying, indicating phenomena out of the ordinary. Outliers are important to identify during EDA for several reasons, including statistical analysis (this was covered in Chapter 4), data quality, and impact to model performance. Additionally, outliers can indicate an interesting event such as in identifying fraud. Methods to detect outliers can vary widely, from simple univariate methods like the Z-score method to complex multivariate methods like the Mahalanobis distance and machine learning–based methods like isolation forests.

While it's important to identify outliers during EDA, the decision on how to handle them requires careful consideration. Depending on the situation, you might decide to keep outliers, remove them, or transform them in some way. Regardless of what you choose, you should always keep in mind the potential impact on your subsequent analyses and results.

Outliers in Regression

Outliers can have a significant impact on regression models due to the nature of these models. Regression models work by estimating the best fit line that minimizes the difference (usually squared difference) between the actual and predicted values. Outliers, being far away from the rest of the data, can exert a disproportionately large influence on the model parameters, causing the regression line to be skewed toward them. This in turn can lead to poor model performance when generalizing to new data.

There are various ways to identify outliers in data. In R, you can use the `boxplot.stats` function, which identifies outliers as data points that are 1.5 times the interquartile range (IQR) above the third quartile or below the first quartile.

In the following R example, I'll use the `boxplot.stats` function and identify `'100'` as the outlier:

```
# Assume df is your dataframe and column is the name of the column
# you want to check for outliers
df <- data.frame(column = c(1, 2, 2, 2, 3, 1, 2, 3, 100))
# The '100' is an outlier

# Get data points considered as outliers
outliers <- boxplot.stats(df$column)$out
print(paste("Outliers: ", paste(outliers, collapse=", ")))

[1] "Outliers:  100"
```

Here is the same approach using Python. In the Python example, `quantile(0.25)` is used to calculate Q1 (the first quartile), and `quantile(0.75)` is used to calculate Q3 (the third quartile). The IQR is the difference between Q3 and Q1. Any data point that falls below Q1 – 1.5IQR or above Q3 + 1.5 IQR is considered an outlier:

```
import pandas as pd
import numpy as np

# Assume df is your DataFrame and 'column' is the name of the column
# you want to check for outliers
df = pd.DataFrame({
    'column': [1, 2, 2, 2, 3, 1, 2, 3, 100]  # The '100' is an outlier
})

Q1 = df['column'].quantile(0.25)
Q3 = df['column'].quantile(0.75)
IQR = Q3 - Q1

# Define bounds for outliers
lower_bound = Q1 - 1.5 * IQR
upper_bound = Q3 + 1.5 * IQR

# Identify the outliers
outliers = df[(df['column'] < lower_bound) | (df['column'] > upper_bound)]
print("Outliers:\n", outliers)

Outliers:
    column
8     100
```

Let's now explore how outliers are addressed in classification.

Outliers in Classification

Outliers can be a concern when creating nonparametric (not statistically based) classification models as well. Although nonparametric models do not make strong assumptions about the underlying distribution of the data, outliers can still influence the model's predictions. The effect of outliers on a nonparametric model will depend on the specific type of model (algorithm) being used.

Classification algorithms can be broadly categorized into parametric and nonparametric based on whether they make assumptions about the underlying distribution or form of the data. Let's consider parametric classification algorithms first:

Logistic regression
: This algorithm assumes a linear relationship between the logic of the response and the predictors.

Linear discriminant analysis (LDA)
: LDA assumes that the predictors are normally distributed and have the same covariance matrix in each class.

Naive Bayes
: This classifier assumes that the features are conditionally independent given the class label. Different types of naive Bayes classifiers (e.g., Gaussian, multinomial) make different assumptions about the distribution of the predictors.

Perceptron
: This is a linear classifier that assumes the classes can be separated by a hyperplane in the feature space.

Let's look at nonparametric options. Nonparametric classification algorithms make no assumption about the distribution or form of the data. Here are some nonparametric algorithms:

Decision trees
: Decision trees learn rules based on the values of the predictors. Decision trees are machine learning models that build decision rules based on predictor values. They recursively split the data into subsets, using predictors to determine the best splits. Each split forms a branch, eventually leading to leaf nodes containing outcome predictions. This process enables decision making based on input feature values.

Random forests
: An ensemble of decision trees.

Support vector machines (SVMs)
: In their basic form (without the kernel trick), SVMs are linear classifiers like the perceptron. However, when used with the kernel trick, they can model complex, nonlinear decision boundaries.

k-nearest neighbors (KNN)
: This algorithm makes predictions for a new instance based on the class labels of its k-nearest neighbors in the feature space.

Neural networks
: Neural networks do not make specific assumptions about the underlying distribution or form of the data.

: Neural networks don't presuppose specific data distributions or forms. Instead, they learn complex patterns and relationships from the data directly, using layers of interconnected neurons. This flexibility enables them to model highly nonlinear relationships without requiring explicit assumptions about the data's underlying structure.

Gradient boosting machines (like XGBoost, LightGBM)
: These are ensembles of decision trees, like random forests. Gradient boosting machines (GBMs) are ensemble models of decision trees, similar to random

forests. However, GBMs sequentially build trees, each correcting errors of the previous ones, thereby enhancing predictive accuracy by focusing on misclassified instances through an iterative learning process.

Nonparametric models like KNN can be sensitive to outliers. An outlier can shift the decision boundary in its vicinity and lead to misclassification, especially for models that rely on distances between data points.

Nonparametric models often have the flexibility to fit complex patterns in the data. This flexibility can lead to overfitting, where the model learns from noise or outliers in the data. As a result, the model may perform poorly on unseen data. Nonparametric models often require more computational resources as they have to consider each data point (or a large subset of data points) for making predictions. Outliers increase the complexity of the data and thus can increase the computational cost of these models.

Just like in parametric models, outliers may indicate errors in data or important but extreme observations. Identifying outliers can provide insights into the data and the problem at hand. That said, it's important to note that not all nonparametric models are affected by outliers in the same way. For instance, decision trees (and by extension, random forests and GBMs), which are also nonparametric models, are generally robust to outliers in the input features since they split the data based on conditions rather than values. Still, when the target variable contains outliers, they can also be affected. As always, understanding the data and the specific model being used is key.

In practice, the decision between using parametric and nonparametric algorithms depends on several factors, including the characteristics of the data and the problem at hand. Parametric algorithms assume a specific form for the underlying data distribution, such as linear regression or logistic regression, and generally have fewer parameters to estimate. This makes them more interpretable and computationally efficient, but they can struggle with complex patterns or nonlinear relationships if the assumed model does not fit the data well. On the other hand, nonparametric algorithms, like decision trees or KNNs, do not make strong assumptions about the data's structure and can adapt to a wider variety of data distributions. While they tend to have higher predictive accuracy, especially with complex or high-dimensional data, they are often less interpretable and can require more data to generalize well. Therefore, the choice between the two involves a trade-off: parametric models offer simplicity and interpretability, while nonparametric models excel in flexibility and predictive power at the cost of increased complexity.

Data Preparation for Modeling

In any data-driven project, the process of data preparation is fundamental to building effective models. This section explores the various stages of data preparation required to ensure that data is suitable for modeling and that the model can generalize well to unseen data. We begin by discussing sampling techniques and the importance of splitting data into training and test sets to properly evaluate model performance. Next, we cover essential steps like data transformation and data formatting, which help make raw data more usable by ensuring consistency and compatibility with algorithms. We also examine how derived attributes (or feature engineering) can create new, meaningful features that improve model accuracy. Additionally, we delve into techniques for data manipulation, such as handling missing values or outliers, before addressing the importance of scaling and normalization to ensure that features contribute equally in modeling. Through these processes, we build a solid foundation for effective and reliable predictive modeling.

Sampling

Sampling is a crucial step in data preparation that directly impacts the quality and performance of machine learning models. It involves selecting a representative subset of the full dataset to use for training and evaluation, especially when dealing with large datasets or imbalanced classes. Effective sampling ensures that the model captures the underlying patterns in the data while maintaining computational efficiency. The key types of sampling techniques used in the preparation process are random, stratified, systematic, cluster, and bootstrap.

Random sampling

In simple random sampling, each data point in the dataset has an equal chance of being selected. This method is straightforward and useful when the data is uniformly distributed and the dataset size is manageable. Random sampling ensures that no bias is introduced during the selection of data points, allowing the sample to represent the population accurately. However, if the dataset is unbalanced or contains rare classes, this technique might not capture those important details, leading to suboptimal models.

Stratified sampling

Stratified sampling ensures that specific subgroups (or strata) of the data are proportionally represented in the sample. This is particularly useful in cases where the dataset is imbalanced, such as when dealing with rare classes in classification problems. By maintaining the distribution of target classes or key features, stratified sampling helps the model learn effectively from both minority and majority classes, reducing bias and improving performance on the minority class during testing.

For example, in a dataset where 90% of the observations belong to one class and only 10% to another, simple random sampling might result in a sample that underrepresents the minority class. Stratified sampling, on the other hand, guarantees that the ratio between classes is preserved in both training and test sets, leading to better generalization.

Systematic sampling

Systematic sampling involves selecting data points at regular intervals from an ordered dataset. For instance, if you need a sample size of 100 from a dataset of 1,000, you could choose every 10th observation. This technique is useful when working with datasets where the data points are evenly distributed or exhibit periodicity. While it is simple and efficient, systematic sampling can introduce bias if there is any underlying pattern in the data that coincides with the sampling interval.

Cluster sampling

Cluster sampling divides the dataset into clusters or groups and then randomly selects some of these clusters to be included in the sample. All data points within the chosen clusters are used in the training process. This technique is particularly useful when the dataset is too large to sample randomly or is distributed across different locations. It is most effective when the clusters themselves are representative of the overall population, but it may introduce bias if the clusters are not diverse.

Bootstrap sampling

Bootstrap sampling is a technique commonly used for model evaluation and variance estimation. It involves sampling data points with replacement, meaning a single data point can appear multiple times in the same sample. Bootstrap sampling is widely used in ensemble methods like bagging (e.g., random forests) to generate multiple training datasets, each slightly different, which improves the model's robustness and reduces overfitting.

Oversampling and undersampling

In datasets with imbalanced classes, where one class significantly outnumbers another, oversampling and undersampling techniques are often used. Oversampling increases the representation of the minority class by duplicating its observations or synthetically generating new samples (e.g., using techniques like SMOTE—synthetic minority oversampling technique). This ensures that the model is not biased toward the majority class.

Undersampling involves reducing the size of the majority class to match the minority class. While it balances the dataset, it can lead to a loss of valuable data, so it's generally used when the dataset is large enough to retain sufficient information for training.

Importance of sampling in model building

The choice of sampling technique can significantly influence the performance of machine learning models. Improper sampling may introduce bias, cause overfitting, or lead to poor generalization on unseen data. On the other hand, effective sampling allows the model to learn from a balanced, representative dataset, leading to better predictions and a more robust model. Sampling is especially important when working with large datasets, where it is often impractical or unnecessary to use the entire dataset for training due to computational constraints. By carefully selecting an appropriate sampling strategy, business analysts can ensure that their models are trained on high-quality data and are capable of performing well in real-world applications.

Training and Testing

The division of data into training and testing sets is essential for building robust and reliable machine learning models. This process helps ensure that the model can generalize to unseen data rather than just memorizing the patterns in the training data.

The training data is the portion of the data used to train the model, where the model learns the underlying patterns and relationships between the features (input) and the target (output). The goal is for the model to find the best mapping from the input data to the output based on these examples. The testing set is a separate portion of the data used to evaluate the model's performance after training. It simulates how the model will perform on unseen data in real-world applications. This is important for measuring the model's ability to generalize to new data and avoid overfitting (when the model performs well on training data but poorly on new data).

A common practice is to split the data into 70–80% for training and 20–30% for testing, but the exact ratio depends on the size of the dataset. The larger the dataset, the smaller the testing set can be, since more data allows for better generalization:

```
from sklearn.model_selection import train_test_split

# Split the data into training (80%) and testing (20%)
X_train, X_test, y_train, y_test = train_test_split(X, y, test_size=0.2,
random_state=42)
```

- `X_train` and `y_train` are the training features and target values.
- `X_test` and `y_test` are the testing features and target values.
- The `random_state` ensures reproducibility, so the same split is generated every time the code runs.

The model learns from the training data by finding patterns and relationships between the features and the target. This process involves adjusting the model parameters (like weights in linear models or decision boundaries in decision trees). The training data must be representative of the overall dataset to avoid introducing bias into the model. If the model is overfitting, it may perform exceptionally well on the training set but fail on new data. This is a sign that the model has learned the noise or specific details of the training data rather than general patterns.

The testing set evaluates how well the model generalizes to new, unseen data. It acts as a stand-in for real-world data that the model has not seen during training. The testing set should be kept separate from the training set and should not be used in the training process. The testing data must also be representative of the problem domain. If it significantly differs from the training data (e.g., in distribution or range), the performance may drop.

The integrity of any analytics output heavily depends on the quality of input data. A substantial part of data science revolves around preparing and cleaning data for analysis, which may involve data manipulation and transformation. Data transformation refers to the process of changing the format, structure, or values of data. Common examples are normalization, standardization, and encoding categorical variables. Data manipulation is the process of changing, adjusting, or organizing data to make it more suitable for analysis. The actions taken for transformation and manipulation depend on the data and the type of analytics project you are working on. Data transformation and manipulation can occur multiple times and iteratively as data is explored and discovered. Data preparation can start with either transformation or manipulation.

Data Transformation

Data transformation is a crucial step in preparing data for machine learning models. It ensures that the data is in the best possible form for a model to learn from. Data transformation can be a format, structure, or context change. Actions to take as part of data transformation will also be impacted from the exploration and discovery of data quality. Data transformation can include formatting, creating derived attributes, and other techniques such as scaling, normalization, and standardization.

Data formatting

Data formatting functions are an integral part of data preparation for machine learning. Some of the basic formatting approaches used are handling missing values, changing the data type, addressing data/time conversions, and manipulating strings. The following list highlights many of the common data formatting use cases and related R and Python functions that can be leveraged:

Missing values
: This includes functions to replace missing values with a specific value, mean, median, or mode. In Python, you can use the `fillna()` function from pandas, and in R, you can use `na.omit()` or `replace_na()` from the tidyr package.

Data type conversion

Functions that convert data types (such as converting a numeric string to a number or vice versa) are essential. In Python, you can use the `astype()` function from pandas, and in R, you can use `as.numeric()`, `as.character()`, `as.factor()`, and similar functions.

Date and time conversion

These functions convert data to a date or time format. In Python, you can use `to_datetime()` from pandas, and in R, you can use `as.Date()` or `strptime()`.

String manipulation

These include functions that manipulate strings, like trimming whitespace, replacing certain characters, or converting to lowercase or uppercase. In Python, you can use `strip()`, `replace()`, or `lower()` from pandas, and in R, you can use `trimws()`, `gsub()`, or `tolower()`.

One-hot encoding. We have learned that regression and classification projects may have different requirements. For instance, with statistical regression, all variables must be numeric. A process called one-hot encoding is often used to convert categorical data variables so they can be provided to machine learning algorithms to improve predictions. It creates new binary columns for each category/label present in the original columns. Therefore, categorical or string data should be encoded to numerical values. In Python, you can use `LabelEncoder()` or `OneHotEncoder()` from sklearn.preprocessing, and in R, you can use `factor()` or `model.matrix()`.

Let's look at an example. One-hot encoding is used to transform categorical variables into binary vectors. Each category becomes a new column, and the value is represented as 1 for the presence of that category and 0 for its absence. This is essential for algorithms that require numeric input, such as regression models. For instance, a feature like a city with values `['New York', 'San Francisco', 'Los Angeles']` would be transformed into three columns, with binary values indicating the presence of each city. Here is an example using Python's preprocessing library:

```
from sklearn.preprocessing import OneHotEncoder

encoder = OneHotEncoder()
X_encoded = encoder.fit_transform(X[['city']])
```

In the these examples, all unique city values would become a column populated by a 1 or a zero.

Binning. Binning is a data preprocessing technique used to group data points by frequency. The original data values that fall into a given small interval, a bin, are replaced by a value representative of that interval. For instance, a binned age feature (`Under 20`, `20-30`, `30-40`, etc.) might be more interpretable than continuous age in some contexts. Binning can assist in handling outliers, improve either sparse or highly granular data, and reduce noise. Binning can be applied to create new categories or to reduce cardinality of categories. Binning transforms continuous features into discrete intervals or "bins." This is useful when you want to reduce the effect of outliers or when a continuous feature is better interpreted as categorical:

```
X['age_binned'] = pd.cut(X['age'], bins=[0, 18, 35, 60, 100],
labels=['Child', 'Young Adult', 'Adult', 'Senior'])
```

For example, instead of using exact age values, you can transform the feature into bins like Child, Young Adult, Adult, and Senior, making the data easier to interpret for some models, as just outlined in the Python example.

Derived attributes

One of the primary goals that an analyst focuses on is determining the features or inputs for the predictive model. Derived attributes, also known as engineered features, are new attributes that are created from existing ones in your dataset. They can provide additional valuable information for your machine learning model that isn't available in the original data. The following list outlines some examples of engineered features. Most of what will be engineered will be based on the problem to be solved. For instance, telecom churn might require the number of customer complaints as a feature, but only a flag that shows a complaint is provided. An engineered feature that counts the number of complaints within a time period would need to be derived:

Ratios
: It is possible to create new attributes by taking the ratio of two existing numerical attributes. For example, in a dataset about houses, you could create a "price per square foot" feature.

Polynomial
: These are features created by raising existing features to an exponent. For example, if your dataset has a feature "x", you could create a new feature "x^2". Polynomial features can help to model relationships that are more complex than simple linear relationships.

Interaction
: These are features that represent the interaction between two or more other features. For example, if you have two binary features "A" and "B", you could create a new feature "A and B" that is 1 when both "A" and "B" are 1, and 0 otherwise.

Temporal
: If data includes dates or timestamps, there are many derived attributes that could be created, like "day of the week", "month of the year", "hour of the day", or "minutes since midnight".

Aggregates
: These are features that summarize multiple data points. For example, you might aggregate transaction data to create features like "average transaction amount" or "number of transactions in the last month".

Text-based
: There are countless derived features that could be created. Some examples include "number of words", "number of unique words", "average word length", or more complex features based on specific words or phrases present in the text.

There are a number of derived features that could be created, and it will be part of EDA and alignment to the problem statement that determines which ones are likely. Also, be careful to avoid creating features that "leak" information from the future (i.e., data that wouldn't actually be available at the time of prediction), as this can lead to overoptimistic performance estimates.

Leaking information from the future occurs when your model inadvertently uses data that wouldn't realistically be available at the time a prediction is made. This gives the model an unfair advantage and results in overly optimistic performance estimates. Let's say you're building a model to predict whether a stock's price will increase tomorrow based on various features. One of the features in your dataset is the stock price at the end of the day. This feature includes the stock's closing price for tomorrow. Since you're trying to predict the price for tomorrow, using the closing price for that day in your training data would be an unrealistic advantage: this is information that wouldn't be available when making the prediction. Why is it a problem?

At the time of prediction, you wouldn't know the stock's closing price for tomorrow, but your model is learning from it, leading to artificially high performance. Instead, your features should only include data up to the current day or any historical data that would be available before making the prediction, such as:

- Today's stock opening price
- Historical stock data (closing prices up to today)
- Technical indicators that are based on past data, like moving averages

By ensuring that your model is only trained on data that would actually be available before the prediction is made, you avoid information leakage and get a more realistic estimate of how the model will perform in the real world.

Other transformation techniques can be applied to improve the data suitability for analysis. Some reasons to apply additional transformation techniques include addressing skewness, high variance, dealing with outliers, and normalizing the data to make the data more suitable for a particular algorithm. Here are a few examples of other transformation techniques:

- Log transformation is a data transformation method in which it replaces each variable `x` with a `log(x)`. The choice of the logarithm base is usually left up to the analyst and depends on the purposes of statistical modeling.
- Power transformations are a family of parametric, monotonic transformations that aim to map data from any form onto a normal distribution.
- Box-Cox is a type of power transformation that's useful for nonnormal dependent variables. The Box-Cox transformation transforms our data so that it closely resembles a normal distribution.

Each of these transformations has its own use cases and is effective in different types of data and models. The choice of transformations depends on the specific requirements of the model and the nature of the input data.

Let's take a further look at an example for log transformation, how it is applied, and how it makes sense in certain scenarios, such as long-term stock market pricing. Suppose you are analyzing the price of a stock over several decades, and the stock has grown significantly in value over time, possibly by several orders of magnitude. For example, let's consider a hypothetical stock whose price increased from $10 in 1990 to $1,000 in 2023. The prices at different points in time might look like this:

- 1990: $10
- 1995: $20
- 2000: $50
- 2010: $200
- 2020: $800
- 2023: $1,000

When dealing with stock prices over a long period, especially when the price grows exponentially, plotting the raw data can show skewed distributions and make it harder to see meaningful patterns. In this case, larger values dominate the plot, and the smaller values seem insignificant. To handle this, a log transformation can be used to normalize the data and bring the values into a more manageable range (see Figure 5-16). For example, applying a log transformation with the natural logarithm (log base e):

- $\log(10) = 2.30$
- $\log(20) = 2.99$
- $\log(50) = 3.91$
- $\log(200) = 5.30$
- $\log(800) = 6.68$
- $\log(1{,}000) = 6.91$

After the log transformation, the stock prices from different time periods are scaled down to be in a more comparable range. This transformation helps highlight relative percentage changes rather than absolute changes. Figure 5-16 shows the following:

- The first plot shows the original stock prices over time, which exhibits exponential growth.

- The second plot displays the log-transformed stock prices, which linearizes the growth, making it easier to analyze trends and percentage changes over time.

This illustrates how log transformation helps normalize data when dealing with exponentially growing values, like long-term stock prices.

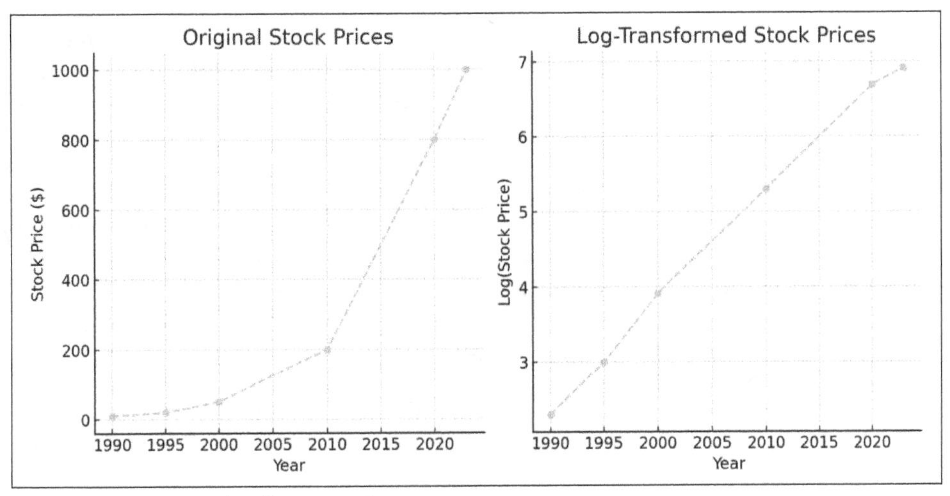

Figure 5-16. Example of log transformation

Scaling, normalization, and standardization

Standardization, normalization, and scaling are three related but distinct concepts in machine learning. They are all preprocessing techniques used to transform data so that it can be more effectively processed by a machine learning algorithm.

Scaling typically refers to adjusting the range of a feature in your dataset. It doesn't assume any specific distribution for the feature and simply adjusts the values to fit within a certain range. The most common method of scaling is min-max scaling, where every feature value is shifted and rescaled so that they end up ranging between 0 and 1. Scaling is particularly important when the features have different units. Min-max scaling transforms features to a fixed range, usually between 0 and 1. The formula for min-max scaling is: $X_{scaled} = X - X_{min} / X_{max} - X_{min}$

The transformation linearly rescales the original values so that the smallest value becomes 0 and the largest value becomes 1, with everything in between proportionally adjusted.

Normalization is sometimes used interchangeably with scaling, particularly min-max scaling. However, in a stricter sense, normalization refers to adjusting the data so that it conforms to a norm, often a unit norm. In other words, normalization scales each input vector to have a length of 1. This is often used when dealing with text data or when calculating similarity.

Standardization is a more specific form of feature scaling. In standardization, the feature is rescaled so that it has a mean of 0 and a standard deviation of 1. This process produces Z-scores for each feature value, giving the output a standard normal distribution. Standardization does not bound values to a specific range as in min-max scaling. The formula for standardization is: $Z = X - \mu / \sigma$.

X is the original value, μ is the mean of the feature, σ is the standard deviation of the feature, which results in Z, which is the standardized value (also called the Z-score). This transformation standardizes the data by centering it around the mean and adjusting the spread so that the data is expressed in terms of standard deviations.

All three techniques adjust the values of numeric features in a dataset to improve the performance of machine learning models. The choice of technique depends on the algorithm used and the nature of the data. For instance, algorithms that calculate distance (like KNN and SVM) or perform gradient descent (like linear regression, logistic regression, and neural networks) often benefit from standardization or scaling, while tree-based algorithms often do not require these preprocessing steps.

Here are some examples of how you might perform scaling, normalization, and standardization in R. For these examples, let's assume we have a dataset `df` with a numeric variable x:

```
# Assume a dataframe df with a numeric variable x
df <- data.frame(x = c(10, 200, 3000, 40000, 500000))
```

Scaling typically focuses on bringing all values into the range [0,1]. Here's how you might do this in R:

```
df$x_scaled <- (df$x - min(df$x)) / (max(df$x) - min(df$x))
```

Normalization means adjusting a distribution to resemble a standard normal distribution (mean 0, standard deviation 1):

```
df$x_normalized <- (df$x - mean(df$x)) / sd(df$x)
```

In some contexts, standardization and normalization are used interchangeably to refer to the process of turning a distribution into a standard normal distribution. So the code example for standardization would be the same as for normalization. However, standardization might refer to transforming a variable to have a specific mean and standard deviation (not necessarily 0 and 1). Here's an example of standardizing a variable to have a mean of 100 and standard deviation of 15:

```
df$x_standardized <- (df$x - mean(df$x)) / sd(df$x) * 15 + 100
```

R packages, such as caret and scale, can provide functions to make these transformations easier. Here's an example of how you might standardize a variable using the scale function:

```
df$x_standardized_scale <- scale(df$x)
# This centers and scales to standard deviation 1
```

And here's how you might perform min-max scaling using the caret package:

```
library(caret)

preproc = preProcess(df, method = c("range"))
df$x_scaled_caret = predict(preproc, newdata = df)$x
```

Let's consider how Python handles these aspects of data preparation. The Python scikit-learn library provides functions to perform scaling, normalization, and standardization. Let's assume we have a Pandas `dataframe df` with a numeric column `x`:

```
import pandas as pd
from sklearn import preprocessing

# Assume a dataframe df with a numeric column 'x'
df = pd.DataFrame({'x': [10, 200, 3000, 40000, 500000]})
```

For scaling, MinMaxScaler can be used as shown here:

```
scaler = preprocessing.MinMaxScaler()
df['x_scaled'] = scaler.fit_transform(df[['x']])
```

In Python, normalization can mean adjusting the data so each value represents its proportion relative to the sum of all values (L1 norm) or so the sum of the squares of the values is 1 (L2 norm). L1 and L2 concepts are explained later in this chapter. Normalizer can be used for this, as shown here:

```
# Normalize using L1 norm
normalizer = preprocessing.Normalizer(norm='l1')
df['x_normalized_l1'] = normalizer.transform(df[['x']])

# Normalize using L2 norm
normalizer = preprocessing.Normalizer(norm='l2')
df['x_normalized_l2'] = normalizer.transform(df[['x']])
```

For standardization, StandardScaler can be used, as shown in this example:

```
standardizer = preprocessing.StandardScaler()
df['x_standardized'] = standardizer.fit_transform(df[['x']])
```

In all of the code snippets, scaling, normalization, and standardization are addressed in Python and R.

Data Manipulation

Data manipulation refers to the process of adjusting data to make it organized and easier to read. These operations include sorting, aggregating, merging, filtering, and more. Data manipulation is used throughout data preparation and is based on the type of analysis that is being done. Common data manipulation techniques include the following:

Filtering
: Filtering is used to select a subset of observations or features that meet certain criteria. For instance, you might want to work with only a particular range of values or exclude outliers.

Sorting
: Sorting rearranges the data in ascending or descending order based on some criteria, which can be useful in various scenarios, such as when wanting to identify the top or bottom values in a dataset.

Aggregation
: Aggregation is used to compute summary statistics about each group, like sum, mean, maximum, minimum, etc. This is commonly used in feature engineering.

Joining
: Joining merges data from different sources based on a common identifier. This is often necessary when working with complex datasets that are split across multiple tables or files. Joining is also used to integrate disparate datasets.

Pivot
: Pivot is used to transform or reshape data. It's commonly used to transform data from long format to wide format.

Discretization
: This is the process of converting continuous variables into discrete ones. This can be useful when you want to create categorical variables that indicate ranges of a continuous variable.

Both R and Python provide several libraries for manipulation. In R, data manipulation tasks can be performed with packages like dplyr. Python offers several libraries for data manipulation, with pandas being the most popular. The following are code samples that show basic functionality, but in reality this could be a complex process and requires a fair amount of thinking and business know-how.

Here is an example in Python:

```
# Import pandas
import pandas as pd

# Load data
df = pd.read_csv('data.csv')

# Sorting
df_sorted = df.sort_values(by='column_name')

# Filtering
df_filtered = df[df['column_name'] > 100]
```

```
# Aggregating
df_agg = df.groupby('column_name').agg({'another_column': 'sum'})
```

And in R:

```
# Load dplyr
library(dplyr)

# Load data
df <- read.csv("data.csv")

# Sorting
df_sorted <- df %>% arrange(column_name)

# Filtering
df_filtered <- df %>% filter(column_name > 100)

# Aggregating
df_agg <- df %>% group_by(column_name) %>%
summarise(sum_column = sum(another_column))
```

Remember, the goal of data manipulation is to create a clean, well-structured, correctly formatted dataset that a machine learning model can learn from effectively. The specific manipulations required will depend on the nature of your data and the model you're using.

Let's walk through an example of data manipulation by sorting, filtering, and aggregating in Python using the popular Titanic dataset. This dataset contains information about passengers on the Titanic, and we'll perform the following operations:

Sorting
: Sort the data by age.

Filtering
: Filter passengers who survived and are female.

Aggregating
: Calculate the mean age of survivors grouped by their passenger class.

Here's how you can do it step-by-step using pandas. First we will load the Titanic dataset:

```
import pandas as pd
import seaborn as sns

# Load the Titanic dataset
df = sns.load_dataset('titanic')

# Show the first few rows to understand the structure of the data
print(df.head())
survived  pclass     sex   age  sibsp  parch     fare embarked  class  \
0         0       3    male  22.0      1      0   7.2500        S  Third
```

```
1         1       1  female  38.0       1       0  71.2833         C  First
2         1       3  female  26.0       0       0   7.9250         S  Third
3         1       1  female  35.0       1       0  53.1000         S  First
4         0       3    male  35.0       0       0   8.0500         S  Third

     who  adult_male deck  embark_town alive  alone
0    man        True  NaN  Southampton    no  False
1  woman       False    C    Cherbourg   yes  False
2  woman       False  NaN  Southampton   yes   True
3  woman       False    C  Southampton   yes  False
4    man        True  NaN  Southampton    no   True
```

Now, let's sort by the age column:

```
# Sort the data by age in ascending order
df_sorted = df.sort_values(by='age', ascending=True)

# Display the first 10 rows of sorted data
print(df_sorted[['age', 'sex', 'survived']].head(10))

      age     sex  survived
803  0.42    male         1
755  0.67    male         1
644  0.75  female         1
469  0.75  female         1
78   0.83    male         1
831  0.83    male         1
305  0.92    male         1
827  1.00    male         1
381  1.00  female         1
164  1.00    male         0
```

Now, we'll filter the dataset to include only female passengers who survived:

```
# Filter the data for female survivors
df_female_survivors = df[(df['sex'] == 'female') & (df['survived'] == 1)]

# Display the filtered data
print(df_female_survivors[['sex', 'age', 'survived']].head())

      sex   age  survived
1  female  38.0         1
2  female  26.0         1
3  female  35.0         1
8  female  27.0         1
9  female  14.0         1
```

Finally, we will group the data by class (passenger class) and calculate the mean age of the survivors in each class:

```
# Group the data by class and calculate the mean age of survivors
df_survivor_age_by_class = df[df['survived'] == 1].groupby('class')['age'].mean()

# Display the result
```

```
print(df_survivor_age_by_class)

class
First     35.368197
Second    25.901566
Third     20.646118
```

This example demonstrates common data manipulation tasks: sorting, filtering, and aggregating using the Titanic dataset in Python.

Data preparation is a crucial aspect of machine learning projects, involving data manipulation and transformation to ensure data quality and suitability for analysis. Techniques such as data formatting, scaling, normalization, and standardization are employed to clean and preprocess the data. Additionally, derived attributes are created to provide valuable insights for predictive models. Both R and Python offer various libraries and functions to perform these data preparation tasks effectively, ensuring the optimal performance of machine learning algorithms.

Selecting and Reducing Features

Features in machine learning are individual measurable properties or characteristics of the phenomena being observed. Simply put, features are the "input" data in the model, which are used to make predictions about the predicted variable. Once we have prepared additional features and cleaned the data, it's crucial to perform feature selection or dimensionality reduction to prevent overfitting. Overfitting occurs when a model learns not only the underlying patterns but also noise and irrelevant details in the training data, leading to poor generalization of new data. By selecting the most important features or reducing the feature space (using methods like PCA or LASSO regression), we can simplify the model, reduce complexity, and improve its performance on unseen data. This helps ensure the model focuses on the most relevant information and avoids overfitting. The end goal of EDA is to have a prepared dataset with the features expected to be used in the modeling phase of the project. Selecting the most relevant features to use is important for several reasons:

- Simplification of models, which improves interpretability
- Shorter training times
- Improved accuracy of the model
- Reducing overfitting on the training data

There are several feature selection methods including filter, wrapper, embedded, and iterative methods that can assist in selecting a subset of relevant features. I'll cover each of these in greater detail throughout the rest of this chapter.

Feature Selection

Feature selection is a critical aspect of machine learning and data analysis, particularly when dealing with high-dimensional datasets. Among the various methods employed for feature selection, filter methods stand out for their simplicity and effectiveness. These methods utilize statistical measures to score each feature based on its relevance to the target variable, allowing for the identification of the most informative features.

Filter methods

Filter methods are methods that apply a statistical measure to assign a scoring to each feature, and then the features are ranked by the score and either selected to be kept or removed from the dataset. Examples include the chi-squared test, information gain, and correlation coefficient scores. Filter methods for feature selection involve selecting relevant features based on their correlation with the target variable. This approach considers each feature individually or in the context of other features, with respect to the target variable. Let's explore a few examples of how you can use filter methods for feature selection in R and Python.

In R, you can use the `cor()` function to calculate the correlation between features and the target variable, and select the features with the highest correlation:

```
# Assume df a data frame and 'target' is the predicted variable

correlations <- cor(df)
cor_with_target <- correlations[, 'target']

# Select top 5 features with highest correlation
selected_features <- head(sort(abs(cor_with_target), decreasing = TRUE), 5)
```

In this R example, absolute values were used to consider both high positive and high negative correlations, as both imply strong relationships with the predicted variable.

In Python, you can use the `SelectKBest` class from the `sklearn.feature_selection` module:

```
from sklearn.feature_selection import SelectKBest, f_classif

# Assume X is your feature set and y is the target variable
# 'k' is the number of top features to select. You can tweak the value of 'k'
# based on your requirements

selector = SelectKBest(f_classif, k=5)
X_new = selector.fit_transform(X, y)

# To get the selected feature names
selected_features = X.columns[selector.get_support()]
```

In this example, f_classif is used as a scoring function for classification tasks. The f_classif function in scikit-learn performs an ANOVA F-test to evaluate the statistical significance of each feature in relation to the target variable in a classification problem, helping to identify which features are most likely to impact the target. If you're performing regression, you might use f_regression instead.

Please remember to replace target, X, y, and df with your own predicted variable and dataset. Additionally, these examples assume that your data is suitable for correlation analysis and does not contain any missing values.

Iterative methods

Iterative methods involve training models over and over with different subsets of features and observing which ones lead to the best-performing model. These methods for feature selection are also known as sequential feature selection. Iterative methods focus on adding or removing features based on a certain criteria until a specific stopping condition is met. Forward selection and backward elimination are two commonly used iterative techniques.

Forward selection is a type of iterative method where we start with having no features in the model. In each iteration, we keep adding the feature that best improves our model until an addition of a new variable does not improve the performance of the model.

Here's the R code using the leaps package:

```
# install and load the necessary packages
install.packages("leaps")
library(leaps)

# load built-in mtcars dataset
data(mtcars)

# Define the target and predictors
x <- mtcars[,3:ncol(mtcars)]
y <- mtcars[,2]

# Run the forward selection
forward <- regsubsets(y~., data=mtcars, nvmax=ncol(x), method="forward")

# Get the results
results <- summary(forward)

# View the best set of variables for each model size
print(results$which)
```

And here's the equivalent Python version using mlxtend:

```
# import the necessary libraries
import pandas as pd
```

```
import seaborn as sns
from mlxtend.feature_selection import SequentialFeatureSelector as SFS
from sklearn.linear_model import LinearRegression

# load built-in mpg dataset
df = sns.load_dataset('mpg').dropna()

# define the target and predictors
X = df.drop('mpg', axis=1)._get_numeric_data() # get the predictors
y = df['mpg'] # target feature

# forward feature selection
sfs = SFS(LinearRegression(),
          k_features=(3,6),
# Change this range according to the number of features in your dataset
          forward=True,
          floating=False,
          scoring='neg_mean_squared_error',
          cv=10)

sfs.fit(X, y)

# print out the selected features.
selected_features= X.columns[list(sfs.k_feature_idx_)]
print(selected_features)
```

```
     cylinders  displacement  horsepower  weight  acceleration  model_year
0            8         307.0       130.0    3504          12.0          70
1            8         350.0       165.0    3693          11.5          70
2            8         318.0       150.0    3436          11.0          70
3            8         304.0       150.0    3433          12.0          70
4            8         302.0       140.0    3449          10.5          70
..         ...           ...         ...     ...           ...         ...
393          4         140.0        86.0    2790          15.6          82
394          4          97.0        52.0    2130          24.6          82
395          4         135.0        84.0    2295          11.6          82
396          4         120.0        79.0    2625          18.6          82
397          4         119.0        82.0    2720          19.4          82

[392 rows x 6 columns]
Index(['cylinders', 'weight', 'model_year'], dtype='object')
```

The `selected_features` variable lists the features that were selected by a feature selection process (likely Sequential Feature Selection, using `sfs.k_feature_idx_`). This method has identified the most important features based on some selection criteria (e.g., maximizing model performance). In this case, the selected features are: `cylinders`, `weight`, and `model_year`.

These three features were found to be the most predictive or relevant to the target variable, according to the feature selection process. The `sfs.k_feature_idx_` refers to the indices of the columns that were selected during the feature selection process.

The selected_features list then maps those indices to the actual column names from X. So after running the feature selection process, the most important features identified are 'cylinders', 'weight', and 'model_year', which are likely the most useful features for the predictive model being built.

Please note that Python doesn't support forward selection out of the box as R does. We need to use the mlxtend package, which isn't included in the standard library but can be installed via pip (pip install mlxtend). The k_features parameter specifies the number of features to select; you can adjust this according to your dataset and specific use case. Finally, always remember to split data into training and validation sets (or use cross-validation) to evaluate the generalizability of the model to new, unseen data.

Wrapper methods

Wrapper methods for feature selection evaluate subsets of variables to determine which subsets will perform better in terms of model prediction. A model is used as a part of the evaluation criteria. The search for the optimal subset can be done in various ways, including sequential forward selection, sequential backward selection, a combination of both (bidirectional), or recursive feature elimination (RFE).

The RFE method is a wrapper method for feature selection. RFE works by recursively removing attributes and building a model on the remaining attributes. It uses the model accuracy to identify which attributes contribute the most to predict the target attribute.

Here is the R version using the caret package:

```
# install the necessary package
install.packages("caret")
library(caret)

# load built-in mtcars dataset
data(mtcars)

# define the target and predictors
x <- mtcars[,3:ncol(mtcars)]
y <- mtcars[,2]

# define the control using a random forest selection function
ctrl <- rfeControl(functions=rfFuncs, method="cv", number=10)

# run the RFE
results <- rfe(x, y, sizes=c(1:10), rfeControl=ctrl)

# display the results
print(results)

Recursive feature selection
```

```
Outer resampling method: Cross-Validated (10 fold)

Resampling performance over subset size:

 Variables   RMSE Rsquared    MAE RMSESD RsquaredSD  MAESD Selected
         1 0.3873   0.8858 0.2238 0.4191    0.21336 0.2420
         2 0.2428   0.9736 0.1766 0.1921    0.04839 0.1359        *
         3 0.2825   0.9618 0.2211 0.2800    0.05443 0.2207
         4 0.3715   0.9715 0.3095 0.2780    0.03834 0.2212
         5 0.3861   0.9739 0.3195 0.2515    0.02824 0.2065
         6 0.3303   0.9736 0.2619 0.2420    0.03026 0.1895
         7 0.3232   0.9788 0.2525 0.2269    0.02206 0.1820
         8 0.3277   0.9772 0.2578 0.2256    0.02065 0.1866
         9 0.3147   0.9758 0.2466 0.2268    0.02513 0.1838

The top 2 variables (out of 2):
   disp, hp
```

And here's the equivalent Python version using sklearn and Seaborn:

```
# import the necessary libraries
import seaborn as sns
from sklearn.feature_selection import RFE
from sklearn.svm import SVR
from sklearn import preprocessing

# load built-in motorcars dataset
df = sns.load_dataset('mpg').dropna()

# define the target and predictors
X = df.drop('mpg', axis=1)._get_numeric_data() # get the predictors
Y = df['mpg'] # target feature

# create a base classifier used to evaluate a subset of attributes
estimator = SVR(kernel="linear")

# create the RFE model and select 3 attributes
selector = RFE(estimator, n_features_to_select=3, step=1)
selector = selector.fit(X, Y)

# print out the ranking
print(selector.ranking_)
[1 3 2 4 1 1]
```

In both code snippets, the output is a ranking of features. In the Python example, `1` indicates a feature that was selected.

Embedded methods

Embedded methods learn which features best contribute to the accuracy of the model while the model is being created. The most common type of embedded feature selection methods are regularization methods. Regularization methods introduce additional constraints into the optimization of a predictive algorithm (such as a regression algorithm) that bias the model toward lower complexity (fewer coefficients). These methods for feature selection combine the benefits of both filter and wrapper methods. They are implemented by algorithms that have their own built-in feature selection methods. Some of the most popular examples of these methods are LASSO and ridge regression, which have built-in penalization functions to reduce overfitting.

Here are examples of using embedded methods for feature selection in R and Python. In R, the glmnet package can be used for LASSO regression:

```
library(glmnet)

# Assume df is a data frame and 'target' is the target variable

# Fit model with the data
fit <- glmnet(as.matrix(df[-'target']), df$target, alpha = 1)

# Get importance of features
importance <- abs(fit$beta[,ncol(fit$beta)])

# Getting the indices of important features
feature_indices <- which(importance > 0)

print("Important feature indices:", feature_indices)
```

In this R example, the `glmnet` function with `alpha = 1` for LASSO regression is used. The coefficients at the last step of the LASSO path (`fit$beta[,ncol(fit$beta)]`) are used to select important features. If running the code, remember to replace `X`, `y`, and `df` with your own data.

Here is an example using real data to demonstrate how to perform LASSO regression using the glmnet package in R, along with the process of selecting important features. We will use the built-in mtcars dataset for the demonstration, where we will predict miles per gallon (`mpg`) using LASSO regression:

```
# Load necessary packages
library(glmnet)

# Load the built-in mtcars dataset
data("mtcars")

# Define the target and feature variables
target <- "mpg"
df <- mtcars

# Convert the data frame to matrix format for glmnet
# Remove the target column from df to use as features
X <- as.matrix(df[, !(names(df) %in% target)])
```

```
y <- df[, target]

# Fit a Lasso model (alpha = 1 for Lasso)
fit <- glmnet(X, y, alpha = 1)

# Get the coefficients for the last model in the lambda sequence
# (the model with the smallest lambda)
importance <- abs(coef(fit)[-1, ncol(fit$beta)])  # Exclude the intercept term

# Get the indices of the important features (where coefficients > 0)
feature_indices <- which(importance > 0)

# Print the important features and their indices
important_features <- names(df)[feature_indices + 1]
# +1 because coef excludes intercept
cat("Important feature indices:", feature_indices, "\n")
Important feature indices: 1 2 3 4 5 6 7 8 9 10
cat("Important features:", important_features, "\n")
Important features: cyl disp hp drat wt qsec vs am gear carb
```

The LASSO regression (L1 regularization) automatically performs feature selection by shrinking the coefficients of less important features to zero. After fitting the model, we extract the important features by checking which coefficients are nonzero. For the mtcars dataset, you will get a list of the most important features (e.g., `hp`, `wt`) based on the LASSO regression model.

In Python, the sklearn library has the LASSO class. LASSO adds a penalty equivalent to the absolute value of the magnitude of coefficients. Here's an example:

```
from sklearn.linear_model import LassoCV

# Assume X is a feature set and y is the target variable

# Fit model with the data
lasso = LassoCV().fit(X, y)

# Get importance of features
importance = np.abs(lasso.coef_)

# Getting the indices of important features
feature_indices = np.where(importance > 0)[0]

print("Important feature indices:", feature_indices)
```

In this example, `LassoCV` is used to fit the data, and then the coefficients are used to select important features. Here is an equivalent example using Python to perform LASSO regression. Since the glmnet package is not available in Python, we will use scikit-learn's implementation of LASSO regression (LASSO class) for feature selection:

```
import numpy as np
import pandas as pd
from sklearn.linear_model import Lasso
from sklearn.preprocessing import StandardScaler

# Load the mtcars dataset
url = "https://raw.githubusercontent.com/selva86/datasets/master/mtcars.csv"
df = pd.read_csv(url, index_col=0)  # Set the car names as the index

# Print column names to confirm the structure
print(df.columns)

# Since 'mpg' is not available, we'll use 'hp' (horsepower) as the target
target = 'hp'

# Drop irrelevant columns (like 'carname', 'fast', and 'cars' if present) to
# focus on numerical data
df_cleaned = df.drop(columns=['carname', 'fast', 'cars'])

# Define the target and features
X = df_cleaned.drop(columns=[target])  # Features (all columns except hp)
y = df_cleaned[target]  # Target (hp)

# Standardize the features to ensure that Lasso regularization treats all
# features equally
scaler = StandardScaler()
X_scaled = scaler.fit_transform(X)

# Fit the Lasso model (alpha controls regularization strength)
lasso = Lasso(alpha=0.1)
lasso.fit(X_scaled, y)

# Get the coefficients from the Lasso model
coefficients = lasso.coef_

# Get the indices of the important features (where coefficients are non-zero)
important_indices = np.where(coefficients != 0)[0]

# Print the important feature indices and their names
important_features = X.columns[important_indices]
print("Important feature indices:", important_indices)
print("Important features:", important_features)
Index(['cyl', 'disp', 'hp', 'drat', 'wt', 'qsec', 'vs', 'am', 'gear', 'carb',
       'fast', 'cars', 'carname'],
      dtype='object')
Important feature indices: [0 1 2 3 4 5 6 7 8]
Important features: Index(['cyl', 'disp', 'drat', 'wt', 'qsec', 'vs',
'am', 'gear', 'carb'], dtype='object')
```

As outlined in the R example, you will get a list of the most important features (e.g., `cyl`, `wt`) based on the LASSO regression model.

Feature Reduction Techniques

Feature reduction and feature selection are both methods used in machine learning for handling high-dimensional data, but they are different in their approach and purpose. Feature selection, as we just explored, is a process involving selecting the most relevant features from the original dataset to use in model construction. The goal of feature selection is to improve model interpretability, reduce overfitting, and improve computational efficiency.

Feature reduction, also known as dimensionality reduction, is a process involving transforming the original high-dimensional data into a lower-dimensional space. The goal of feature reduction is to reduce the dimensionality of the dataset while preserving as much information as possible.

They are used when datasets have a large number of features, which can cause overfitting, increased computational complexity, and decreased model interpretability.

There are different techniques for feature reduction based on the type of model to be created. Both regression and classification models approach feature reduction in different ways. While feature reduction techniques can be used for both regression and classification models, there can be differences in the methods used, mainly because the target variable's nature in regression is continuous, while in classification, it's categorical.

Feature reduction for regression

In regression, feature reduction methods might pay more attention to the linear relationships between features and the target. For example, backward elimination and forward selection in regression use metrics like p-values of coefficients or residual sum of squares to evaluate the importance of features. Several feature reduction techniques are commonly used in regression models, as outlined here:

Backward elimination
: This process involves fitting the model with all features and then iteratively removing the least significant feature (the one with the highest p-value exceeding a certain threshold). The process continues until all insignificant features are removed.

Forward selection
: This is the opposite of backward elimination. You start with no features and keep adding one feature at a time—the one that improves the model the most—until adding new features doesn't lead to improvement.

LASSO regression (L1 regularization)
: LASSO regression not only helps in reducing overfitting but can also be used for feature selection. LASSO can shrink the coefficients of less important features to exactly zero, which effectively eliminates those features.

Ridge regression (L2 regularization)
: Similar to LASSO, ridge regression reduces overfitting. However, it doesn't typically reduce the coefficients to zero but minimizes them, meaning that it includes all features but reduces the impact of less important ones.

Elastic net
: This is a combination of LASSO and ridge regression. It balances between eliminating some features (like LASSO) and keeping all of them (like ridge).

Principal component analysis (PCA)
: This technique is used to reduce linear dimensionality. It works by creating new uncorrelated variables that successively maximize variance. The first principal component accounts for the largest possible variance, and each succeeding component has the highest possible variance given the constraint that it must be orthogonal to (i.e., uncorrelated with) the preceding components.

Linear discriminant analysis (LDA)
: Similar to PCA, LDA also creates linear combinations of features, but it does so in a way that emphasizes the differences between classes. It is a supervised method, meaning that it uses class label information, and is typically used for classification problems.

It is important to note that PCA and LDA are both dimensionality reduction techniques, but they serve different purposes and can result in features that are more difficult to interpret for models.

PCA. PCA is an unsupervised technique that focuses on reducing the dimensionality of the data by finding new features (called principal components) that maximize the variance in the data. It is commonly used to compress high-dimensional data while retaining the most important patterns.

PCA transforms the data into a new coordinate system by finding linear combinations of the original features that capture the greatest variance. The first principal component captures the most variance, the second captures the next most, and so on. The new features (principal components) are combinations of the original features, making it hard to directly understand how they relate to the original variables. This complexity makes it difficult to explain how each original feature influences the outcome since the components are abstract combinations.

LDA. LDA is a supervised technique used for classification. It seeks to reduce dimensionality by finding linear combinations of features that best separate the classes (maximize class separability). LDA maximizes the ratio of between-class variance to within-class variance, ensuring that the resulting components (discriminants) help to distinguish between different classes as much as possible. Like PCA, LDA produces new features that are linear combinations of the original variables. These features are focused on separating classes, but their complex construction can make it hard to directly link back to the original variables, complicating the interpretation of which original features are most important for classifying or predicting outcomes.

In summary, both LDA and PCA generate new features that are combinations of the original ones. While these features improve model performance by reducing dimensionality and noise, they make it harder to interpret the model because the newly created features are less directly linked to the original data attributes, complicating explanations of how specific features influence the outcome.

Feature reduction for classification

In classification, methods like chi-squared test, information gain, or Gini importance are used to assess the correlation between categorical features and categorical target variables. There are several feature reduction techniques that are commonly used in classification models outlined here:

Autoencoders
: These are used for nonlinear dimensionality reduction. An autoencoder is a type of artificial neural network that learns to copy its input to its output. It has an internal (hidden) layer that describes a code used to represent the input, and it is in this layer where dimensionality reduction occurs.

t-distributed stochastic neighbor embedding (t-SNE)
: t-SNE is a nonlinear technique that is particularly good at preserving local structure, making it good for visualization of high-dimensional data.

Uniform Manifold Approximation and Projection (UMAP)
: UMAP is another nonlinear dimensionality reduction technique that has advantages over other techniques in terms of runtime performance, preservation of global structure, and the ability to handle large datasets.

Feature agglomeration
: This technique uses hierarchical clustering to group together features that are closely related.

Choosing which feature reduction technique to use will depend on the specifics of your dataset and the problem you are trying to solve. It is important to note that methods such as autoencoders, t-SNE, UMAP, and feature agglomeration can significantly improve the performance of models by reducing dimensionality and preserving important data structures. However, they also reduce the explainability of features due to the complexity of how new, transformed features are created.

Summary

This chapter has provided an in-depth exploration of data analysis in R and Python, focusing on clustering and unsupervised learning, feature evaluation, outlier identification, data manipulation, transformation, and feature reduction techniques.

You learned about the importance of clustering and unsupervised learning as they allow us to understand the inherent structure of the dataset without relying on predefined labels. Techniques such as k-means and hierarchical clustering, with code examples in both R and Python, were presented.

Then, the section on feature evaluation demonstrated how to assess the importance and relevance of different features in your dataset. This involves understanding each feature's distribution, correlation with other features, and its influence on the output variable. Both statistical and visualization techniques for feature evaluation were explored.

We shifted gears to focus on the identification and handling of outliers. Outliers, which are data points that significantly differ from others in the dataset, were introduced. These can impact the performance of your model, and it's important to leverage techniques for identifying and managing them.

We also learned about data manipulation and transformation and the process of preparing the data for analysis. This process involves data formatting, changing data types, handling missing values, and converting categorical variables to numerical ones. It's also important to consider derived attributes, which involve generating new features from existing ones to capture more complex relationships.

Scaling and standardization are crucial steps in data preprocessing, especially when dealing with features of different scales and units. It's important to understand the differences and when to use each technique.

Feature reduction techniques include methods such as PCA and LDA for dimensionality reduction. They can help mitigate issues related to high-dimensionality, such as overfitting, without losing important information.

Now that you have a strong understanding of how to conduct EDA effectively in both R and Python, let's explore modeling next.

CHAPTER 6

Application and Evaluation of Modeling in R and Python

After the meticulous process of exploratory data analysis, where data is visualized, understood, and preprocessed, we transition into one of the most pivotal stages in the data science life cycle: the modeling phase. This phase is where theoretical knowledge meets practical application. Leveraging the insights gleaned from EDA, data scientists select, design, and train models to predict or classify unknown outcomes. It's a stage where the cleaned and transformed data is fed into algorithms, turning raw information into actionable insights. As we delve deeper into the modeling phase, we'll explore various algorithms, techniques, and best practices to ensure that our models are both accurate and interpretable.

Modeling Steps

Before we dive too deeply into the modeling phase, let's consider the purpose of modeling. Modeling focuses on building and training predictive models using data to make accurate predictions or classifications on new, unseen data. Models can be supervised or unsupervised, and I'll cover both in this section.

Model Selection and Training

One of the first steps in model selection and training is to choose an algorithm. An *algorithm* is a set of well-defined rules or steps that a computer program follows to learn patterns or relationships from data and make predictions or decisions based on that learning. Machine learning algorithms work by analyzing input data (training data) to identify patterns or features, and then they apply this understanding to make predictions or classifications on new, unseen data. For example, in supervised learning, an algorithm learns a mapping from inputs (features) to outputs (labels) by minimizing error between its predictions and the actual results. Common machine learning algorithms include decision trees, linear regression, neural networks, and support vector machines (SVMs). Your choice should be based on the problem you intend to solve (for example, predicting a continuous or categorical result) and the characteristics of your data.

An example of an algorithm could be a decision tree or linear regression. A *decision tree* is a graphical representation and a machine learning algorithm that models decisions and their potential consequences in a tree-like structure. It's used for classification and regression tasks, where it splits data into subsets based on feature conditions to make predictions or decisions. Another example would be linear regression. *Linear regression* is a statistical method and machine learning algorithm used to model the relationship between a dependent variable and one or more independent variables by fitting a linear equation to the observed data. It's primarily used for predicting continuous numeric outcomes and understanding the linear relationship between variables.

Once the algorithm is chosen, the model will be created based on the application of the algorithm against the data. This is done with two steps: training and testing. The training and testing steps are fundamental to the modeling process. They ensure that the model not only fits the data it's trained on but also generalizes well to new, unseen data. Training involves feeding the training data into the chosen algorithm; then the algorithm learns from the data by adjusting internal parameters to minimize the difference between the predictions and the actual outcomes. The goal of this step is to find the model parameters that best fit the training data. The difference between the predictions and the actual outcomes is referred to as *the error or loss*. The goal is to find the model parameters that best fit the training data.

In some cases, a portion of the training data called the validation set is set aside to tune the model hyperparameters and prevent overfitting. *Hyperparameters* in machine learning are settings or configurations that are not learned from the data but are set prior to training a model. They control aspects like model complexity, learning rate, and regularization strength. Hyperparameter tuning involves finding the best combination of these settings to optimize a model's performance. *Overfitting* in machine learning refers to a modeling scenario where a complex model captures noise and random fluctuations in the training data, leading to poor generalization on new, unseen data. Essentially, it means the model has learned the training data too well but struggles to make accurate predictions on fresh data.

Model Evaluation

As a recap from Chapter 5, data for model evaluation is typically divided into three sets: training, validation, and testing. The training set is used to teach the model by allowing it to learn patterns and relationships in the data. The validation set is used during training to fine-tune the model's hyperparameters and ensure it generalizes well, helping to prevent overfitting by providing feedback on performance for unseen data. Finally, the test set is used after the model has been fully trained and optimized, serving as a final, unbiased evaluation of its ability to generalize to new, unseen data. This separation ensures reliable model performance. Different metrics are used to evaluate the model's performance. There are different metrics used for regression and classification, which we'll explore later in this chapter. At this time you would check for overfitting and focus on producing a model that generalizes well (one that doesn't just perform well on the training data).

Before we move on, generalization deserves a quick explanation and example. Imagine you're building a machine learning model to recognize handwritten digits (0–9). You have a dataset of thousands of images of digits, each labeled with the correct number:

- You use 70% of the images to train your model. The model "learns" by analyzing the patterns and shapes of the digits in these images (e.g., how the number 3 looks different from 8).
- Next, you use 15% of the images as a validation set to check how well the model performs on data it hasn't seen during training. If the model starts to memorize the training data (overfitting), its performance will be good on the training set but poor on the validation set. You adjust the model based on these results.
- Finally, you test the model on the remaining 15% of the images, which the model has never seen before. If the model correctly identifies digits in this test set, it means the model has generalized well—it has learned the underlying patterns of the digits rather than just memorizing the training examples. This ability to perform well on unseen data shows the model's generalization capability.

In short, model generalization is the ability of a model to apply what it has learned from the training set to make accurate predictions on new, unseen data (the test set).

Model Optimization

It is important to note that training, testing, and optimization are iterative steps, and the number of times you go through the process depends on when the model is considered optimal. You know you have optimized a predictive model to the point of stopping when further adjustments to the model's hyperparameters or features do not significantly improve its performance on the validation dataset, and the model meets your predefined criteria for accuracy or other relevant metrics. Through each iteration, you then refine the model by adjusting hyperparameters, using different algorithms, or going back to feature engineering. There are techniques that can be used to speed this process along that we will address later in the chapter.

Let's look quickly and review hyperparameter tuning. Imagine you're building a model using a decision tree to classify whether an email is spam or not based on features like the number of links, keywords, etc. For a decision tree, a couple of important hyperparameters include:

Max depth
: This controls how deep the tree can grow (how many splits it can make).

Min samples split
: This controls the minimum number of samples required to split a node.

Initially, you set the max depth of the tree to 5 and min samples split to 2. After training the model, you notice that it's overfitting—it performs well on the training data but poorly on the validation data. To address this, you begin hyperparameter tuning by adjusting the values of max depth and min samples split using a validation set. For example:

- Try max depth = 3 and min samples split = 5.
- Check performance on the validation set.
- Then try max depth = 4 and min samples split = 10.
- Check performance again.

You repeat this process with various combinations of hyperparameters and choose the combination that gives the best performance on the validation set. This fine-tuning helps the model balance between underfitting (too simple) and overfitting (too complex). By tuning hyperparameters like max depth and min samples split, you adjust how the model learns to improve its generalization to new data. This process is crucial because hyperparameters control the behavior of the model, and finding the right values ensures the best possible performance.

Model Deployment

If the model's performance after optimization is satisfactory, it can be deployed to a production environment where it can start taking in new data and making predictions or classifications in real time or in batches. Deploying the model to production involves the following steps:

1. Set up the deployment environment, which may involve configuring servers or cloud services.
2. Deploy the model to a production server or the cloud, making it accessible for predictions.
3. Test the deployed model to ensure it works as expected in the live environment.

After deployment, we are not finished; models require ongoing monitoring and maintenance.

Model Monitoring and Maintenance

Once a model is running in production, it will be necessary to continuously monitor the model's performance in the real world. This step is also referred to as model operations, which we will explore in more detail in Chapter 8. In this step the model is reviewed for performance as new data becomes available. If there are dips in performance, the model is retrained on new data to ensure its relevance and accuracy. Models are supported throughout the life of a model until the model needs to be retired or reaches obsolescence. Now that we have reviewed the modeling steps, let's do a deeper dive into these steps.

Selecting the Right Algorithm

In the realm of business analytics, an algorithm can be thought of as a set of well-defined computational procedures that take some input and produce an output. The efficiency of an algorithm can be critical, especially when dealing with large datasets. Efficient algorithms can save both time and computational resources. There are various types of algorithms, each suited for specific tasks. For example, supervised learning algorithms are used when the outcome variable (label) is known, unsupervised learning algorithms are used when the outcome variable is not known, and reinforcement learning algorithms are used in situations where decisions are made sequentially and the goal is long term, such as game playing or stock trading.

Selecting the right algorithm is the cornerstone of any successful data modeling endeavor. The effectiveness of a machine learning model hinges on its ability to accurately represent complex relationships within the data. An ill-suited algorithm may lead to suboptimal results or misrepresentations of reality. The following list highlights the different algorithm families and the primary use case for each:

Regression algorithms
: Used for predicting a continuous numeric outcome, such as predicting prices or scores. Examples include linear regression and decision trees.

Classification algorithms
: Employed for categorizing data into predefined classes or labels, like spam detection or image recognition. Common algorithms include logistic regression and random forest.

Dimensionality reduction algorithms
: These algorithms help reduce the number of input variables in a dataset while preserving important information. They are useful for visualization and feature selection. Principal component analysis (PCA) is a popular choice.

Neural networks
: Suitable for complex tasks like natural language processing, image recognition, and deep learning problems. Convolutional neural networks (CNNs) are ideal for image tasks, while recurrent neural networks (RNNs) excel in sequence data.

Ensemble algorithms
: Combine multiple models to improve predictive performance and reduce overfitting. Random forest and gradient boosting are widely used ensemble methods.

Anomaly detection algorithms
: Detects rare and unusual patterns or outliers in data, crucial for fraud detection, network security, and quality control. One-class SVMs and isolation forests are examples.

Recommendation algorithms
: Provides personalized recommendations, often seen in recommendation systems for movies, products, or content. Collaborative filtering and matrix factorization are common in this category.

Reinforcement learning algorithms
: Ideal for scenarios where agents learn to take actions to maximize rewards in an environment, such as in autonomous robotics and game playing. Q-learning and deep Q networks (DQNs) are prominent in reinforcement learning.

Choosing the appropriate evaluation metrics is paramount for assessing a model's performance. Accurate evaluation ensures that the model aligns with the specific goals of the analysis, be it classification, regression, clustering, or any other task. The evaluation process provides insights into the model's strengths and weaknesses, aiding in fine-tuning and optimization. We'll explore different types of algorithms as we examine regression and classification modeling scenarios.

Regression

Regression was introduced earlier, but let's do a quick recap. Regression, a fundamental concept in predictive modeling, plays a pivotal role in deciphering relationships between variables and making informed predictions. Regression, in the context of predictive modeling, is a statistical technique used to model the relationship between a dependent variable (often referred to as the target) and one or more independent variables (predictors or features). Its primary objective is to establish a mathematical equation that best describes the dependency between these variables. This equation serves as a predictive model, enabling us to make reasonable predictions about the target variable based on the values of the predictor variables.

Regression analysis is a cornerstone of predictive analytics because it allows us to understand the underlying patterns, trends, and dependencies within data. By fitting a regression model to observed data, we gain insights into how changes in predictor variables influence the target variable. These insights empower decision makers in various fields to make informed choices, optimize processes, and anticipate outcomes with a reasonable degree of certainty.

Common Use Cases

Regression finds widespread application in a multitude of fields, owing to its versatility and interpretability. Some common scenarios where regression analysis proves invaluable include:

Economics
: Regression helps economists understand how changes in factors such as interest rates, inflation, and consumer spending impact economic indicators like GDP and employment rates.

Finance
: Regression models assist in portfolio management, risk assessment, and asset pricing. They help investors and financial institutions make data-driven decisions for better returns and risk management.

Medicine
: Regression analysis is used in medical research to study the relationship between variables like patient age, lifestyle factors, and genetic predisposition in predicting disease outcomes or treatment effectiveness.

Marketing
: Marketers employ regression to determine the impact of advertising expenditures, pricing strategies, and customer demographics on sales and market share.

Environmental science
: Environmental scientists utilize regression to assess the impact of environmental factors on phenomena like climate change, species distribution, and pollution levels.

Social sciences
: In sociology, psychology, and other social sciences, regression helps researchers explore causal relationships between variables like socioeconomic status, education, and well-being.

Regression is a versatile and indispensable tool in predictive modeling, fostering a deeper understanding of relationships within data and facilitating evidence-based decision making across diverse domains.

Linear Regression Equation

Linear regression is widely known and is a simple equation:

$$y = \beta_0 + \beta_1 x + \epsilon$$

Let's walk through each aspect of the equation:

- y is the dependent variable (the variable we are trying to predict).
- x is the independent variable (the input).
- β_0 is the y-intercept.
- β_1 is the slope of the line.
- ϵ represents the error term (the difference between the observed value and the predicted value).

For multiple linear regression, where we have more than one independent variable, the equation extends to:

$$y = \beta_0 + \beta_1 x_1 + \beta_2 x_2 + \ldots + \beta_n x_n + \epsilon$$

Linear regression is a statistical method that models the relationship between a dependent variable and one or more independent variables by fitting a linear equation to the observed data. The goal of linear regression is to find the best fit straight line that accurately predicts the output values within a range:

Dependent and independent variables
: In the context of the equation, y is what we're trying to predict (dependent variable), while x (or x_1, x_2... in multiple regression) is the variable we use to make predictions and is the independent variable.

Coefficients
: The coefficients (β_0, β_1,...) determine the direction and magnitude of the relationship between the independent variables and the dependent variable. For instance, β_1 represents the change in y for a one-unit change in x, holding all other variables constant.

Error term
: In real-world data, there are many other factors at play, which might not be captured by our model. The error term, ϵ, accounts for the variability in y that cannot be explained by x. It's the difference between the observed value and the value predicted by the model.

Linear regression makes several assumptions, including linearity (the relationship between the independent and dependent variables is linear), independence (observations are independent of each other), homoscedasticity (constant variance of the errors), and normality (the errors are normally distributed). Because linear regression makes assumptions about the data, it may lead to inconsequential predictions when these conditions aren't met. In essence, linear regression provides a straightforward way to understand the relationships between variables and make predictions. By finding the line (or hyperplane in multiple dimensions) that best fits the data, we can make informed predictions on new data points.

Linear Regression in R

To wield the power of linear regression in R, the `lm()` function stands as a faithful ally. This function unleashes the capabilities of linear regression by fitting a linear model to your data. Let's dive into a practical example showcasing the application of linear regression in R. The following R code demonstrates how to create a simple linear regression model using the `lm()` function:

```
# Load necessary libraries if not already loaded
# install.packages("ggplot2") # If not already loaded
library(ggplot2) ❶

# Generate sample data
set.seed(123) ❷
x <- rnorm(100, mean = 50, sd = 10) ❸
y <- 2 * x + rnorm(100, mean = 0, sd = 5) ❹

# Create a data frame
data <- data.frame(x, y) ❺

# Fit a linear regression model
linear_model <- lm(y ~ x, data = data) ❻

# Display model summary
summary(linear_model) ❼

Call: ❽
   lm(formula = y ~ x, data = data)

Residuals: ❾
   Min      1Q  Median      3Q     Max
   -9.5367 -3.4175 -0.4375  2.9032 16.4520

Coefficients: ❿
         Estimate Std. Error t value Pr(>|t|)
(Intercept)  0.79778    2.76324   0.289    0.773
x            1.97376    0.05344  36.935   <2e-16 ***
   ---
Signif. codes:  0 '***' 0.001 '**' 0.01 '*' 0.05 '.' 0.1 ' ' 1 ⓫
```

```
Residual standard error: 4.854 on 98 degrees of freedom ⓬
Multiple R-squared:  0.933,  ⓭ Adjusted R-squared:  0.9323 ⓮
F-statistic:  1364 on 1 and 98 DF,  ⓯ p-value: < 2.2e-16 ⓰
```

❶ This line loads the ggplot2 package, which is a popular package in R for data visualization.

❷ `set.seed(123)` sets a seed for random number generation to ensure reproducibility. Every time you run this code, you'll get the same set of random numbers.

❸ x generates 100 random numbers from a normal distribution with a mean of 50 and a standard deviation of 10.

❹ y creates a linear relationship with x (y is approximately twice x) and adds some random noise to it. The noise is from a normal distribution with a mean of 0 and a standard deviation of 5.

❺ This line combines the vectors `x` and `y` into a dataframe named `data`.

❻ This line fits a linear regression model with `y` as the dependent variable and `x` as the independent variable using the dataframe data. The fitted model is stored in the `linear_model` object.

❼ `summary(linear_model)` provides a summary of the linear regression model, including coefficients, residuals, and other statistics.

❽ The call line shows the function call used to fit the model.

❾ The residuals are the differences between the observed values of y and the values predicted by the model. The summary provides the minimum, 1st quartile, median, 3rd quartile, and maximum residuals.

❿ The coefficients section provides details about the model's coefficients:

- `(Intercept)` is the y-intercept of the regression line.
- `x` is the slope of the regression line.
- `Estimate` is the estimated value of the coefficients.
- `Std. Error` is the standard error of the coefficients.
- `t value` is the t-statistic value for testing the hypothesis that the coefficient is equal to zero (no effect).
- `Pr(>|t|)` is the p-value associated with the t-statistic.

⓫ Significance codes provide a quick reference for the significance of each coefficient. In this case, x is highly significant as indicated by ***.

⓬ Residual standard error is the average amount that the response will deviate from the true regression line.

⓭ Multiple R-squared is the proportion of the variance in the dependent variable that is predictable from the independent variable(s). It's a measure of how well the model fits the data.

⓮ Adjusted R-squared is the R-squared value adjusted based on the number of predictors in the model.

⓯ F-statistic tests the hypothesis that all regression coefficients are equal to zero. A large F-statistic indicates that at least some predictors are related to the response.

⓰ p-value is the probability of observing a value as extreme as the F-statistic given that all the regression coefficients are zero. A very small p-value indicates that the predictors are doing a good job in explaining the variation in the response.

In summary, the linear regression model suggests a strong relationship between x and y, with x being a significant predictor of y. The model fits the data well, as indicated by the high R-squared value.

Linear Regression in Python

Here's how you would perform the same operations using Python with the help of the NumPy and statsmodels libraries:

```
# Import necessary libraries
import numpy as np ❶
import statsmodels.api as sm ❷

# Generate sample data
np.random.seed(123) ❸
x = np.random.normal(50, 10, 100) ❹
y = 2 * x + np.random.normal(0, 5, 100) ❺

# Fit a linear regression model
X = sm.add_constant(x)  # Add a constant (intercept) to the model ❻
model = sm.OLS(y, X).fit() ❼

# Display model summary
print(model.summary()) ❽
```

This Python code performs linear regression using the statsmodels library:

❶ numpy is imported as `np` to perform numerical operations.

❷ statsmodels.api is imported as `sm` for performing statistical modeling.

❸ The `np.random.seed(123)` sets a random seed for reproducibility.

❹ `x` is created as an array of 100 random numbers drawn from a normal distribution with mean 50 and standard deviation 10.

❺ `y` is generated as a linear function of `x` with some random noise added to it. This simulates a linear relationship between `x` and `y` with some variability.

❻ `X` is created by adding a constant term `(intercept)` to the `x` values using `sm.add_constant(x)`. This is required because linear regression models typically include an intercept term.

❼ The model is created using `sm.OLS(y, X).fit()`. This line fits a linear regression model to predict `y` based on `X` (including the intercept). The `fit()` function performs the regression and stores the results in the model object.

❽ `model.summary()` generates a summary of the linear regression results. This summary includes statistics such as coefficients, p-values, R-squared, and more. It provides insights into the quality and significance of the model fit.

The code essentially generates random data, fits a linear regression model to it, and then prints out a summary of the regression analysis, including information about the coefficients and their significance. This is a basic example of linear regression modeling for educational purposes. Let's explore the model summary (Figure 6-1).

```
                            OLS Regression Results
==============================================================================
Dep. Variable:                      y   R-squared:                       0.956
Model:                            OLS   Adj. R-squared:                  0.955
Method:                 Least Squares   F-statistic:                     2104.
Date:                Sun, 15 Oct 2023   Prob (F-statistic):           4.83e-68
Time:                        17:28:30   Log-Likelihood:                -299.78
No. Observations:                 100   AIC:                             603.6
Df Residuals:                      98   BIC:                             608.8
Df Model:                           1
Covariance Type:            nonrobust
==============================================================================
                 coef    std err          t      P>|t|      [0.025      0.975]
------------------------------------------------------------------------------
const          0.3194      2.237      0.143      0.887      -4.120       4.759
x1             1.9917      0.043     45.873      0.000       1.906       2.078
==============================================================================
Omnibus:                        5.027   Durbin-Watson:                   1.860
Prob(Omnibus):                  0.081   Jarque-Bera (JB):                5.131
Skew:                          -0.308   Prob(JB):                       0.0769
Kurtosis:                       3.924   Cond. No.                         235.
==============================================================================

Notes:
[1] Standard Errors assume that the covariance matrix of the errors is correctly specified.
```

Figure 6-1. Regression analysis model summary

This Python code provides a much more robust explanation of the linear regression model created:

Dep. variable
: This is the dependent variable, `y`, which is what you're trying to predict or explain.

Model
: Specifies the type of regression model, which in this case is ordinary least squares (OLS).

Method
: The method used to fit the data is "least squares."

Date and time
: The date and time when the regression was run.

No. observations

The number of data points or observations used in the regression.

Df residuals

Degrees of freedom of the residuals. It's the difference between the number of observations and the number of parameters estimated (here, it's 100 observations minus 2 parameters, giving 98 residual degrees of freedom).

Df model

The number of predictors in the model, excluding the constant. In this case, there's one predictor, x_1.

Latex or MathML needed

Represents the proportion of variance in the dependent variable (`y`) explained by the independent variable (`x1`). An R-squared value of 0.956 means 95.6% of the variability in `y` can be explained by x_1. However, high R-squared alone doesn't guarantee that the model fits well; it doesn't account for potential overfitting.

Adj. R-squared

Adjusted R-squared accounts for the number of predictors. Here, since there's only one predictor, the adjusted value (0.955) is very close to R-squared, which is typical when adding predictors does not significantly increase explanatory power.

F-statistic

The F-statistic (2104) tests the overall significance of the model. A high F-statistic indicates that at least one predictor (in this case, `x1`) has a significant relationship with `y`.

Prob (F-statistic)

This is the p-value associated with the F-statistic. A very small value (4.83e-68) suggests that the model is statistically significant, meaning that the predictor (`x1`) is significantly related to the outcome (`y`).

Log-likelihood

This measures how well the model fits the data. It's primarily used in model comparison, with higher values indicating a better fit.

AIC/BIC

Akaike information criterion (AIC = 603.6) and Bayesian information criterion (BIC = 608.8) are used for model selection. Lower values are preferred, but they are most useful when comparing multiple models.

Covariance type

Refers to assumptions made about the covariance of the errors. Nonrobust indicates the standard assumption of homoscedasticity (constant variance).

Coefficients and statistics

Next we examine key residual and model diagnostic tests used to assess the validity and performance of regression models. These diagnostics help determine if the model assumptions—such as normality of residuals, lack of autocorrelation, and multicollinearity—are being met:

coef (x1)
: The coefficient for `x1` is 1.9917, indicating that for every one-unit increase in `x1`, `y` is expected to increase by 1.9917 units.

const (intercept)
: The constant (0.3194) represents the expected value of `y` when `x1` is 0, although the high p-value (0.887) suggests it is not statistically significant.

std err
: The standard error for each coefficient estimates the precision of the coefficient. Lower values, like 0.043 for `x1`, indicate high accuracy.

t
: The t-values test whether the coefficient is significantly different from 0. Larger t-values indicate statistical significance, and for x_1, a t-value of 45.873 shows high significance.

P>|t|
: The p-value for the t-test, where values below 0.05 (e.g., 0.000 for x_1) indicate that the coefficient is significantly different from 0.

[0.025 0.975]
: This is the 95% confidence interval for the coefficient. If 0 falls within the interval, the variable may not be significant. Since the interval for x_1 is (1.906, 2.078), it confirms the significance of the coefficient.

Residual and model diagnostics

Next we examine key residual and model diagnostic tests used to assess the validity and performance of regression models. These diagnostics help determine if the model assumptions—such as normality of residuals, lack of autocorrelation, and multicollinearity—are being met:

Omnibus test
: A test for the normality of residuals, where a value near 0 indicates normal distribution. A p-value of 0.081 suggests that the residuals are not significantly different from normal.

Prob(Omnibus)

This is the p-value for the omnibus test, and in this case, 0.081 means normality cannot be rejected.

Skew

Measures the asymmetry of residuals. A value close to 0 (–0.308) indicates slight skewness, but nothing too concerning.

Kurtosis

Measures the "tailedness" of the distribution of residuals. A kurtosis value of 3.924 is slightly higher than the normal value of 3, indicating heavier tails than a normal distribution.

Durbin-Watson

Tests for autocorrelation in the residuals. A value of 1.860 suggests no significant autocorrelation, which is desirable.

Jarque-Bera (JB)

Another test for normality of residuals. A p-value of 0.0769 similarly suggests that the residuals are not significantly different from normal.

Cond. No

The condition number measures multicollinearity or numerical instability in the model. A value of 235 is higher than 30, which could indicate potential multicollinearity concerns, but it is not extreme.

The examples in R and Python show syntax on how to execute linear regression using one input variable, otherwise known as simple linear regression. In most real-world cases, multiple linear regression is used, where more than one input variable is used for prediction.

Linear Regression Use Case

Multiple linear regression is an extension of simple linear regression, a statistical method used in predictive modeling and data analysis. While simple linear regression involves a single independent variable (predictor) and one dependent variable (outcome), multiple linear regression deals with two or more independent variables to predict a single dependent variable. The equation for multiple linear regression is similar to simple linear regression, but you will see more terms, one for each input variable:

$$Y = \beta_0 + \beta_1 X_1 + \beta_2 X_2 + \beta_3 X_3 + \ldots + \beta_p X_p + \epsilon$$

where:

- Y is the dependent variable.
- β_0 is the intercept (the value of Y when all X variables are zero).
- β_1, β_2, β_3,...,β_p are the coefficients that represent the strength and direction of the relationship between each independent variable and the dependent variable.
- X_1, X_2 ,X_3... X_n are the independent variables.
- ϵ represents the error term, accounting for variability not explained by the model.

Let's explore a dataset to use with multiple linear regression. The "gapminder" dataset is a well-known dataset in the field of analytics and is often used for demonstrating various data analysis techniques. It contains information on various socioeconomic indicators for different countries over multiple years. Here's a description of the key variables in the "gapminder" dataset:

country
: The name of the country

continent
: The continent to which the country belongs (e.g., Africa, Asia, Europe, Americas, Oceania)

year
: The year of data observation

lifeExp
: Life expectancy at birth in years

pop
: Population

gdpPercap
: GDP per capita in US dollars

We will explore this data using R. You may need to install the gapminder library to get started:

```
install.packages("gapminder")

Then load the gapminder library and data:

library("gapminder")
data("gapminder")
```

Then explore the structure:

```
str(gapminder)
tibble [1,704 × 6] (S3: tbl_df/tbl/data.frame)
 $ country  : Factor w/ 142 levels "Afghanistan",..: 1 1 1 1 1 1 1 1 1 1 ...
 $ continent: Factor w/ 5 levels "Africa","Americas",..: 3 3 3 3 3 3 3 3 3 3 ...
 $ year     : int [1:1704] 1952 1957 1962 1967 1972 1977 1982 1987 1992 1997 ...
 $ lifeExp  : num [1:1704] 28.8 30.3 32 34 36.1 ...
 $ pop      : int [1:1704] 8425333 9240934 10267083 11537966 13079460 148803 ...
 $ gdpPercap: num [1:1704] 779 821 853 836 740 ...
```

Now that we are familiar with the data, let's create a linear regression model:

```
# Load the necessary libraries
library(dplyr)
library(lmtest) ❶

# Load the gapminder dataset into a dataframe
data <- gapminder ❷

# Perform multiple linear regression with life expectancy and GDP
model <- lm(lifeExp ~ gdpPercap + pop, data = data) ❸

# Summarize the regression model
summary(model) ❹

Call:
lm(formula = lifeExp ~ gdpPercap + pop, data = data) ❺

Residuals: ❻
    Min      1Q  Median      3Q     Max
-82.754  -7.745   2.055   8.212  18.534

Coefficients: ❼
             Estimate Std. Error t value Pr(>|t|)
(Intercept) 5.365e+01  3.225e-01  166.36  < 2e-16 ***
gdpPercap   7.676e-04  2.568e-05   29.89  < 2e-16 ***
pop         9.728e-09  2.385e-09    4.08 4.72e-05 ***
---
Signif. codes:  0 '***' 0.001 '**' 0.01 '*' 0.05 '.' 0.1 ' ' 1

Residual standard error: 10.44 on 1701 degrees of freedom
Multiple R-squared:  0.3471, Adjusted R-squared:  0.3463 ❽
F-statistic: 452.2 on 2 and 1701 DF,  p-value: < 2.2e-16 ❾

# Test for multicollinearity using VIF (Variance Inflation Factor)
vif(model) ❿

gdpPercap       pop
 1.000656  1.000656
```

1. Load the required libraries: gapminder for the dataset, `dplyr` for data manipulation, and `lmtest` for testing multicollinearity.

2. Load the gapminder dataset into the data variable.

3. Perform multiple linear regression using the `lm` function. In this case, we're predicting `lifeExp` (life expectancy) based on `gdpPercap` (GDP per capita) and `pop` (population).

4. Summarize the regression model using `summary(model)` to see the regression coefficients, R-squared value, and other relevant statistics.

5. `lifeExp` (Life Expectancy) is the dependent variable (response). `gdpPercap` (GDP per capita) and `pop` (Population) are predictor variables (features).

6. Residuals represent the differences between the observed life expectancies and the values predicted by the model. The model's predictions tend to have errors ranging from –82.75 to 18.53 years.

7. The intercept (baseline life expectancy) is approximately 53.65 years. For every one-unit increase in gdpPercap, life expectancy increases by approximately 0.0007676 years (roughly 0.028 days). For every one-unit increase in pop, life expectancy increases by approximately 9.728e-09 years (roughly 0.316 seconds).

8. The model explains some of the variance in life expectancy, as indicated by the Multiple R-squared of 0.3471. This means that approximately 34.71% of the variability in life expectancy can be explained by the predictor variables.

9. The p-value (< 2.2e-16) for the F-statistic indicates that the model is statistically significant, suggesting that at least one of the predictor variables is useful in predicting life expectancy.

10. To check for multicollinearity between the predictor variables (`gdpPercap` and `pop`), we use the Variance Inflation Factor (VIF) test with `vif(model)`. VIF values greater than 10 may indicate multicollinearity.

This model suggests that both GDP per capita (gdpPercap) and Population (pop) are statistically significant predictors of life expectancy, but the impact of population is very small compared to GDP per capita. For a business analyst, this model implies that improving economic conditions (GDP per capita) tends to have a more substantial impact on increasing life expectancy compared to changes in population size. However, other factors not included in this model may also influence life expectancy, and further analysis may be needed to better understand the relationship.

Other Types of Regression

When building predictive models, different types of regression techniques are used depending on the nature of the data and the problem at hand. Three commonly used approaches are polynomial regression, multivariate regression, and time series regression. Each of these methods caters to specific scenarios, from handling nonlinear relationships to predicting multiple outcomes or analyzing time-dependent data. Let's do a brief overview of how each technique works and when it is most applicable.

Polynomial regression

Polynomial regression is a type of regression analysis used when the relationship between the independent variable(s) and the dependent variable is nonlinear. Instead of fitting a straight line, it fits a curve to the data by using polynomial terms (e.g., quadratic or cubic). The model takes the form $y=\beta_0 + \beta_1 x + \beta_2 x^2 + \cdots + \beta_n x^n$, where the degree of the polynomial (n) determines the complexity of the curve. Polynomial regression is useful in cases where data follows a curved trend and linear regression fails to capture the patterns effectively. However, it's important to avoid overfitting by not using a polynomial degree that's too high.

Multivariate regression

This technique is used to predict more than one dependent variable based on one or more independent variables. Unlike multiple linear regression, which predicts a single outcome, multivariate regression models simultaneous relationships between independent variables and multiple outcomes. This is useful when multiple related outcomes need to be predicted together, such as forecasting sales and customer satisfaction based on marketing spend and customer engagement. Multivariate regression allows the model to capture the correlations between the multiple dependent variables, providing a more holistic view of how inputs influence various outcomes.

Time series regression

Time series regression is used to analyze and predict outcomes where data points are ordered chronologically. In this type of regression, the dependent variable is influenced by past values, time-related patterns (like trends or seasonality), or other predictors over time. Time series regression models are widely used in fields like finance, economics, and weather forecasting. Unlike other regression types, time series models often account for temporal dependencies, making use of lag variables or autocorrelation to capture relationships within the data. It's ideal for making predictions based on historical data and identifying trends or cyclical patterns over time.

In addition to these regression approaches, other types of regression techniques, such as LASSO and ridge regression, are used to improve model performance, especially when dealing with multicollinearity or high-dimensional datasets.

LASSO regression

LASSO regression is a form of linear regression that adds an L1 regularization term to the cost function. This penalty term encourages the model to shrink some of the coefficients to exactly zero, effectively performing feature selection by excluding less important variables. This makes LASSO particularly useful when you have many features, as it helps simplify the model and prevent overfitting by keeping only the most relevant features.

Ridge regression

Ridge regression, on the other hand, adds an L2 regularization term to the cost function, which penalizes large coefficients but does not shrink them to zero. Instead of excluding features, ridge regression reduces the magnitude of the coefficients, helping to mitigate overfitting by distributing the effect of multicollinearity across the features. Ridge is useful when all features are believed to contribute to the prediction but you want to regularize or stabilize the model.

Elastic net

An extension of LASSO and ridge regression is elastic net, which combines both L1 and L2 regularization to benefit from both feature selection (like LASSO) and coefficient shrinkage (like ridge). This flexibility makes elastic net a powerful tool when dealing with high-dimensional data or when both feature selection and regularization are needed.

Ridge, LASSO, and elastic net are commonly used when the dataset has many features, some of which may be irrelevant or correlated, and help in preventing overfitting while improving model generalization.

Challenges with Regression Models

It is important to note that linear regression is a statistical learning model and presents challenges with statistical learning. Recall from Chapter 5 that certain data characteristics should be present when choosing linear regression as an algorithm, including:

- The relationship between the independent variable(s) and the dependent variable should be approximately linear. Changes in the independent variable(s) should result in proportional changes in the dependent variable when plotted.
- Observations (data points) should be independent of each other. The value of the dependent variable for one observation should not depend on the value of the dependent variable for another observation.
- The residuals (the differences between observed and predicted values) should be normally distributed. This assumption is more critical for smaller sample sizes.

- If there are multiple independent variables, they should not be highly correlated with each other. High multicollinearity can make it challenging to attribute the variation in the dependent variable to a specific independent variable.
- Outliers, which are data points significantly different from the majority of the data, can have a strong influence on the regression model. Addressing or removing outliers may be necessary.

Even when these characteristics are present, tuning statistical models can be challenging. Some challenges include the following:

Feature selection
: Choosing the right set of features or variables can be tricky. Including irrelevant or redundant features can lead to overfitting, while omitting important ones can result in underfitting.

Model complexity
: Balancing model complexity is crucial. Too many variables or polynomial terms can make the model overly complex, while an overly simple model may not capture underlying patterns.

Regularization
: Linear regression models may require regularization techniques like ridge or LASSO regression to prevent overfitting. Selecting appropriate regularization strengths is a challenge.

Hyperparameters
: Linear regression models may have hyperparameters, such as the learning rate in gradient descent. Tuning these hyperparameters for optimal performance can be nontrivial.

Outliers
: Identifying and handling outliers in the data is essential. Outliers can strongly influence the model's coefficients and predictions.

Multicollinearity
: Dealing with multicollinearity, where predictor variables are highly correlated, requires careful consideration and sometimes feature engineering.

Bias-variance trade-off
: Striking the right balance between bias and variance is crucial. Reducing bias can increase variance and vice versa. Finding the sweet spot is challenging.

Data size
: The amount of data available for training can affect model performance. Small datasets may lead to overfitting, while large datasets may require efficient optimization techniques.

Other Algorithms for Regression

In the realm of data analysis and predictive modeling, regression models are fundamental tools that help in understanding and predicting continuous variables based on other variables in the dataset. While traditional statistical regression, such as linear regression, is widely used due to its simplicity and interpretability, there are numerous other algorithms that can be employed for regression tasks. Each of these algorithms has unique features and is suitable for different types of data and modeling requirements.

The choice of a regression algorithm depends on several factors, including the nature of the relationship between variables (linear or nonlinear), the presence of outliers, the number of features (dimensionality), and the need for robustness and scalability in large datasets. Some algorithms are designed to handle nonlinear relationships and high-dimensional data better than others, while some are more robust to outliers and can automatically detect complex interactions among variables. There are several reasons to consider alternative algorithms besides traditional statistical regression for regression tasks:

Nonlinearity in data
: Traditional linear regression models assume a linear relationship between independent and dependent variables. However, real-world data often exhibits nonlinear relationships. Algorithms like decision trees, random forests, and neural networks can model these complex, nonlinear interactions more effectively.

High dimensionality
: When dealing with datasets with a large number of features (high-dimensional data), traditional regression models can suffer from overfitting, where the model learns noise in the training data rather than the actual relationships. Algorithms like ridge regression, LASSO regression, and principal component regression include regularization techniques that help manage high dimensionality and reduce overfitting.

Robustness to outliers
: Some regression algorithms are more robust to outliers than others. For instance, linear regression can be heavily influenced by outliers, while tree-based methods like random forests are generally more resistant to them.

Handling of categorical variables
: Traditional regression models require numerical input, so categorical variables need to be transformed (e.g., one-hot encoding). Some algorithms, like decision trees and their ensemble methods (random forests, gradient boosting machines), can natively handle categorical variables without extensive preprocessing.

Complex interactions and polynomials
: Some algorithms can automatically capture complex interactions and polynomial terms. For example, neural networks and SVMs with nonlinear kernels can model complex relationships without manually creating interaction or polynomial features.

Scalability and performance
: In cases where datasets are extremely large, some algorithms scale better than traditional regression methods. Machine learning techniques like stochastic gradient descent are designed to handle very large datasets efficiently.

Flexibility and customization
: Advanced algorithms often provide more options for customization and tuning. You can adjust various hyperparameters to optimize performance for specific types of data or problems.

Predictive performance
: Ultimately, the choice of algorithm is often driven by the goal of maximizing predictive performance. In many real-world applications, machine learning models outperform traditional statistical models in terms of accuracy and other performance metrics.

When building predictive models, linear regression is often a starting point. However, depending on the complexity of the data and the specific task at hand, alternative algorithms may provide better performance. These alternatives, such as decision trees, XGBoost, and SVMs, each have strengths that make them suitable for different types of problems, including nonlinear relationships, high-dimensional datasets, and classification tasks. Understanding when to use each algorithm is crucial for optimizing model accuracy and efficiency. Popular alternatives to linear regression and the conditions under which each algorithm performs best include the following:

Decision trees
: Best for nonlinear relationships and data with many categorical variables. Handles missing data well.

Random forest
: Suitable for complex, high-dimensional data. It reduces overfitting by averaging multiple decision trees.

XGBoost
: Great for large datasets and structured data. It's highly efficient and performs well with limited tuning.

SVMs
: Effective for smaller datasets with a clear margin of separation. Works well for classification tasks.

k-nearest neighbors (KNN)
Good for low-dimensional data and small datasets but sensitive to noise.

Each algorithm has its strengths and weaknesses, and the choice of which to use depends on the specifics of the data and the problem at hand. It's often beneficial to try multiple models and compare their performance to find the best solution for a particular regression task. Choosing the correct algorithm will be explored more in Chapter 7.

Decision Trees for Regression

In this section, we delve into the world of decision trees specifically tailored for regression analysis. Decision trees, known for their versatility in predictive modeling, offer a distinct approach when applied to regression tasks. Here, we explore the fundamental concepts, present a practical Python code example utilizing the DecisionTreeRegressor from the scikit-learn library, and bring the results to life through visualization using powerful tools like Matplotlib or Seaborn.

Distinguishing regression trees from classification trees

Decision trees, often associated with classification tasks, are also potent tools in the realm of regression analysis. While classification trees aim to categorize data points into discrete classes or labels, regression trees take a different route. Their primary objective is to model and predict continuous numeric values. In essence, regression trees allow us to explore and understand the relationships between input variables and continuous outcomes, making them indispensable for various predictive modeling scenarios.

Let's consider a practical example and harness the power of decision trees for regression in Python. Our vehicle for this exploration is the `DecisionTreeRegressor` class (see Figure 6-2) from the scikit-learn library, a go-to choice for machine learning enthusiasts and practitioners:

```
# Import necessary libraries
import numpy as np ❶
import matplotlib.pyplot as plt ❷
from sklearn.tree import DecisionTreeRegressor ❸

# Generate sample data
rng = np.random.RandomState(1) ❹
X = np.sort(5 * rng.rand(80, 1), axis=0) ❺
y = np.sin(X).ravel() ❻
y[::5] += 3 * (0.5 - rng.rand(16)) ❼

# Fit regression model using a decision tree
regr = DecisionTreeRegressor(max_depth=5) ❽
regr.fit(X, y) ❾
```

```
# Predict
X_test = np.arange(0.0, 5.0, 0.01)[:, np.newaxis] ⑩
y_pred = regr.predict(X_test) ⑪

# Plot the results
plt.figure() ⑫
plt.scatter(X, y, s=20, edgecolor="black", c="darkorange", label="data") ⑬
plt.plot(X_test, y_pred, color="cornflowerblue", linewidth=2, label="prediction") ⑭
plt.xlabel("data")
plt.ylabel("target")
plt.title("Decision Tree Regression")
plt.legend() ⑮
plt.show() ⑯
```

This code snippet demonstrates the use of a decision tree for regression tasks using Python's NumPy, Matplotlib, and sklearn libraries:

① NumPy is used for numerical operations.

② matplotlib.pyplot is used for plotting graphs.

③ DecisionTreeRegressor from sklearn.tree is a machine learning model for regression.

④ `rng = np.random.RandomState(1)` initializes a random number generator with a fixed state for reproducibility.

⑤ `X = np.sort(5 * rng.rand(80, 1), axis=0)` generates 80 random numbers, multiplies them by 5, and sorts them. This array X represents the feature data.

⑥ `y = np.sin(X).ravel()` calculates the sine of each number in X to create the target variable y.

⑦ `y[::5] += 3 * (0.5 - rng.rand(16))` adds noise to every 5th element in `y` to make the data more realistic.

⑧ `regr = DecisionTreeRegressor(max_depth=5)` creates a decision tree regressor model with a maximum depth of 5.

⑨ `regr.fit(X, y)` fits the model to the data `X` and target `y`.

⑩ `X_test = np.arange(0.0, 5.0, 0.01)[:, np.newaxis]` creates test data `X_test` from 0.0 to 5.0 with a step of 0.01.

⑪ `y_pred = regr.predict(X_test)` uses the fitted model to predict the target values `y_pred for X_test`.

⓬ `plt.figure()` initializes a new figure for plotting.

⓭ `plt.scatter(...)` plots the original data (`X` versus `y`) as a scatter plot.

⓮ `plt.plot(...)` plots the predicted data (`X_test` versus `y_pred`) as a line plot.

⓯ `plt.xlabel("data"), plt.ylabel("target"), plt.title(...), plt.legend()` adds labels, title, and legend to the plot.

⓰ `plt.show()` displays the plot.

The chart in Figure 6-2 depicts the original data points as a scatter plot and the predictions from the decision tree regressor as a line plot. The x-axis represents the feature data (X), and the y-axis represents the target variable (y). The scatter plot (in dark orange) shows the actual data with some noise added. The line plot (in cornflower blue) shows the predictions made by the decision tree. The line is likely to be piecewise constant or step-like because decision trees make predictions based on the mean of the target values in each leaf node.

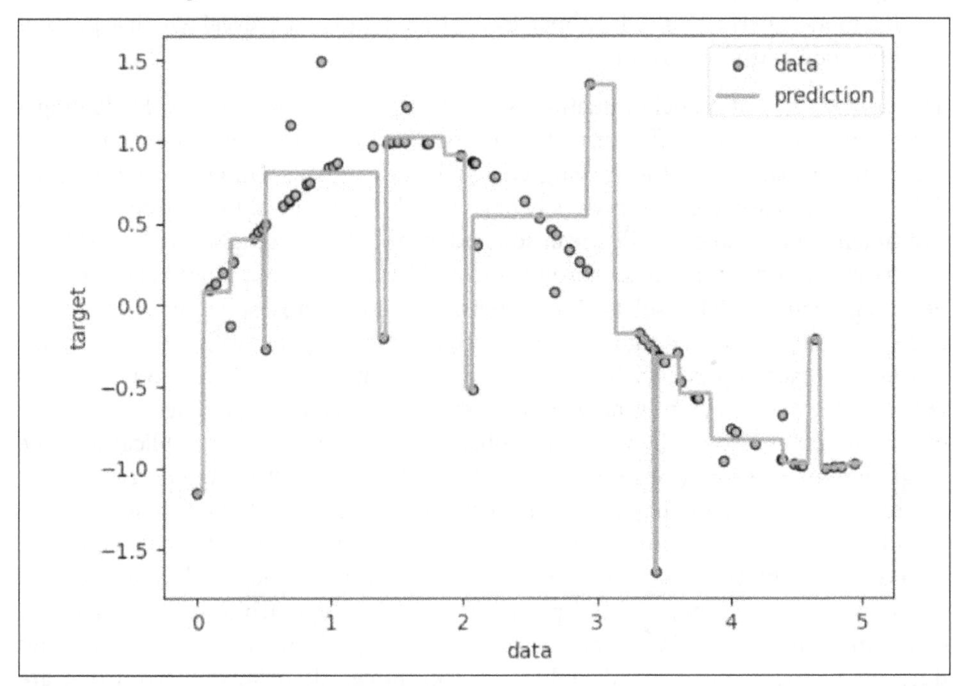

Figure 6-2. Decision tree regression results

Linear Regression Evaluation

Model evaluation is a critical step in the data modeling process. It helps us assess how well our machine learning models are performing and whether they are making accurate predictions. In this section, we'll discuss the importance of evaluating model performance and explore common evaluation metrics for both regression and classification tasks in R and Python.

Evaluating model performance is essential to ensure that our models are making accurate and reliable predictions. It helps us:

Assess the model's predictive power
: By evaluating a model's performance, we can determine how well it generalizes to unseen data.

Compare different models
: Model evaluation allows us to compare multiple models and choose the one that performs best for a specific task.

Identify areas for improvement
: If a model's performance is subpar, evaluation metrics can reveal which aspects of the model need improvement.

The primary role of model evaluation is to find the most robust model. Evaluating a regression model using different metrics is essential due to the multifaceted nature of model performance and the varying characteristics of data. Different metrics shed light on various aspects of a model's behavior, making it crucial for a comprehensive assessment. For instance, while mean squared error (MSE) is sensitive to outliers and emphasizes larger errors, mean absolute error (MAE) provides a more balanced view of average errors. Relying solely on one metric can lead to misleading interpretations; a low MSE might indicate a good fit, but it could also suggest overfitting. This multifaceted approach is particularly important when comparing different models, as certain metrics may inherently favor one model type over another. The relevance of different metrics also varies with the problem context. In financial applications, the magnitude of an error (captured by MSE) might be more critical, whereas in customer service, the frequency of errors (better represented by MAE) could be more significant. Additionally, using a range of metrics helps assess the robustness and generalization capability of the model, ensuring it performs well across different scenarios and is not just optimized for a specific metric. This approach is also beneficial for communicating with stakeholders, as different groups might prefer or understand different metrics. Business stakeholders, for example, often favor metrics that are easily interpretable and directly related to business outcomes. In summary, employing a variety of metrics for evaluating regression models ensures a thorough understanding of their performance, aligns evaluation with specific problem requirements, and caters to diverse stakeholder needs.

Table 6-1 contains a list of commonly used evaluation metrics for regression models along with their definitions and applications. Each of these metrics provides different insights into the performance of a regression model and is chosen based on the specific context and requirements of the analysis.

Table 6-1. Regression metrics

Metric	Definition	Use	Interpretation
Mean absolute error (MAE)	The average of the absolute differences between the predicted values and the actual values. It measures the average magnitude of errors in a set of predictions, without considering their direction.	MAE is useful for understanding the average error magnitude but does not give information about the error direction.	Lower MAE indicates better model performance.
Mean squared error (MSE)	The average of the squares of the errors between the predicted and actual values. It emphasizes larger errors by squaring them.	MSE is helpful in highlighting large errors (due to the squaring of each term) and is widely used, but it can be more sensitive to outliers.	Lower MSE means better performance, but it's sensitive to outliers.
Root mean squared error (RMSE)	The square root of the mean squared error. It scales the MSE to the error units and is one of the most commonly used metrics.	RMSE is useful for interpreting the average error in the same units as the response variable, making it more interpretable than MSE.	Lower RMSE suggests better accuracy.
Mean absolute percentage error (MAPE)	The average of the absolute percentage errors between the predicted and actual values. It provides an understanding of prediction accuracy in terms of percentage.	MAPE is useful for comparing the accuracy of prediction models on a percentage scale, but it can be misleading for data with zero or near-zero values.	Lower MAPE indicates better performance, but it's sensitive to zero values.
R-squared (R^2)	The proportion of the variance in the dependent variable that is predictable from the independent variables. It ranges from 0 to 1.	R^2 is used to measure the strength of the relationship between the model and the dependent variable. However, it can be misleading with nonlinear relationships or in the presence of irrelevant variables.	Values closer to 1 indicate a better fit.
Adjusted R-squared	A modified version of R^2 that adjusts for the number of predictors in the model. Unlike R^2, it penalizes the model for adding variables that do not improve the model.	Adjusted R^2 is more accurate for comparing models with a different number of independent variables.	A higher value indicates a more efficient model with fewer predictors.
Mean bias error (MBE)	The average of the differences between predicted and actual values. It indicates the average bias in the predictions.	MBE is used to check if a model tends to underpredict or overpredict the values. A positive value indicates overprediction, while a negative value indicates underprediction.	Ideally, MBE should be close to zero, indicating minimal bias in the model's predictions.

Metric	Definition	Use	Interpretation
Root mean squared logarithmic error (RMSLE)	Similar to RMSE, but the logarithm of the predicted and actual values is taken before calculating the square root of the mean squared error.	RMSLE is particularly useful when the dataset has not been scaled, and you want to penalize underestimates more than overestimates.	Lower RMSLE indicates better model performance, and it is particularly useful when predicting targets that span multiple orders of magnitude, minimizing the effect of large differences.
Akaike information criterion (AIC)	A measure of the quality of a statistical model for a given set of data. It deals with the trade-off between the goodness of fit of the model and the complexity of the model.	AIC is used for model selection. It is useful when comparing different models on the same dataset.	Lower AIC suggests a better model.

Model evaluation in R

Let's consider a simple linear regression model in R using the mtcars dataset, which is built into R. This dataset contains various attributes of cars. We'll predict the miles per gallon (mpg) based on the displacement (disp). After building the model, we'll evaluate it using various metrics:

```
# Load necessary library
library(ggplot2)

# Load the mtcars dataset
data(mtcars)

# Build a linear regression model
model <- lm(mpg ~ disp, data=mtcars)

# Summary of the model
summary(model)
Call:
lm(formula = mpg ~ disp, data = mtcars)

Residuals:
    Min      1Q  Median      3Q     Max
-4.8922 -2.2022 -0.9631  1.6272  7.2305

Coefficients:
             Estimate Std. Error t value Pr(>|t|)
(Intercept) 29.599855   1.229720  24.070  < 2e-16 ***
disp        -0.041215   0.004712  -8.747 9.38e-10 ***
---
Signif. codes:  0 '***' 0.001 '**' 0.01 '*' 0.05 '.' 0.1 ' ' 1

Residual standard error: 3.251 on 30 degrees of freedom
Multiple R-squared:  0.7183, Adjusted R-squared:  0.709
```

```
F-statistic: 76.51 on 1 and 30 DF,  p-value: 9.38e-10

# Predictions
predictions <- predict(model, mtcars)

# Metrics
# Mean Absolute Error (MAE)
mae <- mean(abs(predictions - mtcars$mpg))

# Mean Squared Error (MSE)
mse <- mean((predictions - mtcars$mpg)^2)

# Root Mean Squared Error (RMSE)
rmse <- sqrt(mse)

# R-squared
rsq <- summary(model)$r.squared

# Print metrics
cat("MAE:", mae, "\nMSE:", mse, "\nRMSE:", rmse, "\nR-squared:", rsq)
MAE: 2.605473
MSE: 9.911209
RMSE: 3.148207
R-squared: 0.7183433
```

In this example, we used the mtcars dataset to build a linear regression model, predicting mpg based on disp. The `summary(model)` function provides an overview of the model's coefficients and statistical significance.

Model evaluation in Python

Using the same example with the mtcars dataset, here is the same example, but in Python:

```
import pandas as pd
import seaborn as sns
from sklearn.linear_model import LinearRegression
from sklearn.metrics import mean_squared_error, r2_score
import numpy as np

# Load the mtcars dataset
mtcars = sns.load_dataset('mpg').dropna()

# Prepare the data
X = mtcars[['displacement']]  # Predictor variable
y = mtcars['mpg']  # Response variable

# Build a linear regression model
model = LinearRegression()
model.fit(X, y)

# Predictions
```

```
predictions = model.predict(X)

# Metrics
# Mean Absolute Error (MAE)
mae = np.mean(np.abs(predictions - y))

# Mean Squared Error (MSE) and Root Mean Squared Error (RMSE)
mse = mean_squared_error(y, predictions)
rmse = np.sqrt(mse)

# R-squared
rsq = r2_score(y, predictions)

# Print metrics
print(f"MAE: {mae:.2f}")
print(f"MSE: {mse:.2f}")
print(f"RMSE: {rmse:.2f}")
print(f"R-squared: {rsq:.2f}")
MAE: 3.51
MSE: 21.37
RMSE: 4.62
R-squared: 0.65
```

In this Python example, we used the mtcars dataset to perform a linear regression, predicting mpg based on disp. The LinearRegression model from sklearn.linear_model was used for this purpose.

The MAE gives us an average of the absolute errors between our predictions and actual values, which is useful for understanding the average magnitude of our errors without considering their direction. The MSE and RMSE provide similar insights but give more weight to larger errors, which can be particularly important if large errors are more undesirable in the application context. The RMSE is more interpretable in the units of the target variable (mpg in this case) as it is on the same scale as the target variable.

The R-squared value indicates the proportion of variance in the dependent variable (mpg) that is predictable from the independent variable (disp). It gives an idea of how well the unseen data is likely to be predicted by the model, with a higher R-squared value indicating a better fit.

These metrics collectively offer a comprehensive view of the model's performance. MAE and RMSE provide insights into the average error magnitude, while MSE emphasizes larger errors. R-squared offers a measure of how well the model explains the variability in the data. Together, they provide a balanced approach to evaluating the model's accuracy and effectiveness.

Classification

Classification algorithms are pivotal tools, orchestrating the categorization of data into predefined classes or labels. These algorithms discern patterns and structures within data, enabling machines to make informed decisions without explicit human intervention. From filtering emails as spam or not spam to diagnosing diseases based on patient records, classification algorithms play an integral role in business analytics.

Classification algorithms are about categorizing data into predefined classes or groups based on the input features of the data. They operate by learning from a set of data points whose class labels are known and then applying this learned knowledge to classify new, unseen data points. Here's a step-by-step breakdown of how they generally work:

Training phase

- Input data: The algorithm starts with a training dataset, where the class labels of each data point are known.
- Learning: The algorithm learns the relationship between the features of the data and its class labels. This process is called "fitting" the model to the data.

Model evaluation

- Once trained, the model's performance is evaluated using a separate dataset, known as the validation or test dataset.
- Metrics such as accuracy, precision, recall, and F1-score are used to gauge how well the model is likely to perform on unseen data.

Prediction phase

- New data: When new data points come in without class labels, the trained model predicts their classes based on what it has learned.
- Output: The algorithm assigns each new data point to one of the predefined classes.

Classification

- Binary classification: Deals with two classes (e.g., spam or not spam, sick or healthy).
- Multiclass classification: Each instance is assigned to one and only one class from a set of multiple classes (e.g., classifying a set of images into "cat," "dog," or "bird").
- Multilabel classification: Each instance can belong to multiple classes simultaneously. For example, in text classification, a document might be labeled as both "sports" and "technology" if it covers topics related to both.

Different algorithms use different techniques to classify data. For instance, decision trees split the data based on feature values, SVMs find the hyperplane that best separates the classes, and neural networks use layers of mathematical nodes to make decisions. KNN classifies a data point based on how its neighbors are classified.

As more data becomes available or as the underlying data distribution changes, classification models might need to be retrained or fine-tuned to maintain or improve their accuracy.

In essence, classification algorithms in data science are about understanding the patterns and relationships within labeled data and leveraging this understanding to categorize new, unlabeled data. They form the backbone of many modern systems, from email filters to medical diagnosis tools, making them indispensable in our data-driven world.

Common Use Cases

Classification models are a cornerstone of machine learning, widely used across various industries and applications. These models categorize data into predefined classes or groups based on input features, making them essential for tasks requiring discrete outputs. Here are some common use cases for classification models:

Customer segmentation
: In marketing and sales, classification models help segment customers into distinct groups based on purchasing behavior, demographics, or engagement levels. This segmentation enables businesses to tailor marketing strategies and personalize customer experiences.

Fraud detection
: Financial institutions use classification models to identify potentially fraudulent transactions. By analyzing patterns in transaction data, these models can flag unusual activities that deviate from a user's typical behavior, helping prevent fraud and financial losses.

Healthcare diagnostics
: In healthcare, classification models assist in diagnosing diseases by analyzing medical images, patient history, and laboratory results. For instance, models can classify medical images to detect conditions like cancer or heart disease, aiding in early diagnosis and treatment planning.

Spam filtering
: Email services use classification algorithms to distinguish between spam and legitimate emails. By analyzing email content, sender information, and user behavior, these models effectively filter out unwanted emails, enhancing user experience.

Sentiment analysis

In social media monitoring and customer feedback analysis, classification models determine the sentiment of text data, categorizing opinions into positive, negative, or neutral. This application is crucial for brand monitoring and understanding customer satisfaction.

Credit scoring

Banks and credit agencies use classification models to assess the creditworthiness of individuals. By analyzing credit history, income, and other financial indicators, these models classify individuals into different risk categories, influencing lending decisions.

Quality control in manufacturing

Classification models can be used to classify products as defective or nondefective based on visual inspection or sensor data, ensuring consistent product quality and reducing waste.

Predictive maintenance

Classification models predict equipment failures in industrial settings by analyzing sensor data. This predictive maintenance helps in scheduling repairs and reducing downtime.

Natural language processing (NLP)

In NLP, classification models are used for tasks like language detection, topic classification, and intent recognition in chatbots and virtual assistants. Sentiment analysis is another use case.

Image and video recognition

In computer vision, classification models identify objects, faces, or actions in images and videos. This technology is used in various applications, from security surveillance to autonomous vehicles. This may go beyond a simple classification, as it also may detect which areas in the image correspond to a specific target.

These use cases demonstrate the versatility and impact of classification models in machine learning. By automating the process of categorizing data, they enable more efficient, accurate, and insightful decision making across diverse domains.

Classification Algorithms

Classification algorithms are a subset of machine learning algorithms used for categorizing or classifying data into a defined number of classes. These algorithms are essential for tasks where the output is discrete or categorical, such as identifying whether an email is spam or not, or predicting whether a loan applicant is low or high risk.

Here are some common classification algorithms along with their definitions:

Logistic regression
: A classification algorithm used for binary classification problems. It estimates the probability that a given input point belongs to a certain class.

Decision trees
: A decision tree splits the data into subsets based on the value of input features. Each node in the tree represents a feature, each branch represents a decision rule, and each leaf node represents the outcome.

Random forest
: An ensemble method that operates by constructing a multitude of decision trees at training time and outputting the class that is the mode of the classes of the individual trees. It's known for its high accuracy, handling of large datasets, and ability to manage thousands of input variables without variable deletion.

SVMs
: A set of supervised learning methods used for classification and regression. In classification tasks, SVMs aim to find the hyperplane that best separates the classes in the feature space.

Naive Bayes
: A collection of classification algorithms based on Bayes's theorem. It's "naive" because it assumes independence between predictors in a dataset. Particularly useful for large datasets and performs well with categorical input variables compared to numerical variables.

KNN
: A simple, instance-based learning algorithm. KNN stores all available cases and classifies new cases by a majority vote of its k neighbors. The output is a class membership.

Neural networks
: Inspired by the structure and functions of the brain, neural networks are a set of algorithms, modeled loosely after the human brain, designed to recognize patterns. They interpret sensory data through a kind of machine perception, labeling, or clustering raw input.

Gradient boosting machines (GBM)
: An ensemble technique that builds the model in a stage-wise fashion. It constructs new models that correct the errors made by previous models and combines them for the final prediction.

Adaptive boosting (AdaBoost)
: A boosting technique used as an ensemble method in machine learning. It works by weighting the instances in the dataset by how easy or difficult they are to classify, allowing the algorithm to pay more or less attention to them in the construction of subsequent models.

Extreme gradient boosting (XGBoost)
: An efficient and scalable implementation of gradient boosting machines, known for its speed and performance. It is widely used in machine learning competitions and practical applications.

Each of these algorithms has its strengths and weaknesses, and the choice of algorithm often depends on the specific requirements of the task, such as the size and nature of the dataset, the accuracy required, the training time available, and the number of features.

Classification in R

Let's consider logistic regression for binary classification. Logistic regression is often used for binary classification tasks, where the outcome is either 0 or 1 (e.g., spam or not spam). For example:

```
# Load necessary library
library(MASS)

# Using built-in dataset 'Pima.tr' from MASS package for demonstration
# This dataset is on diabetes in Pima Indian Women
data("Pima.tr")

# Logistic regression model
model <- glm(type ~ ., data = Pima.tr, family = binomial())

# Predicting values (as probabilities)
predictions <- predict(model, Pima.tr, type = "response")

# Convert probabilities to binary outcome (0 or 1)
predicted_classes <- ifelse(predictions > 0.5, 1, 0)

# Printing first few predictions
print(head(predicted_classes))

1 2 3 4 5 6
0 1 0 1 0 0
```

In this example, notice the following:

- We use the Pima.tr dataset from the MASS package, which contains data for diabetes classification.
- A logistic regression model is created using the `glm` function with the binomial family, indicating a binary outcome.
- The model is then used to make predictions. The `predict` function returns probabilities, which we convert into binary outcomes (1 if the probability is greater than 0.5; otherwise 0).

Decision trees are popular for classification as they are easy to interpret. The rpart package in R is commonly used for decision trees:

```
# Load necessary libraries
library(rpart)

# Using the built-in iris dataset
data(iris)

# Creating a decision tree model
# Here, we are predicting the species of iris based on other measurements
model <- rpart(Species ~ ., data = iris, method = "class")

# Predicting the species
predictions <- predict(model, iris, type = "class")

# Printing first few predictions
print(head(predictions))

  1      2      3      4      5      6
setosa setosa setosa setosa setosa setosa
Levels: setosa versicolor virginica

# Decision Tree Example in R

# Load necessary libraries
library(rpart)
library(caret)  # for train/test split

# Using the built-in iris dataset
data(iris)

# Set seed for reproducibility
set.seed(123)

# Splitting the data into 80% training and 20% testing
trainIndex <- createDataPartition(iris$Species, p = 0.8, list = FALSE)
trainData <- iris[trainIndex, ]
testData <- iris[-trainIndex, ]
```

```
# Creating a decision tree model on the training data
# We are predicting the species of iris based on other measurements
model <- rpart(Species ~ ., data = trainData, method = "class")

# Predicting the species on the test data
predictions <- predict(model, testData, type = "class")

# Printing the first few predictions on the test data
print(head(predictions))
     1      2      6     16     23     34
setosa setosa setosa setosa setosa setosa
Levels: setosa versicolor virginica
```

In this example:

- The Iris dataset is used. It is a classic dataset for classification tasks and contains various measurements of iris flowers and their species.
- A seed is set for reproducibility. This ensures that any random processes, such as data splitting, yield the same results each time.
- The data is split into two parts: 80% is used for training the model, and 20% is reserved for testing. This ensures that we can evaluate the model's performance on unseen data.
- A decision tree model is created using the training data. The model is designed to predict the species of the iris flower based on measurements like sepal length, petal width, etc.
- The trained model is then used to predict the species of the flowers in the test data.
- The first few predictions are printed to see how the model performed on the test data.

These are basic examples to get started with classification in R. The actual implementation in a real-world scenario would involve additional steps like data preprocessing, feature selection, model validation, and tuning.

Classification in Python

The context around classification is the same in Python. The following example shows the same case used in R: we are classifying the species of the iris flow based on flower characteristics:

```
# Import necessary libraries ❶
from sklearn.tree import DecisionTreeClassifier
from sklearn.datasets import load_iris
from sklearn.model_selection import train_test_split
from sklearn.metrics import accuracy_score
```

```
# Load the Iris dataset ❷
iris = load_iris()
X = iris.data
y = iris.target

# Split the dataset into training and testing sets ❸
X_train, X_test, y_train, y_test = train_test_split(X, y, test_size=0.2,
  random_state=42)

# Create a decision tree classifier ❹
clf = DecisionTreeClassifier()

# Train the classifier on the training data ❺
clf.fit(X_train, y_train)

# Predict the species ❻
predictions = clf.predict(X_test)

# Convert predicted numbers to class labels using iris.target_names
predicted_classes = iris.target_names[predictions]

# Printing first few predictions (show class description instead of numbers) ❼
print("Predicted class labels:")
print(predicted_classes[:5])

# Calculate accuracy on the test set ❽
accuracy = accuracy_score(y_test, predictions)
print(f"Accuracy: {accuracy * 100:.2f}%")
Predicted class labels:
['versicolor' 'setosa' 'virginica' 'versicolor' 'versicolor']
Accuracy: 100.00%
```

❶ Import scikit-learn's DecisionTreeClassifier, load_iris to load the Iris dataset, train_test_split for data splitting, and accuracy_score for evaluating accuracy.

❷ Load the Iris dataset using `load_iris()`, which provides the feature data (X) and target labels (y).

❸ Split the dataset into training and testing sets using train_test_split. We reserve 20% of the data for testing, and the random seed ensures reproducibility.

❹ Create a decision tree classifier using `DecisionTreeClassifier()`.

❺ Train the classifier on the training data using `clf.fit(X_train, y_train)`.

❻ Make predictions on the test data using `clf.predict(X_test)`.

❼ Print the first few predictions to check the model's performance.

❽ Calculate the accuracy of the model on the test set using accuracy_score and print the accuracy percentage.

In both Python and R, classification modeling involves similar steps and libraries for building predictive models. First, both languages require importing relevant libraries such as scikit-learn in Python and rpart in R to facilitate classification tasks. Next, they load datasets, split data into training and testing sets, and select appropriate machine learning algorithms, like decision trees, to build classification models. Training the models on the training data and making predictions on the test data are common steps in both languages. Finally, accuracy assessment using metrics like accuracy_score in Python or confusionMatrix in R helps evaluate model performance.

Python and R share similarities in their approach to classification modeling, making it easy for users familiar with one language to adapt to the other. These languages prioritize ease of use and have extensive libraries for data manipulation, preprocessing, and model evaluation, making them versatile tools for classification tasks in data science and machine learning. So far, we have explored creating, training, and testing classification models; the next section will address classification evaluation.

Classification Use Case: Telecom Churn

To demonstrate classification using multiple algorithms, we will use a common business problem: predicting churn in the telecommunications industry. Telecom churn refers to the phenomenon where customers stop using a telecom provider's services, switching to a competitor or ceasing the service altogether. Churn is a critical metric in the telecommunications industry because retaining existing customers is often more cost-effective than acquiring new ones.

Here is a data dictionary of the dataset we will be using outlined in Table 6-2. This dataset was synthetically generated using common publicly available datasets.

Table 6-2. Churn data dictionary

Column name	Data type	Description
customerID	string	Unique identifier for each customer
genfer	categorical	Gender of the customer (Male, Female)
SeniorCitizen	numeric	Indicates if the customer is a senior citizen (1: Yes, 0: No)
Partner	categorical	Whether the customer has a partner (Yes, No)
Dependents	categorical	Whether the customer has dependents (Yes, No)
tenure	numeric	Number of months the customer has stayed with the company
PhoneService	categorical	Whether the customer has phone service (Yes, No)
MultipleLines	categorical	Whether the customer has multiple lines (Yes, No, No phone service)
InternetService	categorical	Type of internet service (DSL, Fiber optic, No)

Column name	Data type	Description
OnlineSecurity	categorical	Whether the customer has online security (Yes, No, No internet service)
OnlineBackup	categorical	Whether the customer has online backup (Yes, No, No internet service)
DeviceProtection	categorical	Whether the customer has device protection (Yes, No, No internet service)
TechSupport	categorical	Whether the customer has tech support (Yes, No, No internet service)
StreamingTV	categorical	Whether the customer has streaming TV service (Yes, No, No internet service)
StreamingMoves	categorical	Whether the customer has streaming movie service (Yes, No, No internet service)
Contract	categorical	Type of contract (Month-to-month, One year, Two year)
PaperlessBilling	categorical	Whether the customer uses paperless billing (Yes, No)
PaymentMethod	categorical	Method of payment (Electronic check, Mailed check, Bank transfer, etc.)
MonthlyCharges	numeric	Monthly charges for the customer
TotalCharges	numeric	Total charges incurred by the customer
Churn	categorical	Whether the customer has churned (Yes, No)

Examples using a decision tree and logistic regression were explored to illustrate the coding approaches. Now we will explore the telecom churn case using multiple classification algorithms. Both Python and R examples will be discussed, and we will start with Python and the data preparation needed for modeling. Classification evaluation of all models will be addressed in "Classification Evaluation" on page 207.

Python example

Our Python example for telecom churn starts with data preparation:

```
# Import necessary libraries
import pandas as pd
from sklearn.model_selection import train_test_split
from sklearn.preprocessing import LabelEncoder, StandardScaler

# Load the dataset

import pandas as pd
from sklearn.model_selection import train_test_split
from sklearn.preprocessing import LabelEncoder, StandardScaler

# Load the dataset
file_path = '/content/sample_data/synthetic_telecom_churn_data.csv'
# Update with actual path
data = pd.read_csv(file_path)

# Handle missing values in 'TotalCharges'
(convert to numeric and fill missing with median)
data['TotalCharges'] = pd.to_numeric(data['TotalCharges'], errors='coerce')
data['TotalCharges'].fillna(data['TotalCharges'].median(), inplace=True)

# Drop 'customerID' column
data.drop('customerID', axis=1, inplace=True)
```

```
# Convert 'Churn' column (target variable) to binary labels
label_encoder = LabelEncoder()
data['Churn'] = label_encoder.fit_transform(data['Churn'])
# 'Yes' -> 1, 'No' -> 0

# Encode all other categorical columns using Label Encoding
categorical_columns = data.select_dtypes(include=['object']).columns
for column in categorical_columns:
    data[column] = label_encoder.fit_transform(data[column])

# Split the data into features (X) and target (y)
X = data.drop('Churn', axis=1)
y = data['Churn']

# Standardize the numeric features
scaler = StandardScaler()
X_scaled = pd.DataFrame(scaler.fit_transform(X), columns=X.columns)

# Split the data into training (60%), temporary (40%) sets
X_train, X_temp, y_train, y_temp = train_test_split(X_scaled, y, test_size=0.4,
random_state=42, stratify=y)

# Split the temporary set into validation (50% of temp) and test (50% of temp)
X_val, X_test, y_val, y_test = train_test_split(X_temp, y_temp, test_size=0.5,
random_state=42, stratify=y_temp)

# Check the shape of the resulting datasets
print(f"Training data shape: {X_train.shape}, Validation data shape:
{X_val.shape}, Testing data shape: {X_test.shape}")
data = pd.read_csv(file_path)

# Handle missing values in 'TotalCharges'
(convert to numeric and fill missing with median)
data['TotalCharges'] = pd.to_numeric(data['TotalCharges'], errors='coerce')
data['TotalCharges'].fillna(data['TotalCharges'].median(), inplace=True)

# Drop 'customerID' column
data.drop('customerID', axis=1, inplace=True)

# Convert 'Churn' column (target variable) to binary labels
label_encoder = LabelEncoder()
data['Churn'] = label_encoder.fit_transform(data['Churn'])
# 'Yes' -> 1, 'No' -> 0

# Encode all other categorical columns using Label Encoding
categorical_columns = data.select_dtypes(include=['object']).columns
for column in categorical_columns:
    data[column] = label_encoder.fit_transform(data[column])

# Split the data into features (X) and target (y)
X = data.drop('Churn', axis=1)
```

```
y = data['Churn']

# Standardize the numeric features
scaler = StandardScaler()
X_scaled = pd.DataFrame(scaler.fit_transform(X), columns=X.columns)

# Split the data into training (60%), temporary (40%) sets
X_train, X_temp, y_train, y_temp = train_test_split(X_scaled, y, test_size=0.4,
random_state=42, stratify=y)

# Split the temporary set into validation (50% of temp) and test (50% of temp)
X_val, X_test, y_val, y_test = train_test_split(X_temp, y_temp, test_size=0.5,
random_state=42, stratify=y_temp)

# Check the shape of the resulting datasets
print(f"Training data shape: {X_train.shape}, Validation data shape:
{X_val.shape}, Testing data shape: {X_test.shape}")
Training data shape: (60000, 19), Validation data shape: (20000, 19),
Testing data shape: (20000, 19)
Training data shape: (60000, 19), Validation data shape: (20000, 19),
Testing data shape: (20000, 19)
```

This code is focused on preparing a dataset for machine learning by handling missing values, encoding categorical variables, scaling features, and splitting the dataset into training, validation, and test sets. Here's a breakdown of the key steps:

Import libraries
: The necessary libraries (pandas, train_test_split, LabelEncoder, and StandardScaler) from sklearn are imported. These are used for data manipulation, encoding, scaling, and splitting the data.

Load the dataset
: The dataset is read from a CSV file (*synthetic_telecom_churn_data.csv*) into a Pandas DataFrame.

Handle missing values
: The column TotalCharges is converted to numeric (since it might contain non-numeric values), and any missing values in this column are replaced with the median value.

Drop irrelevant columns
: The customerID column is dropped as it's not relevant for the model (likely just a unique identifier).

Encode the target variable
: The target variable Churn is converted from categorical values ('Yes' or 'No') into binary labels (1 for 'Yes' and 0 for 'No') using LabelEncoder.

Encode other categorical variables
: The remaining categorical columns are encoded into numerical values using LabelEncoder. Each unique category is converted into a corresponding integer.

Separate features and target
: The data is split into features (X, all columns except Churn) and the target variable (y, the Churn column).

Standardize numeric features
: The features are scaled using StandardScaler to standardize the data, which transforms the numeric columns so that they have a mean of 0 and a standard deviation of 1. This is especially important for algorithms sensitive to feature scales.

Split the dataset
: The dataset is split into the training set (60% of the data) using train_test_split. Temporary set (40% of the data), which is then further split into the validation set (20% of the original data, half of the temporary set) and the test set (20% of the original data, other half of the temporary set). This is done using the stratify parameter to ensure that the Churn class proportions are preserved in each split.

Output the data shapes
: The shapes of the resulting training, validation, and test sets are printed to ensure they were split correctly.

In summary, the code preprocesses the telecom churn dataset by handling missing values, encoding categorical variables, and standardizing numerical features. The code splits the data into training, validation, and test sets, preserving the class distribution in each split for better model evaluation.

So now that the data is loaded and ready to go, the next Python code examples are going to show how to initialize and train the model, predict and evaluate on the validation set, and then print out the evaluation metrics. Example 6-1 demonstrates the steps using the random forest algorithm.

Example 6-1. Random forest

```
from sklearn.ensemble import RandomForestClassifier ❶
from sklearn.metrics import classification_report, accuracy_score ❷

# Initialize and train Random Forest model
rf_model = RandomForestClassifier(n_estimators=100, random_state=42)  ❸
rf_model.fit(X_train, y_train) ❹

# Predict and evaluate on validation set
y_pred_rf = rf_model.predict(X_val) ❺
print("Random Forest Evaluation:")
print(classification_report(y_val, y_pred_rf)) ❻
```

```
print("Accuracy:", accuracy_score(y_val, y_pred_rf)) ❼
```

```
Random Forest Evaluation:
              precision    recall  f1-score   support

           0       0.73      1.00      0.84     14599
           1       0.20      0.00      0.00      5401

    accuracy                           0.73     20000
   macro avg       0.46      0.50      0.42     20000
weighted avg       0.59      0.73      0.62     20000

Accuracy: 0.7292
```

This code initializes, trains, and evaluates a random forest classifier using the scikit-learn library. Here's a detailed explanation of each part:

❶ Import RandomForestClassifier, the random forest model from scikit-learn's ensemble module that builds an ensemble of decision trees.

❷ Import classification_report and accuracy_score. classification_report generates a detailed performance report for classification tasks, including metrics like precision, recall, F1-score, and support. accuracy_score computes the accuracy of the model, which is the ratio of correctly predicted instances to the total number of instances.

❸ RandomForestClassifier initializes a random forest classifier. `n_estimators=100` specifies that the random forest will use 100 decision trees. `random_state=42` sets a seed to ensure that the randomness involved in training the model is reproducible. This ensures consistent results across runs.

❹ The `fit()` method trains the random forest model on the training dataset: `X_train` is the feature matrix from the training set. `y_train` is the corresponding target labels from the training set.

❺ The `predict()` method is used to make predictions on the validation set (`X_val`), generating a predicted set of labels (`y_pred_rf`).

❻ `classification_report(y_val, y_pred_rf)` compares the actual validation labels (`y_val`) with the predicted labels (`y_pred_rf`) and generates a performance report, including precision, which is the proportion of true positives among the instances labeled as positive, and recall, which is the proportion of true positives among all actual positive instances. F1-score is the harmonic mean of precision and recall, giving a single performance metric, and support is the number of true instances for each label. We will explore the evaluation results from the used algorithms in "Classification Evaluation" on page 207.

7. `accuracy_score(y_val, y_pred_rf)` calculates the accuracy of the model, which is the proportion of correct predictions compared to the total number of instances in the validation set.

In summary, the code imports necessary libraries and initializes a random forest classifier with one hundred trees and trains it on the training data. Then the trained model is used to make predictions on the validation data. Next the model is evaluated by generating a classification report and calculating the overall accuracy of the model on the validation set. Last, the classification report provides a detailed breakdown of the model's performance, while the accuracy score gives an overall performance metric.

Examples 6-2 through 6-8 have the same code steps. The only difference will be the classifier used. We will see examples that cover SVM, naive Bayes, KNN, neural network, GBM, AdaBoost, and XGBoost. These algorithms were described in "Classification Algorithms" on page 187.

Example 6-2. SVMs

```
from sklearn.svm import SVC

# Initialize and train SVM model
# kernel = 'linear': Specifies the kernel type. This can be changed to
# #'radial' or 'polynomial' depending on your dataset and problem.
svm_model = SVC(kernel='linear')
svm_model.fit(X_train, y_train)

# Predict and evaluate on validation set
y_pred_svm = svm_model.predict(X_val)
print("SVM Evaluation:")
print(classification_report(y_val, y_pred_svm))
print("Accuracy:", accuracy_score(y_val, y_pred_svm))
```

Example 6-3. Naive Bayes

```
from sklearn.naive_bayes import GaussianNB

# Initialize and train naive Bayes model
nb_model = GaussianNB()
nb_model.fit(X_train, y_train)

# Predict and evaluate on validation set
y_pred_nb = nb_model.predict(X_val)
print("Naive Bayes Evaluation:")
```

```
print(classification_report(y_val, y_pred_nb))
print("Accuracy:", accuracy_score(y_val, y_pred_nb))
```

Example 6-4. KNN

```
from sklearn.neighbors import KNeighborsClassifier

# Initialize and train KNN model
knn_model = KNeighborsClassifier(n_neighbors=5)
knn_model.fit(X_train, y_train)

# Predict and evaluate on validation set
y_pred_knn = knn_model.predict(X_val)
print("KNN Evaluation:")
print(classification_report(y_val, y_pred_knn))
print("Accuracy:", accuracy_score(y_val, y_pred_knn))
```

Example 6-5. Neural networks (MLP)

```
from sklearn.neural_network import MLPClassifier

# Initialize and train Neural Network model
mlp_model = MLPClassifier(hidden_layer_sizes=(100,100), max_iter=300,
random_state=42)
mlp_model.fit(X_train, y_train)

# Predict and evaluate on validation set
y_pred_mlp = mlp_model.predict(X_val)
print("Neural Network Evaluation:")
print(classification_report(y_val, y_pred_mlp))
print("Accuracy:", accuracy_score(y_val, y_pred_mlp))
```

Example 6-6. Gradient boosting machines (GBM)

```
from sklearn.ensemble import GradientBoostingClassifier

# Initialize and train GBM model
gbm_model = GradientBoostingClassifier(random_state=42)
gbm_model.fit(X_train, y_train)

# Predict and evaluate on validation set
y_pred_gbm = gbm_model.predict(X_val)
print("GBM Evaluation:")
print(classification_report(y_val, y_pred_gbm))
print("Accuracy:", accuracy_score(y_val, y_pred_gbm))
```

Example 6-7. AdaBoost (adaptive boosting)

```
from sklearn.ensemble import AdaBoostClassifier

# Initialize and train AdaBoost model
ada_model = AdaBoostClassifier(n_estimators=100, random_state=42)
ada_model.fit(X_train, y_train)

# Predict and evaluate on validation set
y_pred_ada = ada_model.predict(X_val)
print("AdaBoost Evaluation:")
print(classification_report(y_val, y_pred_ada))
print("Accuracy:", accuracy_score(y_val, y_pred_ada))
```

Example 6-8. XGBoost (extreme gradient boosting)

```
from xgboost import XGBClassifier

# Initialize and train XGBoost model
xgb_model = XGBClassifier(use_label_encoder=False, eval_metric='logloss',
random_state=42)
xgb_model.fit(X_train, y_train)

# Predict and evaluate on validation set
y_pred_xgb = xgb_model.predict(X_val)
print("XGBoost Evaluation:")
print(classification_report(y_val, y_pred_xgb))
print("Accuracy:", accuracy_score(y_val, y_pred_xgb))
```

We just reviewed examples that outline how to create a model, train, and complete predictions for SVM, naive Bayes, KNN, neural networks, GBM, AdaBoost, and XGBoost in Python. Now let's review the same examples in R. Keep in mind, we will review the evaluation of each model in "Classification Evaluation" on page 207.

First, we introduce the code in R to prepare the data for modeling:

```
library(caret)
library(dplyr)
library(randomForest) ❶

# Load the dataset
data <- read.csv("path_to_your_dataset/synthetic_telecom_churn_data.csv") ❷

# Handle missing values in 'TotalCharges'
data$TotalCharges <- as.numeric(as.character(data$TotalCharges))
data$TotalCharges[is.na(data$TotalCharges)] <- median(data$TotalCharges, na.rm =
TRUE) ❸

# Drop 'customerID' column
data <- data %>% select(-customerID) ❹
```

```
# Convert 'Churn' column (target variable) to a factor (binary)
data$Churn <- factor(data$Churn, levels = c("No", "Yes"), labels = c(0, 1)) ❺

# Encode other categorical columns using dummy variables, excluding the 'Churn'
# column
dummy_data <- dummyVars(Churn ~ ., data = data)
data_preprocessed <- predict(dummy_data, newdata = data) ❻

# Convert the preprocessed data into a data frame and add the 'Churn' column back
data_preprocessed <- as.data.frame(data_preprocessed)
data_preprocessed$Churn <- data$Churn  # Add the 'Churn' column back as factor ❼

# Split the data into training (60%), temporary (40%) sets
set.seed(42)
trainIndex <- createDataPartition(data_preprocessed$Churn, p = 0.6, list = FALSE)
trainData <- data_preprocessed[trainIndex, ]
tempData <- data_preprocessed[-trainIndex, ] ❽

# Now split tempData into validation (50% of tempData) and test (50% of tempData)
valIndex <- createDataPartition(tempData$Churn, p = 0.5, list = FALSE)
validationData <- tempData[valIndex, ]
testData <- tempData[-valIndex, ] ❾

# Ensure 'Churn' is a factor in train, validation, and test sets
trainData$Churn <- as.factor(trainData$Churn)
validationData$Churn <- as.factor(validationData$Churn)
testData$Churn <- as.factor(testData$Churn) ❿

# Prepare features (X) and target (y) for Random Forest
X_train <- trainData %>% select(-Churn)
y_train <- trainData$Churn

X_val <- validationData %>% select(-Churn)
y_val <- validationData$Churn ⓫
```

Let's explore what this code does in more detail:

❶ Load the necessary libraries. `caret` is used for machine learning tasks, including data partitioning, model training, and evaluation. `dplyr` is used for data manipulation (e.g., selecting or transforming columns). `randomForest` is used to train a random forest model.

❷ Load the dataset from a CSV file into the data object.

❸ Handle missing values in `'TotalCharges'`. The TotalCharges column is converted to numeric, and missing values (NA) are replaced with the median of the column. This handles any missing or invalid data in the TotalCharges column.

❹ Drop the `'customerID'` column. The customerID column is dropped using dplyr's `select()` function, as it is likely irrelevant for model training.

❺ Convert `'Churn'` to a factor. The Churn column (the target variable) is converted to a binary factor with levels `'No'` and `'Yes'` mapped to 0 and 1, respectively, preparing it for classification.

❻ Encode categorical columns as dummy variables. The `dummyVars()` function from the caret library creates dummy variables (one-hot encoding) for all categorical columns in the dataset except Churn. The `predict()` function applies this transformation to the dataset, converting categorical features into numeric dummy variables.

❼ Add the `'Churn'` column back. After dummy encoding, the transformed dataset is converted into a dataframe, and the Churn column (which was not encoded) is added back as the target variable.

❽ Split the data into training (60%) and temporary (40%) sets. The dataset is split into a training set (60%) and a temporary set (40%) using `createDataPartition()` from the caret library. The partition is stratified to maintain the proportion of the Churn target class in both sets.

❾ Split the temporary set into validation and test sets (50% each). The temporary set is further split into validation (50% of temporary) and test sets (50% of temporary). This allows for model tuning and testing.

❿ Ensure `'Churn'` is a factor. The Churn column is ensured to be treated as a factor in the training, validation, and test sets, ensuring correct behavior during model training.

⓫ Prepare features `X` and `y` for the model. The dataset is split into features (`X_train`, `X_val`) and target (`y_train`, `y_val`) variables. The Churn column is removed from the feature matrices.

As a recap, the code preprocesses a telecom churn dataset by handling missing values, dummy encoding categorical variables, and dropping irrelevant columns. The code splits the data into training, validation, and test sets. The code is structured for robust model training and evaluation, ensuring that categorical variables are properly encoded and the target variable is treated as a factor for classification.

Now, let's look at Example 6-9.

Example 6-9. Random forest

```
# Load necessary libraries
library(caret)
library(randomForest) ❶

# Step 1: Train the Random Forest model using the preprocessed data
set.seed(42)
rf_model <- randomForest(x = X_train, y = y_train, ntree = 100) ❷

# Step 2: Predict on the validation set
rf_pred <- predict(rf_model, newdata = X_val) ❸

# Step 3: Evaluate the model using confusion matrix
confusionMatrix(rf_pred, y_val) ❹
```

❶ Load the randomForest and caret libraries. randomForest is used to build the random forest model, and caret provides useful functions for data splitting and evaluation.

❷ Use the `randomForest()` function to train a random forest classifier on the training data. `x = X_train`: the features of the training data, excluding the target variable Churn. `y = y_train`: the target labels (Churn), which we are trying to predict. `ntree = 100` specifies the number of decision trees in the Random Forest. Here, we are using one hundred trees. `set.seed(42)` ensures reproducibility by setting a random seed, making sure that the random processes produce the same results every time.

❸ After the model is trained, use `predict()` to predict the labels for the validation set (`X_val`). `newdata = X_val` is the feature set of the validation data for which predictions are made.

❹ To evaluate the results using a confusion matrix we can add an additional step:

```
# Step 3: Evaluate the model using confusion matrix
confusionMatrix(rf_pred, y_val)

Confusion Matrix and Statistics

                  Reference
Prediction     0      1
                0 14585  5397
                1    14     4

                           Accuracy : 0.7294
                                 95% CI : (0.7232, 0.7356)
        No Information Rate : 0.73
        P-Value [Acc > NIR] : 0.5669
```

```
                                   Kappa : -3e-04

       Mcnemar's Test P-Value : <2e-16

                            Sensitivity : 0.9990410
                            Specificity : 0.0007406
                         Pos Pred Value : 0.7299069
                         Neg Pred Value : 0.2222222
                             Prevalence : 0.7299500
                         Detection Rate : 0.7292500
                   Detection Prevalence : 0.9991000
                      Balanced Accuracy : 0.4998908

                       'Positive' Class : 0
```

The `confusionMatrix()` function compares the predicted values (`rf_pred`) with the actual validation labels (`y_val`). This provides key metrics like accuracy, precision, recall, F1-score, and the confusion matrix itself, which shows true positives, true negatives, false positives, and false negatives. "Classification Evaluation" on page 207 will explore this further. Let's look at the code examples for the other approaches, starting with SVMs (Examples 6-10 through 6-16).

Example 6-10. SVM

```
# Load necessary libraries
library(caret)
library(e1071)  # For SVM

# Step 1: Train the SVM model
svm_model <- svm(x = X_train, y = y_train, kernel = 'linear')

# Step 2: Predict on the validation set
svm_pred <- predict(svm_model, newdata = X_val)

# Step 3: Evaluate the model using confusion matrix
confusionMatrix(svm_pred, y_val)
```

Example 6-11. Naive Bayes

```
# Load necessary libraries
library(caret)
library(e1071)  # For naive Bayes

# Step 1: Train the naive Bayes model
nb_model <- naiveBayes(x = X_train, y = y_train)

# Step 2: Predict on the validation set
nb_pred <- predict(nb_model, newdata = X_val)
```

```
# Step 3: Evaluate the model using confusion matrix
confusionMatrix(nb_pred, y_val)
```

Example 6-12. KNN

```
# Load necessary libraries
library(caret)

# Step 1: Train the KNN model
knn_model <- train(x = X_train, y = y_train, method = "knn", tuneLength = 5)
#tuneLength = 5: Automatically tunes the value of k

# Step 2: Predict on the validation set
knn_pred <- predict(knn_model, newdata = X_val)

# Step 3: Evaluate the model using confusion matrix
confusionMatrix(knn_pred, y_val)
```

Example 6-13. Neural network

```
# Load necessary libraries
library(caret)
library(nnet)  # For neural network

# Step 1: Train the Neural Network model
nn_model <- train(x = X_train, y = y_train, method = "nnet", tuneLength = 5,
trace = FALSE)

# Step 2: Predict on the validation set
nn_pred <- predict(nn_model, newdata = X_val)

# Step 3: Evaluate the model using confusion matrix
confusionMatrix(nn_pred, y_val)
```

Example 6-14. GBM

```
library(gbm)

# Load necessary libraries
library(caret)
library(gbm)  # For GBM

# Step 1: Train the GBM model
gbm_model <- train(x = X_train, y = y_train, method = "gbm", verbose = FALSE)

# Step 2: Predict on the validation set
gbm_pred <- predict(gbm_model, newdata = X_val)

# Step 3: Evaluate the model using confusion matrix
confusionMatrix(gbm_pred, y_val)
```

Example 6-15. AdaBoost

```
# Load necessary libraries
library(caret)
library(adabag)  # For AdaBoost

# Step 1: Train the AdaBoost model
adaboost_model <- train(x = X_train, y = y_train, method = "AdaBoost.M1")

# Step 2: Predict on the validation set
adaboost_pred <- predict(adaboost_model, newdata = X_val)

# Step 3: Evaluate the model using confusion matrix
confusionMatrix(adaboost_pred, y_val)
```

Example 6-16. XGBoost

```
# Load necessary libraries
library(caret)
library(xgboost)  # For XGBoost

# Step 1: Train the XGBoost model
xgb_model <- train(x = X_train, y = y_train, method = "xgbTree", tuneLength = 5)

# Step 2: Predict on the validation set
xgb_pred <- predict(xgb_model, newdata = X_val)

# Step 3: Evaluate the model using confusion matrix
confusionMatrix(xgb_pred, y_val)
```

We have reviewed the code to complete classification using multiple algorithms in Python and R. Let's review how classification models are evaluated.

Classification Evaluation

Classification model evaluation is a critical aspect of the data science and machine learning process. It involves assessing the performance and effectiveness of models designed to classify data into different categories or classes. The goal of classification model evaluation is to determine how well a model can make accurate predictions on unseen data and identify any potential issues or areas for improvement.

The process of evaluating classification models typically begins with the selection of appropriate evaluation metrics, such as accuracy, precision, recall, F1-score, and area under the ROC curve (AUC-ROC) depending on the specific problem and objectives (these will be defined in the next two sections). These metrics provide insights into different aspects of model performance, such as overall correctness, the ability to identify relevant cases, and the trade-off between precision and recall.

Furthermore, classification model evaluation often involves techniques like cross-validation, which helps estimate a model's performance on diverse subsets of the data to ensure robustness. Additionally, visualizations, such as confusion matrices and ROC curves, are frequently used to gain a deeper understanding of how well the model is performing across various categories.

In this context, classification model evaluation serves as a critical checkpoint in the machine learning workflow, enabling data scientists and analysts to make informed decisions about model selection, fine-tuning, and deployment, ultimately ensuring that models meet the desired objectives and provide valuable insights for decision making.

Metrics

Classification metrics are crucial for assessing the performance of machine learning models that work with categorical data. They provide insights into various aspects of model accuracy, such as the ability to correctly identify classes, handle false positives, and balance trade-offs between precision (correct positive predictions) and recall (ability to capture all actual positives).

Each metric serves a different purpose depending on the classification problem. Data scientists select the most relevant metrics based on the specific goals of their tasks to fine-tune models and ensure optimal performance. This careful evaluation helps in improving the model's predictive capabilities. Most of the metrics used for evaluation are derived from a confusion matrix.

Confusion matrix

A confusion matrix is a fundamental tool used in classification model evaluation to assess the performance of a machine learning model, especially in binary or multi-class classification tasks. It provides a clear and detailed summary of the model's predictions compared to the actual ground truth. A confusion matrix typically consists of four main components (Figure 6-3):

True positives (TP)
: These are instances where the model correctly predicted the positive class (e.g., correctly identifying individuals with a disease).

True negatives (TN)
: These are instances where the model correctly predicted the negative class (e.g., correctly identifying individuals without a disease).

False positives (FP)
: These are instances where the model incorrectly predicted the positive class when it should have been negative (e.g., incorrectly diagnosing a healthy individual as having a disease, also known as a type I error).

False negatives (FN)
: These are instances where the model incorrectly predicted the negative class when it should have been positive (e.g., failing to diagnose a diseased individual, also known as a type II error).

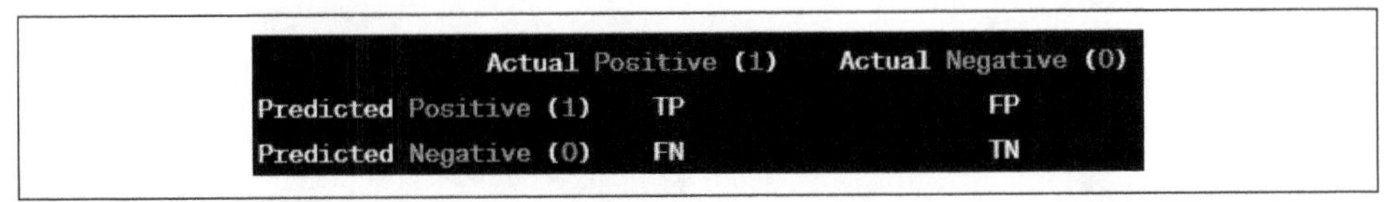

	Actual Positive (1)	Actual Negative (0)
Predicted Positive (1)	TP	FP
Predicted Negative (0)	FN	TN

Figure 6-3. Example of a confusion matrix

Now, let's see how the confusion matrix is used in classification model evaluation:

Accuracy
: Accuracy is calculated as (TP + TN) / (TP + TN + FP + FN). It measures the overall correctness of the model by considering both true positives and true negatives. However, it may not be suitable for imbalanced datasets.

Precision
: Precision is calculated as TP / (TP + FP). It evaluates the accuracy of the positive predictions made by the model. High precision indicates a low false positive rate.

Recall (Sensitivity)
: Recall is calculated as TP / (TP + FN). It measures the model's ability to identify all relevant instances of the positive class. High recall indicates a low false negative rate.

F1-Score
: The F1-Score is the harmonic mean of precision and recall. It balances precision and recall, making it a suitable metric when there is a trade-off between false positives and false negatives.

Specificity
: Specificity is calculated as TN / (TN + FP). It assesses the model's ability to correctly identify negative instances. High specificity indicates a low false positive rate.

By examining the values in the confusion matrix and calculating these metrics, you can gain valuable insights into how well your classification model is performing and make informed decisions about model improvements or adjustments, such as tuning thresholds or selecting different algorithms.

Model evaluation in R

The following code highlights classification using the iris data, including the evaluation step:

```
# Load necessary libraries
library(caret)
library(e1071) ❶

# Load the Iris dataset and convert it to binary classification
data(iris) ❷
iris$Species <- factor(ifelse(iris$Species == "setosa", "setosa", "non-setosa"))

# Split the dataset into training and testing sets
set.seed(123)
trainIndex <- createDataPartition(iris$Species, p = 0.8,
                                  list = FALSE,
                                  times = 1)
data_train <- iris[ trainIndex,]
data_test  <- iris[-trainIndex,]  ❸

# Build a Support Vector Machine (SVM) classifier
svm_model <- svm(Species ~ ., data = data_train) ❹

# Make predictions on the test set
predictions <- predict(svm_model, newdata = data_test) ❺

# Create a confusion matrix
conf_matrix <- table(Actual = data_test$Species, Predicted = predictions) ❻

# Print the confusion matrix
print(conf_matrix) ❼

       Predicted
Actual       non-setosa setosa
  non-setosa         20      0
  setosa              0     10

# Calculate accuracy
accuracy <- sum(diag(conf_matrix)) / sum(conf_matrix)
cat("Accuracy:", accuracy, "\n") ❽

Accuracy: 1
```

❶ Load two necessary libraries, caret and e1071, which are used for data preprocessing and SVM modeling.

❷ Load the Iris dataset using `data(iris)`. Then, it is converted into a binary classification problem. The original three species classes are reduced to two classes: `"setosa"` and `"non-setosa"` using `ifelse`. This prepares the data for binary classification.

❸ Split the dataset into training and testing sets. The `set.seed(123)` ensures reproducibility. `createDataPartition` is used to create an 80-20 split (80% training, 20% testing) based on the `Species` variable.

❹ Build an SVM classifier (`svm_model`) using the `Species` as the target variable and all other variables as predictors.

❺ The model is used to make predictions on the test set (`data_test`) using the `predict` function.

❻ Create a confusion matrix (`conf_matrix`) to evaluate the model's performance. It compares the actual species labels (`Actual`) from the test set with the predicted labels (`Predicted`).

❼ Print the confusion matrix to the console using `print(conf_matrix)`.

❽ Finally, calculate accuracy by summing the diagonal elements of the confusion matrix (correct predictions) and dividing it by the total number of predictions. The result is printed to the console.

Model evaluation in Python

The following code includes a classification example using SVM in Python:

```
# Import necessary libraries
from sklearn import datasets
from sklearn.model_selection import train_test_split
from sklearn.svm import SVC
from sklearn.metrics import confusion_matrix, accuracy_score ❶

# Load the Iris dataset and convert it to binary classification
iris = datasets.load_iris()
X = iris.data ❷
y = [1 if label == 0 else 0 for label in iris.target]  # Setosa vs.
non-Setosa ❸

# Split the dataset into training and testing sets
X_train, X_test, y_train, y_test = train_test_split(X, y, test_size=0.2,
random_state=42) ❹

# Build a Support Vector Machine (SVM) classifier
svm_model = SVC(kernel='linear', C=1)
svm_model.fit(X_train, y_train) ❺

# Make predictions on the test set
predictions = svm_model.predict(X_test) ❻

# Create a confusion matrix
```

```
conf_matrix = confusion_matrix(y_test, predictions) ❼

# Print the confusion matrix
print("Confusion Matrix:")
print(conf_matrix) ❽

# Calculate accuracy
accuracy = accuracy_score(y_test, predictions) ❾
print("Accuracy:", accuracy) ❿
Confusion Matrix:
[[20  0]
 [ 0 10]]
Accuracy: 1.0
```

❶ Import the necessary libraries, including datasets from sklearn to load the Iris dataset, train_test_split from sklearn.model_selection to split the dataset, SVC from sklearn.svm to build an SVM classifier, and confusion_matrix and accuracy_score from sklearn.metrics to evaluate the model.

❷ Load the Iris dataset, which contains flower measurements (features) for different species of Iris. The data (`X`) consists of the measurements, and the labels (`y`) indicate the species.

❸ Convert the problem into a binary classification task. The goal is to classify whether a flower is Setosa (class 1) or not Setosa (class 0). A list comprehension is used to transform the original labels into this binary format.

❹ Split the dataset into two parts: the training set (80% of the data) is used to train the model, and the testing set (20% of the data) is used to evaluate the model. The split is done randomly, but `random_state=42` ensures that the results are reproducible.

❺ Build an SVM classifier with a linear kernel. The regularization parameter `C=1` controls the trade-off between a smooth decision boundary and classifying the training examples correctly. The model is then trained using the training data (`X_train, y_train`)

❻ After training, the SVM model is used to predict the class labels for the test data (`X_test`).

❼ Create a confusion matrix to assess how well the model is performing. It shows the counts of true positives, true negatives, false positives, and false negatives based on the test data predictions.

❽ Print the confusion matrix to give a summary of how many samples were correctly or incorrectly classified.

❾ Calculate the accuracy of the model by comparing the predicted labels with the true labels from the test data. Accuracy is the ratio of correctly classified instances to the total instances.

❿ Finally, print the calculated accuracy to give a measure of the overall performance of the model.

In the provided code examples for both R and Python, the accuracy rate may appear as 100% due to the specific data split and the simplicity of the binary classification task we are working on, which is distinguishing between the `"setosa"` and `"non-setosa"` iris flowers. This high accuracy rate can occur for a few reasons:

Data split
: The dataset is split into a training set and a testing set. If the data split happens to separate the two classes (setosa and non-setosa) very cleanly in both sets, it can lead to a high accuracy because the model may correctly predict all instances in the testing set.

Model simplicity
: In both examples, we are using a linear SVM classifier. Linear SVMs are known for their simplicity and effectiveness when data is linearly separable, as might be the case here.

Data separability
: The Iris dataset is a well-known dataset in machine learning, and the "setosa" class is often linearly separable from the other classes. This means that a simple linear classifier like SVM can achieve high accuracy for this specific task.

Small dataset
: The Iris dataset is relatively small, with only 150 instances. In some cases, smaller datasets can lead to overfitting, especially when the data is relatively simple and separable.

It's important to note that achieving 100% accuracy on a classification task is not common in real-world scenarios. In more complex datasets with overlapping classes and noise, models are unlikely to achieve perfect accuracy.

While the provided examples demonstrate the basics of classification and evaluation, they may not reflect the typical performance of models in more complex problems.

Calculating classification metrics in Python

In addition to the metrics discussed, there are two others worth mentioning: micro- and macro-averages.

Micro-averaging aggregates the contributions of all classes to compute the overall precision, recall, and F1 score. It treats every prediction equally, regardless of the class. In other words, it calculates the metrics globally by counting the total true positives, false positives, and false negatives across all classes. Micro-average first sums the true positives, false positives, and false negatives for all classes, and then calculates precision, recall, and F1 based on these aggregated values. Micro-average is useful when you want to evaluate the overall performance of the model, without giving more weight to classes with more instances. It's often used when the class distribution is imbalanced.

Macro-averaging calculates precision, recall, and F1 score for each class independently and then averages these metrics. Unlike micro-averaging, it gives equal weight to each class, regardless of the number of samples in that class. Macro-average calculates the precision, recall, and F1 score for each class individually and then takes the unweighted mean of these values. You would use the macro-average when you want to measure the model's performance equally across all classes, especially if you are interested in how the model performs on each class, irrespective of class distribution (i.e., it's beneficial when dealing with imbalanced datasets).

Let's define our classification metrics and then calculate them based on the following confusion matrix in Python:

Precision
: Precision measures the accuracy of positive predictions made by the model. It answers the question, "Of all instances predicted as positive, how many were actually correct positives?"

Recall (Sensitivity)
: Recall measures the ability of the model to correctly identify all relevant instances. It answers the question, "Of all actual positive instances, how many were correctly predicted as positive?"

F1 Score
: The F1 score is the harmonic mean of precision and recall, providing a balance between the two metrics. It is particularly useful when there is an uneven class distribution.

Micro-average metrics
: Micro-average metrics consider the overall performance across all classes. They are calculated by aggregating the counts of true positives, false positives, and false negatives across all classes.

Macro-average metrics
: Macro-average metrics calculate the average of individual class metrics and offer an overall view of model performance without considering class imbalances.

Here is the confusion matrix to calculate those metrics:

```
# Confusion matrix
confusion_matrix = [[30, 0, 0],
                    [0, 25, 3],
                    [0, 2, 20]]

# Calculate precision, recall, and F1 score for each class
precision_class_0 = confusion_matrix[0][0] / (confusion_matrix[0][0] +
confusion_matrix[1][0] + confusion_matrix[2][0])
recall_class_0 = confusion_matrix[0][0] / (confusion_matrix[0][0] +
confusion_matrix[0][1] + confusion_matrix[0][2])
f1_score_class_0 = 2 * (precision_class_0 * recall_class_0) /
(precision_class_0 + recall_class_0)

precision_class_1 = confusion_matrix[1][1] / (confusion_matrix[1][1] +
confusion_matrix[0][1] + confusion_matrix[2][1])
recall_class_1 = confusion_matrix[1][1] / (confusion_matrix[1][1] +
confusion_matrix[1][0] + confusion_matrix[1][2])
f1_score_class_1 = 2 * (precision_class_1 * recall_class_1) /
(precision_class_1 + recall_class_1)

precision_class_2 = confusion_matrix[2][2] / (confusion_matrix[2][2] +
confusion_matrix[0][2] + confusion_matrix[1][2])
recall_class_2 = confusion_matrix[2][2] / (confusion_matrix[2][2] +
confusion_matrix[2][0] + confusion_matrix[2][1])
f1_score_class_2 = 2 * (precision_class_2 * recall_class_2) /
(precision_class_2 + recall_class_2)

# Calculate micro-average and macro-average metrics
total_true_positives = sum([confusion_matrix[i][i] for i in range(3)])
total_false_positives = sum([sum(confusion_matrix[i]) - confusion_matrix[i][i]
for i in range(3)])
total_false_negatives = sum([sum(confusion_matrix[i]) - confusion_matrix[i][i]
for i in range(3)])

micro_precision = total_true_positives / (total_true_positives +
total_false_positives)
micro_recall = total_true_positives / (total_true_positives +
total_false_negatives)
micro_f1_score = 2 * (micro_precision * micro_recall) / (micro_precision +
micro_recall)

macro_precision = (precision_class_0 + precision_class_1 + precision_class_2) /
3
macro_recall = (recall_class_0 + recall_class_1 + recall_class_2) / 3
macro_f1_score = (f1_score_class_0 + f1_score_class_1 + f1_score_class_2) / 3
```

```
# Print all variables
print("Confusion Matrix:")
print(confusion_matrix)
print("\nClass 0 Precision:", precision_class_0)
print("Class 0 Recall:", recall_class_0)
print("Class 0 F1 Score:", f1_score_class_0)
print("\nClass 1 Precision:", precision_class_1)
print("Class 1 Recall:", recall_class_1)
print("Class 1 F1 Score:", f1_score_class_1)
print("\nClass 2 Precision:", precision_class_2)
print("Class 2 Recall:", recall_class_2)
print("Class 2 F1 Score:", f1_score_class_2)
print("\nMicro-Average Precision:", micro_precision)
print("Micro-Average Recall:", micro_recall)
print("Micro-Average F1 Score:", micro_f1_score)
print("\nMacro-Average Precision:", macro_precision)
print("Macro-Average Recall:", macro_recall)
print("Macro-Average F1 Score:", macro_f1_score)
Confusion Matrix:
[[30, 0, 0], [0, 25, 3], [0, 2, 20]]

Class 0 Precision: 1.0 ❶
Class 0 Recall: 1.0 ❷
Class 0 F1 Score: 1.0 ❸

Class 1 Precision: 0.9259259259259259 ❹
Class 1 Recall: 0.8928571428571429 ❺
Class 1 F1 Score: 0.9090909090909091 ❻

Class 2 Precision: 0.8695652173913043 ❼
Class 2 Recall: 0.9090909090909091 ❽
Class 2 F1 Score: 0.888888888888889 ❾

Micro-Average Precision: 0.9375 ❿
Micro-Average Recall: 0.9375 ⓫
Micro-Average F1 Score: 0.9375 ⓬

Macro-Average Precision: 0.9318303811057435 ⓭
Macro-Average Recall: 0.933982683982684 ⓮
Macro-Average F1 Score: 0.9326599326599326 ⓯
```

Now, let's interpret the metrics:

❶ Precision for Class 0 is 1.0, indicating that all positive predictions for Class 0 were correct (no false positives for Class 0).

❷ Precision for Class 1 is approximately 0.926, signifying that about 92.6% of positive predictions for Class 1 were correct (some false positives for Class 1).

❸ Precision for Class 2 is approximately 0.870, showing that around 87.0% of positive predictions for Class 2 were correct (some false positives for Class 2).

❹ Recall for Class 0 is 1.0, signifying that all actual instances of Class 0 were correctly identified by the model (no false negatives for Class 0).

❺ Recall for Class 1 is approximately 0.893, indicating that about 89.3% of actual Class 1 instances were correctly identified by the model (some false negatives for Class 1).

❻ Recall for Class 2 is approximately 0.909, showing that about 90.9% of actual Class 2 instances were correctly identified by the model (some false negatives for Class 2).

❼ The F1 score for Class 0 is 1.0, representing a perfect balance between precision and recall for this class.

❽ The F1 score for Class 1 is approximately 0.909, showing a good balance between precision and recall for this class.

❾ The F1 score for Class 2 is approximately 0.889, indicating a good balance between precision and recall for this class.

❿ Micro-average precision is 0.9375, representing an overall precision score across all classes. It indicates that, on average, about 93.75% of positive predictions across all classes were correct.

⓫ Micro-average recall is 0.9375, signifying an overall recall score across all classes. On average, about 93.75% of actual instances across all classes were correctly identified.

⓬ Micro-average F1 score is 0.9375, providing an overall balance between precision and recall across all classes.

⓭ Macro-average precision is approximately 0.932, calculating the average precision score across all classes. It provides an overall view of precision performance.

⓮ Macro-average recall is approximately 0.934, calculating the average recall score across all classes. It offers an overall view of recall performance.

⓯ Macro-average F1 score is approximately 0.933, providing an average balance between precision and recall across all classes.

Collectively, these metrics provide a comprehensive evaluation of the classification model's performance, considering precision, recall, and F1 score for each class as well as micro- and macro-averages across all classes. The values indicate how well the model is performing in terms of correctly classifying instances and balancing precision and recall.

Classification Use Case Evaluation

A telecom use case was provided earlier in the chapter to demonstrate the different modeling approaches for classification. In this section, the confusion matrices for each of the models will be examined. The following are the prediction results of each model depicted using the confusion matrix:

```
Random Forest Evaluation:
              precision    recall  f1-score   support

           0       0.73      1.00      0.84     14599
           1       0.20      0.00      0.00      5401

    accuracy                           0.73     20000
   macro avg       0.46      0.50      0.42     20000
weighted avg       0.59      0.73      0.62     20000

Accuracy: 0.7292

SVM Evaluation:
              precision    recall  f1-score   support

           0       0.73      1.00      0.84     14599
           1       0.00      0.00      0.00      5401

    accuracy                           0.73     20000
   macro avg       0.36      0.50      0.42     20000
weighted avg       0.53      0.73      0.62     20000

Accuracy: 0.72995
Naive Bayes Evaluation:
              precision    recall  f1-score   support

           0       0.73      1.00      0.84     14599
           1       0.00      0.00      0.00      5401

    accuracy                           0.73     20000
   macro avg       0.36      0.50      0.42     20000
weighted avg       0.53      0.73      0.62     20000

Accuracy: 0.72995

KNN Evaluation:
              precision    recall  f1-score   support

           0       0.73      0.87      0.80     14599
           1       0.27      0.13      0.18      5401
```

```
    accuracy                           0.67     20000
   macro avg       0.50      0.50      0.49     20000
weighted avg       0.61      0.67      0.63     20000

Accuracy: 0.672
Neural Network Evaluation:
              precision    recall  f1-score   support

           0       0.73      0.80      0.76     14599
           1       0.27      0.20      0.23      5401

    accuracy                           0.64     20000
   macro avg       0.50      0.50      0.49     20000
weighted avg       0.60      0.64      0.62     20000

Accuracy: 0.63805
GBM Evaluation:
              precision    recall  f1-score   support

           0       0.73      1.00      0.84     14599
           1       0.50      0.00      0.00      5401

    accuracy                           0.73     20000
   macro avg       0.62      0.50      0.42     20000
weighted avg       0.67      0.73      0.62     20000

Accuracy: 0.72995
AdaBoost Evaluation:
              precision    recall  f1-score   support

           0       0.73      1.00      0.84     14599
           1       0.55      0.00      0.00      5401

    accuracy                           0.73     20000
   macro avg       0.64      0.50      0.42     20000
weighted avg       0.68      0.73      0.62     20000

Accuracy: 0.73
XGBoost Evaluation:
              precision    recall  f1-score   support

           0       0.73      0.99      0.84     14599
           1       0.27      0.01      0.03      5401

    accuracy                           0.72     20000
   macro avg       0.50      0.50      0.43     20000
weighted avg       0.60      0.72      0.62     20000

Accuracy: 0.7231
```

Most models demonstrate significant bias toward the majority class (nonchurn, 0):

High Precision and Recall for Class 0
: All models performed well in predicting the majority class (nonchurners). The precision and recall for class 0 is quite high, typically near 0.73 to 1.00.

Poor Performance for Class 1 (Churners)
: Across all models, the recall for class 1 (churn) is extremely low, close to zero, meaning the models were unable to correctly identify churners. Precision for class 1 is also very low, indicating a significant imbalance issue. In most cases, the F1-score is 0.00 for class 1, meaning the models failed to predict this class properly.

Accuracy Is Misleading
: While most models show an accuracy around 0.73, this is due to the imbalanced nature of the dataset. Since the dataset contains more nonchurners, the models are biased toward predicting the majority class correctly, making accuracy an unreliable metric.

One unaddressed issue in the data preparation was a class imbalance. From the data profiling report (Figure 6-4), there are substantially more Class 0 than Class 1.

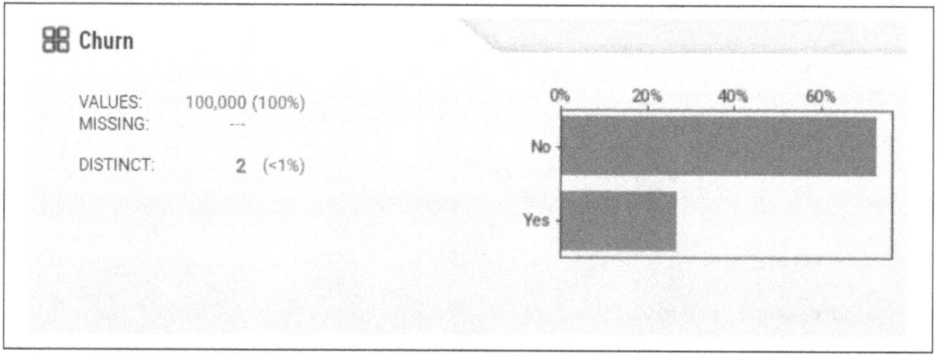

Figure 6-4. Churn class imbalance

SMOTE (reviewed in Chapter 5) is a popular method used to address class imbalance in datasets. It works by generating synthetic samples for the minority class, thereby increasing its representation in the dataset without simply duplicating existing examples. SMOTE achieves this by interpolating between existing minority class instances to create new, similar instances, which helps the model learn a more balanced decision boundary. This can lead to improved performance, particularly in classification tasks where the minority class is often underrepresented and overlooked. By balancing the class distribution, SMOTE reduces bias toward the majority class and enhances the model's ability to predict the minority class more accurately. However, it is important to evaluate the effectiveness of SMOTE using metrics like precision, recall, and F1-score, as it may not always improve performance depending on the nature of the data.

To address the imbalance issue, SMOTE is applied to the data preparation phase directly after the initial data split and before the split to include the validation set. The data preparation code has been modified to show this in Figure 6-5.

```
# Import necessary libraries
import pandas as pd
from sklearn.model_selection import train_test_split
from sklearn.preprocessing import LabelEncoder, StandardScaler
from imblearn.over_sampling import SMOTE  # Import SMOTE from imbalanced-learn

# Load the dataset
file_path = '/content/sample_data/synthetic_telecom_churn_data.csv' # Update with actual path
data = pd.read_csv(file_path)

# Handle missing values in 'TotalCharges' (convert to numeric and fill missing with median)
data['TotalCharges'] = pd.to_numeric(data['TotalCharges'], errors='coerce')
data['TotalCharges'].fillna(data['TotalCharges'].median(), inplace=True)

# Drop 'customerID' column
data.drop('customerID', axis=1, inplace=True)

# Convert 'Churn' column (target variable) to binary labels
label_encoder = LabelEncoder()
data['Churn'] = label_encoder.fit_transform(data['Churn'])  # 'Yes' -> 1, 'No' -> 0
```

```
# Encode all other categorical columns using Label Encoding
categorical_columns = data.select_dtypes(include=['object']).columns
for column in categorical_columns:
    data[column] = label_encoder.fit_transform(data[column])

# Split the data into features (X) and target (y)
X = data.drop('Churn', axis=1)
y = data['Churn']

# Standardize the numeric features
scaler = StandardScaler()
X_scaled = pd.DataFrame(scaler.fit_transform(X), columns=X.columns)

# Split the data into training (60%), temporary (40%) sets
X_train, X_temp, y_train, y_temp = train_test_split(X_scaled, y, test_size=0.4, random_state=42, stratify=y)

# Apply SMOTE to the training set only
smote = SMOTE(random_state=42)
X_train_smote, y_train_smote = smote.fit_resample(X_train, y_train)

# Verify class distribution after SMOTE
print(f"Original training set class distribution:\n{y_train.value_counts()}")
print(f"Resampled training set class distribution after SMOTE:\n{y_train_smote.value_counts()}")
```

```
# Split the temporary set into validation (50% of temp) and test (50% of temp)
X_val, X_test, y_val, y_test = train_test_split(X_temp, y_temp, test_size=0.5, random_state=42, stratify=y_temp)

# Check the shape of the resulting datasets
print(f"Training data shape: {X_train_smote.shape}, Validation data shape: {X_val.shape}, Testing data shape: {X_test.shape}")
```

```
Original training set class distribution:
Churn
0    43798
1    16202
Name: count, dtype: int64
Resampled training set class distribution after SMOTE:
Churn
0    43798
1    43798
Name: count, dtype: int64
Training data shape: (87596, 19), Validation data shape: (20000, 19), Testing data shape: (20000, 19)
```

Figure 6-5. SMOTE added to the data preparation step (you can see the image at full size in color online (https://oreil.ly/mban_0605png))

Once the data preparation has been run to include SMOTE, the results show that both classes are now balanced, with 43,798 records for each class. The next step is to initialize, train, and test the models using the balanced classes. Here are the new confusion matrices:

```
Random Forest Evaluation:
              precision    recall  f1-score   support
           0       0.73      0.99      0.84     14599
           1       0.25      0.01      0.01      5401
    accuracy                           0.73     20000
   macro avg       0.49      0.50      0.43     20000
weighted avg       0.60      0.73      0.62     20000
Accuracy: 0.72595

SVM Evaluation:
             precision   recall  f1-score  support
          0       0.73     0.40      0.51    14599
          1       0.27     0.60      0.37     5401
    accuracy                         0.45    20000
   macro avg       0.50     0.50      0.44    20000
weighted avg       0.60     0.45      0.48    20000
Accuracy: 0.45125

KNN Evaluation:
             precision    recall  f1-score   support

           0      0.73      0.52      0.61     14599
           1      0.27      0.48      0.34      5401
    accuracy                          0.51     20000
   macro avg      0.50      0.50      0.47     20000
weighted avg      0.60      0.51      0.54     20000
 Accuracy: 0.5077

Naive Bayes Evaluation:
             precision   recall  f1-score   support

           0      0.73     0.55      0.63     14599
           1      0.27     0.45      0.34      5401
     accuracy                         0.52     20000
   macro avg      0.50     0.50      0.48     20000
weighted avg      0.61     0.52      0.55     20000
Accuracy: 0.52195

Neural Network Evaluation:
             precision   recall  f1-score   support
            0     0.73     0.67      0.70     14599
           1      0.27     0.33      0.29      5401
     accuracy                         0.58     20000
   macro avg      0.50     0.50      0.50     20000
weighted avg      0.60     0.58      0.59     20000
Accuracy: 0.57795

GBM Evaluation:
             precision   recall  f1-score   support
```

```
           0       0.73      1.00      0.84     14599
           1       0.00      0.00      0.00      5401
    accuracy                           0.73     20000
   macro avg       0.36      0.50      0.42     20000
weighted avg       0.53      0.73      0.62     20000
 Accuracy: 0.72995

AdaBoost Evaluation:
              precision    recall  f1-score   support
           0       0.73      0.99      0.84     14599
           1       0.29      0.01      0.02      5401
    accuracy                           0.73     20000
   macro avg       0.51      0.50      0.43     20000
weighted avg       0.61      0.73      0.62     20000
Accuracy: 0.7268

XGBoost Evaluation:
              precision    recall  f1-score   support
           0       0.73      0.99      0.84     14599
           1       0.29      0.01      0.02      5401
    accuracy                           0.73     20000
   macro avg       0.51      0.50      0.43     20000
weighted avg       0.61      0.73      0.62     20000
Accuracy: 0.72635
```

The results are certainly different after SMOTE was applied. Let's look at the results before and after in Table 6-3.

Table 6-3. Comparison of results before and after applying SMOTE

Method	Before SMOTE	After SMOTE
Random forest	• Precision (Class 1): 0.20 • Recall (Class 1): 0.00 (almost no churners detected) • Accuracy: 0.7292 • The model is heavily biased toward class `0` (nonchurners), correctly predicting almost all nonchurners but failing to detect churners (class `1`). The recall for class `1` is almost zero, meaning it's not identifying the minority class at all.	• Precision (Class 1): 0.25 • Recall (Class 1): 0.01 • Accuracy: 0.72595 • After SMOTE, the model slightly improves in detecting churners, but the recall is still very low (only 1%). The model still favors the majority class, but SMOTE helped only a bit by increasing the recall marginally.
SVM	• Precision (Class 1): 0.00 • Recall (Class 1): 0.00 • Accuracy: 0.72995 • The SVM completely fails to identify any churners before SMOTE, predicting only nonchurners.	• Precision (Class 1): 0.27 • Recall (Class 1): 0.60 • Accuracy: 0.45125 • SMOTE significantly improves recall (60%) for churners, meaning the model can now identify many churners, but the overall accuracy drops significantly. This suggests that the SVM starts overpredicting churners but at the expense of lower accuracy.

Method	Before SMOTE	After SMOTE
Naive Bayes	• Precision (Class 1): 0.00 • Recall (Class 1): 0.00 • Accuracy: 0.72995 • Similar to SVMs, naive Bayes predicts only the majority class, failing to detect any churners.	• Precision (Class 1): 0.27 • Recall (Class 1): 0.45 • Accuracy: 0.52195 • After SMOTE, naive Bayes improves recall for churners but still performs poorly overall. Accuracy drops, but the model is now at least detecting a decent proportion of churners (45% recall).
KNN	• Precision (Class 1): 0.27 • Recall (Class 1): 0.13 • Accuracy: 0.672 • KNN has a small amount of success in predicting churners before SMOTE, but recall remains low.	• Precision (Class 1): 0.27 • Recall (Class 1): 0.48 • Accuracy: 0.5077 • SMOTE improves recall for class 1, but accuracy drops. KNN is now predicting more churners but at the expense of precision and overall accuracy.
Neural network (MLP)	• Precision (Class 1): 0.27 • Recall (Class 1): 0.20 • Accuracy: 0.63805 • The neural network performs similarly to KNN before SMOTE, detecting some churners but with low recall and accuracy.	• Precision (Class 1): 0.27 • Recall (Class 1): 0.33 • Accuracy: 0.57795 • After SMOTE, recall improves, but accuracy remains relatively low. The neural network is performing better at detecting churners but is still biased toward the majority class.
Gradient boosting machine (GBM)	• Precision (Class 1): 0.50 • Recall (Class 1): 0.00 • Accuracy: 0.72995 • GBM fails to detect any churners before SMOTE but has a high precision for the few times it does predict churners.	• Precision (Class 1): 0.00 • Recall (Class 1): 0.00 • Accuracy: 0.72995 • After SMOTE, GBM still fails to detect churners, and performance remains biased toward the majority class.
AdaBoost	• Precision (Class 1): 0.55 • Recall (Class 1): 0.00 • Accuracy: 0.73 • AdaBoost has high precision but fails to recall any churners before SMOTE.	• Precision (Class 1): 0.29 • Recall (Class 1): 0.01 • Accuracy: 0.7268 • After SMOTE, AdaBoost's recall improves slightly, but it still performs poorly at detecting churners, as seen by its low recall and precision.
XGBoost	• Precision (Class 1): 0.27 • Recall (Class 1): 0.01 • Accuracy: 0.7231 • XGBoost performs similarly to other models before SMOTE, with very low recall for churners.	• Precision (Class 1): 0.29 • Recall (Class 1): 0.01 • Accuracy: 0.72635 • After SMOTE, XGBoost shows little improvement in detecting churners, with the recall remaining very low.

SMOTE improves recall for the minority class in many models (particularly SVMs, naive Bayes, and KNN), but at the cost of reduced precision and accuracy. The improvement in recall means these models are better at identifying churners. Accuracy typically drops after SMOTE because the models are now predicting more churners, but they often misclassify nonchurners as churners, resulting in lower overall accuracy. Some models like GBM and AdaBoost fail to benefit from SMOTE, indicating that these algorithms might not work well with oversampling techniques or require different handling of imbalanced data. SVMs, KNN, and naive Bayes show promise after SMOTE, but further tuning of hyperparameters (using GridSearchCV or RandomizedSearchCV) might help improve both precision and recall for the minority class. For example, you could tune the number of neighbors in KNN, the kernel in SVMs, and priors in naive Bayes. Chapter 7 explores hyperparameter tuning.

It is possible to evaluate models by shifting focus from accuracy to precision, recall, and F1-score for class 1 (churners) since churn prediction often places more importance on correctly identifying churners rather than nonchurners. It is possible to use an ROC-AUC curve as alternative metrics for evaluation. The ROC curve is a graphical representation used to evaluate the performance of a binary classification model. It plots the True Positive Rate (Sensitivity) against the False Positive Rate (1 - Specificity) across different threshold values. The curve helps assess how well the model distinguishes between the two classes. A model with a perfect classification would have a curve that rises steeply toward the top-left corner.

AUC is a single scalar value derived from the ROC curve that quantifies the overall ability of the model to differentiate between the positive and negative classes. AUC values range between 0 and 1. A value of 1 indicates a perfect model, while 0.5 suggests a random classifier with no discriminatory power.

Figure 6-6 shows the random forest model's confusion matrix and ROC curve. Even though the accuracy for the model is 72.5%, this result simply measures the proportion of correct predictions. If the dataset is imbalanced (i.e., if one class, such as "no churn," dominates), the model can predict the majority class most of the time and still achieve high accuracy, even if it is not effectively identifying the minority class (e.g., "churn"). AUC-ROC (area under the ROC curve) measures the model's ability to distinguish between the two classes (in this case, churn versus no churn). An AUC of 0.50 indicates that the model cannot discriminate between the classes better than random guessing, meaning it is failing to rank or prioritize instances of the positive class (e.g., churn) versus the negative class.

In cases of class imbalance, it's essential to consider both accuracy and AUC (or other metrics like precision, recall, and F1-score) to properly assess the performance of a model. A high accuracy combined with a low AUC can indicate a misleading model that only works well for the majority class but fails to correctly classify the minority class.

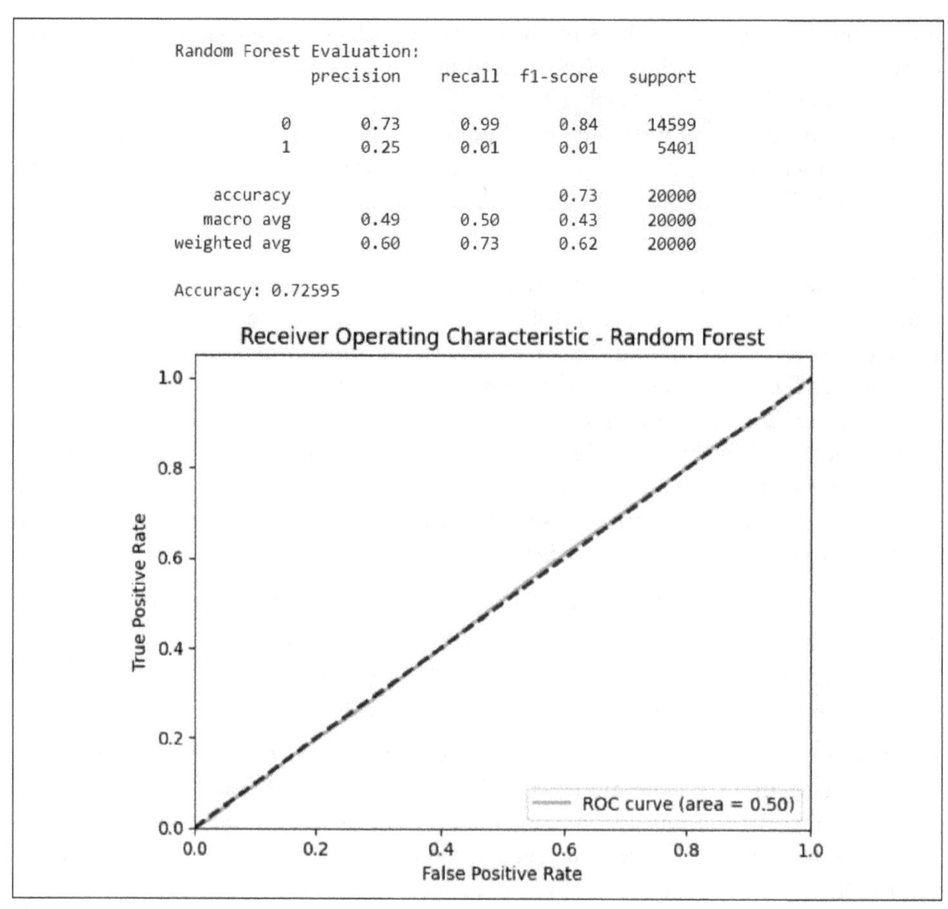

```
Random Forest Evaluation:
              precision    recall  f1-score   support

           0       0.73      0.99      0.84     14599
           1       0.25      0.01      0.01      5401

    accuracy                           0.73     20000
   macro avg       0.49      0.50      0.43     20000
weighted avg       0.60      0.73      0.62     20000

Accuracy: 0.72595
```


Figure 6-6. ROC curve for RF churn

Summary

We've journeyed through the intricate landscape of regression and classification models using both R and Python. The fundamental concepts of both approaches were introduced while providing examples of the practical implementation of multiple algorithms, each with its unique strengths and applications. The hands-on coding sections in R and Python demonstrated how to effectively implement models, offering step-by-step code explanations. This practical approach was designed to not only enhance your understanding but also equip you with the skills to apply these models to real-world datasets.

A pivotal aspect of this chapter was the emphasis on evaluating models. We thoroughly explored the use of a confusion matrix as a tool for evaluation, highlighting its ability to provide a nuanced view of model performance. We discussed how accuracy, precision, recall, and the F1 score, derived from the confusion matrix, are critical for assessing the effectiveness of classification models.

This chapter has been a comprehensive guide to understanding and applying regression and classification models in R and Python. From theoretical concepts to practical implementation and evaluation, we have covered a broad spectrum of topics essential for anyone looking to delve into the world of regression and classification models. Next, we'll explore choosing algorithms.

CHAPTER 7

Modeling and Algorithm Choice

Selecting the appropriate algorithm for building a machine learning model is an important decision. Whatever is chosen will influence the outcome and effectiveness of your predictive analytics projects, so it's important to understand the nuances of the options that are available. This chapter goes into the process of choosing the right machine learning algorithm, taking into account the nature of the data, the problem to be solved, and the constraints of the computational environment. Whether you're working with structured data for regression analysis, unstructured data for deep learning applications, or another analytical area, understanding the strengths, weaknesses, and peculiarities of various algorithms is crucial.

This chapter will guide you through the considerations of algorithm choices, including decision trees, support vector machines (SVMs), neural networks, and ensemble methods. It's designed to provide you with the knowledge needed to make informed decisions that align with project goals, data characteristics, and performance requirements.

Algorithms

A machine learning algorithm is a computational method used to enable a computer to learn from data, identify patterns, and make decisions with minimal human intervention. As we have learned, machine learning is about constructing algorithms that can learn from and make predictions or decisions based on input data. These algorithms improve their accuracy over time by processing more data and adjusting strategies to perform better, mimicking the way humans learn from experience.

As discussed in Chapters 5 and 6, machine learning algorithms are broadly categorized into three main types based on their learning style. These categories are supervised, unsupervised, and reinforcement learning.

Machine learning algorithms utilize various techniques including math, statistics, and logic. Each technique has its own strengths and weaknesses, making it more suitable for certain types of problems and datasets.

The choice of algorithm depends on several factors, including:

- The size, quality, and nature of the data
- The task to be performed (e.g., prediction, classification, clustering)
- The complexity of the model
- The computational resources available
- The interpretability of the model output

Machine learning algorithms are also the foundation of AI systems and are increasingly used in a wide range of applications, from email filtering and speech recognition to medical diagnosis and stock market analysis. The ability to automatically learn and improve from experience makes machine learning a key driver of innovation and efficiency across industries.

Algorithm Criteria

Selecting the optimal algorithm for an analytics project is a multilayered process that requires a comprehensive understanding of various factors that can significantly influence the success of a project. Business analysts play a prominent role in algorithm selection to ensure the specific needs and objectives of their modeling projects are met. The algorithm selection process starts with the foundational step of accurately defining the problem type. This could be a classification, regression, clustering, or another form of machine learning task.

Model interpretation and its relevance in business contexts is another important aspect, where explaining predictions and decisions to stakeholders is often as critical as the predictions themselves. The balance between model accuracy and the practicalities of training and prediction speed also need to be examined. This is a trade-off analysts must consider when efficiency and performance need to be balanced. A key process in optimizing algorithm performance is hyperparameter tuning. Business analysts will need to understand this process as well as strategies for managing the size of the dataset, which can range from small, easily manageable sets to big data scenarios requiring specialized approaches. The interaction of features within the model will also be addressed, as understanding and engineering these interactions can unlock significant predictive power. Let's start exploring algorithm selection by understanding the problem type.

Problem Type

Selecting an algorithm for a modeling project is fundamentally contingent upon the type of problem you're aiming to solve. Each class of machine learning problem (classification, regression, clustering, or dimensionality reduction) has a suite of algorithms best suited to its specific nature. When selecting an algorithm for a modeling project, it's important to recognize that you may not always need to choose just one model. Depending on the complexity of the problem, a chain or sequence of models may be more effective in addressing different aspects of the prediction task. For example, you might start with a regression model to predict the number of applications received in a month, followed by a classification model to assess the probability of default for each application. This layered approach allows you to break down complex problems into more manageable tasks, each handled by the model best suited to its specific objective. By combining multiple models, you can better capture nuances in the data and enhance overall predictive accuracy across different stages of the process. Understanding that machine learning problems often require a combination of models helps ensure that you're addressing each problem type effectively, whether it involves prediction, classification, or another goal.

Understanding the characteristics of these problem types is the first step in navigating the vast landscape of machine learning algorithms, so let's explore each one individually in the sections that follow.

Classification problems

In classification, the goal is to predict a discrete label for given inputs. Accordingly, algorithms like logistic regression, decision trees, SVMs, and neural networks often come to the forefront. Classification can have two (binary classification) or more (multiclass) classes.

For binary classification, logistic regression is a straightforward and efficient choice, providing not only class predictions but also probabilities that offer insights into the model's confidence. There are also other methods to consider. Decision trees are highly interpretable and can easily handle categorical features, but they are prone to overfitting. In contrast, SVMs are powerful in high-dimensional spaces, making them ideal for complex datasets. For more complex or nonlinear decision boundaries, ensemble methods like random forests or gradient boosting machines (XGBoost, LightGBM) and deep learning models can provide higher accuracy at the cost of increased computational complexity and reduced interpretability.

Regression problems

When predicting a continuous outcome, regression algorithms are employed. Linear regression is the simplest and most interpretable option, effective for problems with a linear relationship between the input features and the target variable. However, when dealing with nonlinear relationships, polynomial regression, decision trees for regression, and ensemble regressors like gradient boosting regressors can be more appropriate. Neural networks, especially those with deep architectures, have the flexibility to model complex, nonlinear interactions between features but require substantial data and computational power.

Clustering problems

Clustering algorithms are used when the goal is to group data points into clusters based on similarity. K-means is a popular choice for its simplicity and efficiency, suitable for a wide range of applications, though it assumes clusters to be spherical and equally sized. Hierarchical clustering allows for the discovery of nested clusters, providing a more detailed data structure, which is useful for hierarchical data. DBSCAN is another algorithm that excels in identifying clusters of varying shapes and sizes, proving robust against outliers.

Dimensionality reduction

When dealing with high-dimensional data, dimensionality reduction algorithms like principal component analysis (PCA) and t-distributed stochastic neighbor embedding (t-SNE) are invaluable for visualization and to improve model efficiency by reducing the number of input variables. t-SNE is a machine learning algorithm used for dimensionality reduction, particularly well-suited for the visualization of high-dimensional datasets by mapping them to a lower-dimensional space while preserving the relative distances between points. PCA is particularly effective for linear dimensionality reduction, while t-SNE provides superior visualizations of high-dimensional data by preserving the local structure of the data in a lower-dimensional space.

Ultimately, selecting the right algorithm involves balancing these considerations against the specific requirements and constraints of your modeling project. Experimentation, coupled with cross-validation, provides a pragmatic approach to evaluating different algorithms' performance on the dataset, resulting in the most effective solution for the problem type. Experimentation is the process of systematically applying different algorithms, tuning parameters, or using various data preprocessing techniques to build models that best predict or classify data, with the goal of identifying the most effective approach for a given problem. Cross-validation is a technique to assess how well a model will generalize to an independent dataset. It involves dividing the data into several parts, training the model on some parts and testing it on the remaining ones, and then averaging the results to estimate the model's performance.

Interpretable Models

Model interpretability is an important consideration in many fields, particularly in industries like finance and healthcare, where decisions need to be explained to stakeholders or must comply with regulatory requirements. Interpretable models allow humans to understand the decision-making process of the algorithm, fostering trust and enabling insightful analysis of the model's behavior. There are some algorithms that are referred to as "black box models." This refers to a type of algorithm or model that operates with little to no transparency regarding its internal workings or decision-making process. Users can see the input and output of the model but have limited understanding of how or why the model arrives at a particular decision. This term often applies to complex models like deep neural networks, where the exact relationships between inputs and outputs are difficult to interpret or explain. When model interpretability is a priority, a business analyst will need to select algorithms that enhance transparency and understanding.

Linear models

Both linear regression for continuous outcomes and logistic regression for classification problems are classic examples of interpretable models. Their simplicity lies in the direct relationship between input features and the model's output, where each coefficient represents the change in the output variable for a one-unit change in the corresponding input variable, holding all other variables constant. The actual model is an equation of coefficients where it is clear how feature values are associated with predictions. The following example in Python shows the linear regression equation, where 0.60 is the coefficient, meaning this would be the change of the output variable based on the value of x:

```
from sklearn.linear_model import LinearRegression
import numpy as np

# Example dataset
# X represents the independent variable
X = np.array([[1], [2], [3], [4], [5]])
# y represents the dependent variable
y = np.array([2, 4, 5, 4, 5])

# Create and fit the model
model = LinearRegression().fit(X, y)

# Get the slope (m) and the intercept (c) of the line
slope = model.coef_[0]
intercept = model.intercept_

# Display the equation of the model
print(f"The equation of the model is: y = {slope:.2f}x + {intercept:.2f}")

# Making a prediction
```

```
# For example, predicting y for x = 6
predicted_value = model.predict([[6]])
print(f"The predicted value for x = 6 is: {predicted_value[0]:.2f}")

The equation of the model is: y = 0.60x + 2.20
The predicted value for x = 6 is: 5.80
```

In summary, both linear and logistic regression offer clear and interpretable models, where the relationship between input features and outcomes is straightforward, making it easy to understand how changes in input variables affect predictions.

Decision trees

Decision trees are another class of highly interpretable models, offering a visual representation of the decision-making process through a tree-like structure of nodes (decisions) and branches (consequences). They are particularly useful for binary classification and regression tasks. Trees can be easily explained to nontechnical stakeholders by tracing the path from the root to a leaf to understand the sequence of decisions leading to a prediction. However, when trees become too deep, they may overfit and lose some interpretability, which brings ensemble methods into play. The following Python code creates a basic decision tree using the Iris dataset. This is depicted in Figure 7-1, showing the nodes and decisions:

```
from sklearn.datasets import load_iris
from sklearn.tree import DecisionTreeClassifier, plot_tree
import matplotlib.pyplot as plt

# Load the iris dataset
iris = load_iris()
X = iris.data
y = iris.target

# Create and fit the decision tree model
clf = DecisionTreeClassifier(random_state=123)
clf.fit(X, y)

# Plot the decision tree
plt.figure(figsize=(20,10))
plot_tree(clf, filled=True, feature_names=iris.feature_names,
class_names=iris.target_names, rounded=True)
plt.title("Decision Tree of Iris Dataset")
plt.show()
```

Let's look at a decision tree for the telecom churn example. See Figure 7-2, which depicts the decision tree to predict churn. The tree is hard to interpret as a diagram, so the rules are listed after the figure.

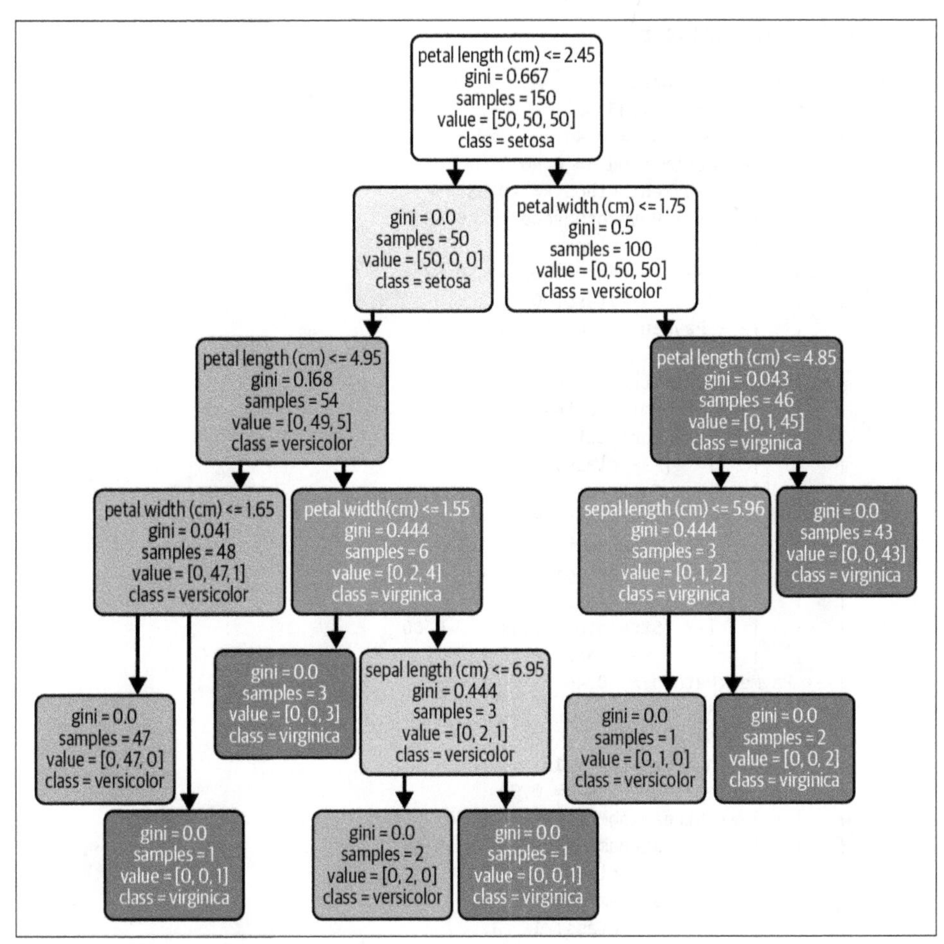

Figure 7-1. Decision tree for Iris dataset

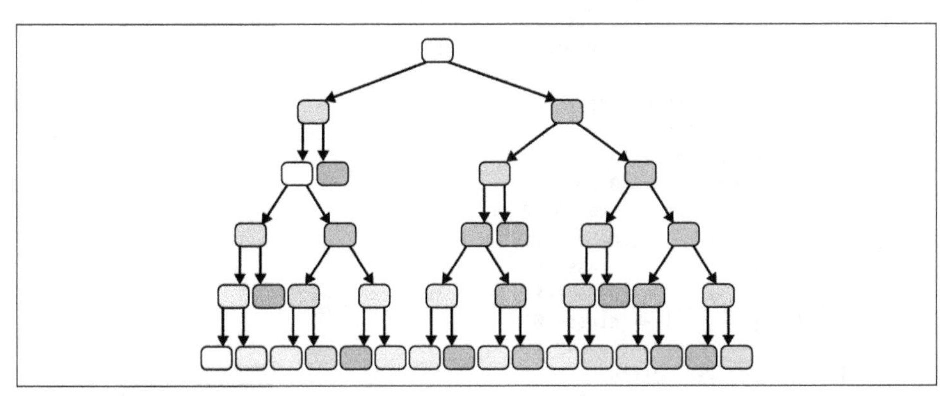

Figure 7-2. Conceptual decision tree structure

Here are the rules for the tree:

```
Decision tree structure:
|--- PaymentMethod <= 1.33
|   |--- PaymentMethod <= 0.40
|   |   |--- PaymentMethod <= 0.40
|   |   |   |--- PaymentMethod <= -0.53
|   |   |   |   |--- PaymentMethod <= -0.53
|   |   |   |   |   |--- class: 1
|   |   |   |   |--- PaymentMethod >  -0.53
|   |   |   |   |   |--- class: 0
|   |   |   |--- PaymentMethod >  -0.53
|   |   |   |   |--- class: 1
|   |   |--- PaymentMethod >  0.40
|   |   |   |--- StreamingMovies <= -1.13
|   |   |   |   |--- Contract <= 1.66
|   |   |   |   |   |--- class: 0
|   |   |   |   |--- Contract >  1.66
|   |   |   |   |   |--- class: 0
|   |   |   |--- StreamingMovies >  -1.13
|   |   |   |   |--- StreamingMovies <= -0.00
|   |   |   |   |   |--- class: 1
|   |   |   |   |--- StreamingMovies >  -0.00
|   |   |   |   |   |--- class: 0
|   |--- PaymentMethod >  0.40
|   |   |--- class: 1
|--- PaymentMethod >  1.33
|   |--- StreamingMovies <= 1.13
|   |   |--- StreamingMovies <= 0.00
|   |   |   |--- StreamingMovies <= -0.00
|   |   |   |   |--- StreamingMovies <= -1.13
|   |   |   |   |   |--- class: 0
|   |   |   |   |--- StreamingMovies >  -1.13
|   |   |   |   |   |--- class: 1
|   |   |   |--- StreamingMovies >  -0.00
|   |   |   |   |--- DeviceProtection <= 1.24
|   |   |   |   |   |--- class: 0
|   |   |   |   |--- DeviceProtection >  1.24
|   |   |   |   |   |--- class: 0
|   |   |--- StreamingMovies >  0.00
|   |   |   |--- class: 1
|   |--- StreamingMovies >  1.13
|   |   |--- Contract <= 0.42
|   |   |   |--- Contract <= -0.82
|   |   |   |   |--- OnlineSecurity <= 1.39
|   |   |   |   |   |--- class: 0
|   |   |   |   |--- OnlineSecurity >  1.39
|   |   |   |   |   |--- class: 0
|   |   |   |--- Contract >  -0.82
|   |   |   |   |--- class: 1
|   |   |--- Contract >  0.42
|   |   |   |--- Contract <= 0.43
|   |   |   |   |--- TotalCharges <= 0.40
```

```
|   |   |   |   |   |--- class: 0
|   |   |   |   |--- TotalCharges >  0.40
|   |   |   |   |   |--- class: 0
|   |   |   |--- Contract >  0.43
|   |   |   |   |--- Contract <= 1.66
|   |   |   |   |   |--- class: 1
|   |   |   |   |--- Contract >  1.66
|   |   |   |   |   |--- class: 0
```

At first glance, it may not be clear what we are able to determine from the tree. From the top to the bottom of the decision tree rules, the features are listed in the order of importance with the associated decision rules.

The decision tree shows how the model makes decisions by splitting the data based on feature values at each level. Here's a breakdown of key parts of the decision tree:

Top split: `PaymentMethod <= 1.33`
: The first split in the tree is based on the `PaymentMethod` feature, which indicates that how customers pay (e.g., electronic check, mailed check, credit card, etc.) is a crucial factor in determining whether they will churn. This feature appears in several nodes and subnodes, showing its strong influence in the decision-making process.

Subsequent nodes
: The tree continues to split on `PaymentMethod`, and then moves to other features like `StreamingMovies`, `Contract`, and `DeviceProtection`. The repetition of `PaymentMethod` in the tree highlights its high predictive power.

  The nodes that follow use features such as `StreamingMovies` and `Contract` to further distinguish between churners and nonchurners. For example, if a customer streams movies and has a certain contract type, it influences whether they are likely to churn.

  The tree also includes splits based on `TotalCharges` and `OnlineSecurity` in some branches, although these are less frequent, indicating they have less influence on the model's decision making.

Ensemble models

While generally considered less interpretable due to their complexity, certain ensemble methods like random forests and gradient boosting machines (GBMs) offer a balance between interpretability and performance. Techniques such as feature importance scores help identify which variables are most influential in predicting the target variable, providing insights into the model's decision-making process. Additionally, tools like SHapley Additive exPlanations (SHAP) and local interpretable model-agnostic explanations (LIME) can break down predictions to show the impact of each feature.

SHAP is a method for interpreting machine learning models by breaking down predictions into the contributions of each feature, based on cooperative game theory's Shapley values. SHAP assigns an importance value (SHAP value) to each feature by computing the average marginal contribution of the feature across all possible combinations of other features. This provides a global and local interpretability framework for understanding how each feature impacts the prediction for both individual instances and across the entire model. SHAP is model-agnostic and can be used to explain complex models like neural networks and ensemble methods, making it a powerful tool for gaining insights into black-box models.

LIME is an interpretability technique that explains individual predictions of any machine learning model by approximating the original model with a simpler, interpretable model locally around the instance being predicted. It works by generating perturbations of the input data and training a surrogate interpretable model, such as a linear model or decision tree, on these perturbed samples. LIME then uses this surrogate model to explain the prediction of the original black-box model for a specific instance, providing insight into which features contributed the most to the prediction. LIME is particularly useful for understanding local behaviors of complex models.

Generalized additive models

Generalized additive models (GAMs) extend linear models by allowing nonlinear relationships between each feature and the target, using smooth functions. They maintain a level of interpretability by preserving the ability to independently analyze each feature's effect on the response variable. GAMs provide a good compromise between the flexibility of nonlinear models and the interpretability of linear models. The following Python example shows how a GAM model could be reviewed. It does the following:

- Generates a synthetic dataset with two features, where the response variable `y` is a nonlinear function of the first feature and linearly dependent on the second feature, with some added noise
- Fits a GAM to the data, using a spline (`s`) for the first feature to capture its nonlinear effect and a linear term (`f`) for the second feature
- Plots the partial dependence of the model on the first feature, showing how the response variable is expected to change with this feature, holding all other features constant

- Prints the model's intercept and coefficients for the nonlinear and linear terms, providing insight into how each feature influences the response variable:

```
from pygam import LinearGAM, s, f
import numpy as np
import matplotlib.pyplot as plt

# Generating synthetic data
np.random.seed(42)
X = np.random.rand(100, 2)  # 100 samples, 2 features
y = np.sin(2 * np.pi * X[:, 0]) + np.random.normal(0, 0.2, 100) + X[:, 1]

# Fitting a GAM with a spline term for the first feature and a linear term
for the second
gam = LinearGAM(s(0) + f(1)).fit(X, y)

# Plotting the partial dependence for the first feature
fig, ax = plt.subplots()
XX = gam.generate_X_grid(term=0)
ax.plot(XX[:, 0], gam.partial_dependence(term=0, X=XX))
ax.plot(XX[:, 0], gam.partial_dependence(term=0, X=XX, width=.95)[1], c='r',
ls='--')
plt.title('Partial Dependence of Feature 0')

# Adding a caption below the plot
caption = "This plot shows the partial dependence of the first feature with
95% confidence interval."
fig.text(0.5, -0.05, caption, ha='center', fontsize=10)

# Adjust the layout to make space for the caption
plt.tight_layout()
plt.show()

# Explaining the model
print("Model Explanation:")
print("Coefficients:", gam.coef_)
```

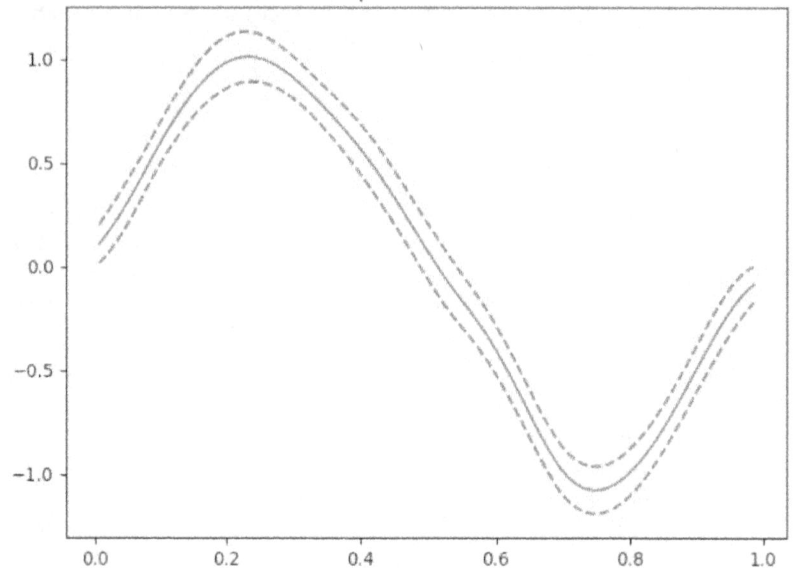

This plot shows the partial dependence of the first feature with 95% confidence interval.

```
Model Explanation:
Coefficients: [-4.14173547e-02  7.86252085e-02  3.80470544e-01  7.336006 ...
  9.65016876e-01  1.04131299e+00  9.36730903e-01  7.48749240e-01
  5.39801783e-01  2.42044348e-01 -7.19989362e-02 -2.87286552e-01
 -6.34001954e-01 -1.02401311e+00 -1.11239341e+00 -9.71490674e-01
 -6.84093543e-01 -3.11193137e-01 -4.30228588e-02 -2.22778167e-02
 -2.04951266e-15 -9.66941105e-16 -9.01487202e-16 -1.23087850e-15
 -5.12687644e-16 -3.17643126e-17  1.00758008e-15  2.03555094e-16
 -2.47715218e-16  2.63338628e-16 -6.90117702e-17  9.63203708e-17
 -4.06257155e-16 -7.74952941e-16  3.09032262e-16 -2.49217163e-16
 -1.10133108e-17 -5.73984861e-17 -4.14632422e-16  1.67919266e-16
  4.22527674e-16 -3.81928638e-17  5.90990053e-17  1.40527531e-16
  5.57565006e-17 -2.65277355e-01 -3.54985743e-01 -3.67534703e-01
 -2.73328912e-01 -3.08646320e-01 -1.64104075e-01 -2.74196549e-01
 -1.70595684e-01 -1.99462505e-01 -1.25419848e-01 -2.96083855e-01
 -3.21482812e-01 -2.77091641e-01 -2.09652206e-01 -2.13736314e-01
 -1.31241316e-01 -1.70003604e-01  7.76867706e-03 -1.37887696e-01
 -1.34183974e-01  5.59026524e-16 -1.00361684e-01 -5.09419956e-16
 -1.56132301e-01 -7.82432882e-02  1.03607339e-01 -1.86632758e-02
  4.14670856e-02  1.17665206e-01 -1.28336731e-02 -1.40841962e-16
  2.67752838e-01  4.22036517e-16  8.65095159e-02  1.58211358e-16
  3.65310539e-01  1.18504126e-01  7.70677272e-02  2.63964513e-01
  3.51606387e-01  2.24195731e-01  5.04379735e-01 -2.27779289e-17
  5.18228273e-01  2.93432604e-01  3.68920612e-01  5.15493384e-01
  3.22445905e-01  2.85068021e-01  3.92919696e-01  0.00000000e+00
  0.00000000e+00  0.00000000e+00  0.00000000e+00  0.00000000e+00
```

```
0.00000000e+00  0.00000000e+00  0.00000000e+00  0.00000000e+00
0.00000000e+00  0.00000000e+00  0.00000000e+00  0.00000000e+00
0.00000000e+00  0.00000000e+00  0.00000000e+00  0.00000000e+00
0.00000000e+00  0.00000000e+00  0.00000000e+00  0.00000000e+00
0.00000000e+00  0.00000000e+00  0.00000000e+00  0.00000000e+00
4.63163174e-01]
```

The coefficient output from the GAM includes a large number of values, many of which are quite small or even zero. Here's what the results generally mean and how you can interpret them:

- The coefficients for a GAM represent the contribution of each spline function or linear component to the target variable. In this case, the model uses splines to capture the nonlinear relationship between the features and the target.
- For nonlinear features, GAM uses splines, so there are multiple coefficients for each feature. These coefficients together form the smooth curve that describes the relationship between a feature and the target variable.
- For Feature 0, the coefficients may represent how the smooth term (nonlinear) for this feature behaves, and for Feature 1, the coefficients represent its linear impact.
- Coefficients close to zero (like those after 0.00000000e+00) mean that those terms do not contribute significantly to the prediction.
- The positive values (like 9.65016876e-01 or 1.04131299e+00) indicate that as the feature value increases, the target variable also increases.
- The negative values (like -4.14173547e-02) indicate a negative relationship between that feature and the target variable.
- Many of the coefficients toward the end of the list are exactly zero (e.g., 0.00000000e+00). These likely correspond to terms that are not included or are irrelevant in the model.
- The partial dependence plot generated earlier would provide a clear picture of how Feature 0 (modeled as a spline) influences the target variable. This is where we can visually interpret the nonlinear relationship between the feature and the target.
- If the coefficients for Feature 0 have a mix of positive and negative values, it indicates a nonlinear relationship between the feature and the target. This is expected in a spline-based model, where the relationship is more complex than linear.
- The linear feature (Feature 1) will have a single coefficient, and its value tells you whether the relationship between Feature 1 and the target is positive or negative.

The model suggests that Feature 0 has a nonlinear relationship with the target, which is reflected in the varying coefficients for the spline terms. Feature 1, modeled linearly, shows how each unit increase in that feature affects the target (likely in a straightforward manner).

The partial dependence plot and the coefficients offer insights into these relationships, helping to understand how each feature contributes to the model's predictions. The next step would be to examine the plot for nonlinearities and see if the model aligns with real-world expectations.

When prioritizing model interpretability, it's crucial to consider the trade-offs with model complexity and accuracy. Highly interpretable models may not always capture complex patterns in the data as effectively as more sophisticated algorithms. Therefore, the choice of algorithm should align with the project's objectives, the stakeholders' needs for explanation, and the regulatory environment. Engaging domain experts in the modeling process can also enhance the interpretability and relevance of the model's outputs. With the growing availability of tools and techniques to enhance the interpretability of even complex models, the focus has shifted toward making machine learning more transparent and accountable across all domains.

Prediction Accuracy

When the primary objective of an analytics project is to maximize prediction accuracy, the selection of the right algorithm becomes focused on understanding the complexities of the data and the trade-offs among various modeling approaches. High prediction accuracy is critical in applications where the cost of errors is high, such as disease diagnosis, fraud detection, or financial forecasting. Multiple algorithms, when used to create models, can have a higher degree of accuracy. Examples include neural networks, ensemble methods, and SVMs.

Complexity and capacity

Algorithms with greater complexity and capacity, such as deep learning neural networks, have the potential to model intricate patterns and interactions within the data that simpler models might miss. Deep learning, in particular, excels in areas like image recognition, natural language processing, and any domain where the input data is high-dimensional and the relationships between variables are nonlinear. However, these models require large amounts of data to train effectively and avoid overfitting.

Ensemble methods

Ensemble methods, such as random forests, GBMs, and stacking, combine the predictions of multiple models to improve accuracy. By aggregating the predictions of several models, ensemble methods can often achieve higher accuracy than any single model alone. Random forests, for instance, build upon the simplicity of decision trees but enhance prediction accuracy through bagging, which reduces variance without increasing bias. Similarly, GBMs improve upon this by sequentially correcting errors of the trees, focusing on difficult cases to boost performance.

SVMs

For datasets where the boundary between classes is not clear, SVMs can be particularly effective. They work by finding the hyperplane that best separates different classes in the feature space, with the capability to handle linear and nonlinear separations through the use of kernel functions. SVMs are powerful for classification and regression tasks but may require careful tuning of parameters and kernel choice.

Feature engineering

Beyond the choice of algorithm, the accuracy of predictions can be significantly influenced by the quality of input features. Techniques such as feature engineering, which involves creating new features from existing data, and feature selection, which focuses on eliminating irrelevant or redundant features, can greatly enhance model performance. Algorithms that inherently perform feature selection, like LASSO regression or tree-based models, can be advantageous in scenarios with high-dimensional data.

Training Speed

The training speed of an algorithm is crucial in environments where model development cycles are fast-paced or where computational resources are limited. Rapid training allows for quicker iterations, enabling data scientists and machine learning engineers to experiment with different models, tune hyperparameters more efficiently, and accelerate the overall development process. This aspect becomes particularly important during the exploratory phase of a project, in situations requiring frequent model updates, or when working with very large datasets. Several key considerations for optimizing training speed without compromising the effectiveness of the model include algorithmic efficiency, data size and quality, parallelization, distributing computing, and others.

Algorithmic efficiency

Some algorithms are inherently faster to train due to their simpler mathematical operations. For instance, linear models like linear regression and logistic regression typically train faster than more complex models like deep neural networks or ensemble methods such as GBMs and random forests. Simpler algorithms are particularly advantageous in early stages of model exploration or when the dataset size is not prohibitively large.

Data size and quality

The volume and quality of the training data directly impact training speed. Working with smaller, well-preprocessed datasets can significantly reduce training time. Techniques such as feature selection to remove irrelevant or redundant features and data sampling methods like stratified sampling or using minibatches can make training processes more efficient while maintaining a representative dataset.

Parallelization and distributed computing

Leveraging parallel processing and distributed computing can dramatically reduce training time for algorithms that support these capabilities. Decision tree–based algorithms, such as those used in random forests and GBM, can often be parallelized across multiple cores or machines. Similarly, deep learning frameworks are well optimized for GPU acceleration, which can offer substantial speed improvements over CPU-based training.

Algorithm selection based on problem complexity

The complexity of the problem at hand should guide the selection of the algorithm. For simpler problems, starting with more straightforward and less computationally intensive algorithms may yield satisfactory results much faster. More complex or high-dimensional problems might necessitate advanced algorithms, but exploring dimensionality reduction techniques or feature engineering could allow for the use of faster-training models.

Early stopping

Implementing early stopping during training can prevent overfitting and reduce training time. By monitoring the model's performance on a validation set and stopping the training process once performance ceases to improve or starts to degrade, resources are conserved for other experiments or iterations.

Prediction Speed

Optimizing for prediction speed is a priority in scenarios where decisions need to be made rapidly, such as real-time fraud detection, dynamic pricing models, or high-frequency trading algorithms. In these scenarios, the ability of a machine learning model to quickly make accurate predictions based on new data can be just as critical as the precision of the predictions themselves. Selecting the right algorithm to maximize prediction speed without impacting accuracy includes reviewing model complexity, number of dimensions, computational resources, and batch processing.

Model complexity

Generally, simpler models predict faster. Algorithms like linear regression, logistic regression, and decision trees tend to have faster prediction times compared to more complex models like deep learning neural networks or ensemble methods like GBM and random forests. The trade-off between model complexity and prediction speed must be carefully managed; simpler models may not capture complex patterns as effectively but can offer significant advantages in terms of speed.

Dimensionality reduction

High-dimensional datasets can significantly slow down prediction times. Techniques for dimensionality reduction, such as PCA or autoencoders in deep learning, can reduce the number of features the model needs to consider, thereby speeding up prediction times. Reducing the input size not only accelerates the prediction process but can also help in mitigating overfitting, leading to models that generalize better.

Model training and serving architecture

The infrastructure on which models are trained and served can have a substantial impact on prediction speed. Leveraging optimized libraries and hardware acceleration (e.g., GPUs for neural networks) can enhance computational efficiency. Additionally, deploying models in a production environment using technologies like model serving frameworks (e.g., TensorFlow Serving, TorchServe) can further optimize prediction latency.

Algorithm computational speed

Some algorithms are inherently faster at making predictions. For example, k-nearest neighbors (KNN) can be slow in prediction because it requires computing the distance between the input sample and each training instance. In contrast, models that learn a direct mapping from inputs to outputs, such as linear models or shallow neural networks, can make predictions more quickly.

Model pruning and quantization

Techniques like pruning, which removes unnecessary weights or neurons from neural networks, and quantization, which reduces the precision of the numbers used in the model, can significantly decrease the computational burden during inference. These techniques are especially relevant for deploying models on edge devices with limited computational resources.

Hyperparameter Tuning

Hyperparameter tuning is the process of optimizing the settings of the algorithm parameters that are not learned from data, with the goal of determining the right combination that provides the best model performance. Since different algorithms have varying sensitivities with hyperparameters, the selection of an algorithm influences the approach and efficiency of hyperparameter tuning. The following areas need to be considered.

Algorithm complexity and hyperparameter space

Algorithms with a large number of hyperparameters or those whose performance is highly sensitive to hyperparameter values might offer higher potential for optimization but require more extensive tuning efforts. For example, deep learning models and ensemble methods like GBM have complex hyperparameter spaces, making them more challenging but potentially more rewarding to tune. In contrast, simpler models with fewer hyperparameters, such as linear regression or decision trees, can be easier and faster to tune but might offer limited performance improvements.

Automated hyperparameter tuning tools

The availability and effectiveness of automated hyperparameter tuning tools should be considered. Tools like grid search, random search, and Bayesian optimization, as well as more advanced AutoML solutions, can automate the search for optimal hyperparameters. The choice of algorithm might influence the suitability of these tools; for instance, models with high-dimensional hyperparameter spaces might benefit more from Bayesian optimization, which is more efficient in exploring large spaces, compared to grid search, which can be computationally prohibitive.

Computational resources

The computational cost of hyperparameter tuning is directly tied to the algorithm's training speed and the complexity of the hyperparameter space. Algorithms that are computationally intensive to train require more resources for thorough hyperparameter tuning. It's essential to balance the expected performance gains against the available computational budget. For projects with limited resources, prioritizing algorithms that are relatively efficient to train and tune may be necessary.

Tuning versus performance trade-off

Some algorithms might achieve acceptable performance with default or slightly adjusted hyperparameters, reducing the need for extensive tuning. In cases where rapid development cycles are critical, it might be beneficial to opt for algorithms that are less sensitive to hyperparameter changes. For projects where the highest possible accuracy is a priority, choosing more complex algorithms and comprehensive hyperparameter tuning might be justified.

Sensitivity analysis

Before committing to extensive tuning, performing a sensitivity analysis can help identify which hyperparameters are most influential on model performance. This approach allows for focusing tuning efforts on the parameters that matter most, improving efficiency. Algorithms that offer clear documentation and understanding of their hyperparameters' roles can facilitate more effective sensitivity analysis.

Cross-validation strategy

The strategy used for evaluating model performance during hyperparameter tuning, such as k-fold cross-validation, can assist in obtaining reliable tuning results. The choice of algorithm might affect the feasibility of certain evaluation strategies, especially for algorithms that require long training times. In such cases, techniques like stratified sampling or using a validation set instead of full cross-validation can save time and resources.

Example of hyperparameter tuning

Going back to the telecom churn example with the decision tree, here is an example of hyperparameter tuning with GridSearchCV. GridSearchCV is a technique used in machine learning to find the optimal hyperparameters for a model. It exhaustively tests all possible combinations of a specified parameter grid by cross-validating the model for each combination. This helps in identifying the best-performing hyperparameters based on evaluation metrics like accuracy, precision, or others. Essentially, it automates the process of hyperparameter tuning, ensuring the best model configuration is found efficiently:

```
# Import necessary libraries
from sklearn.tree import DecisionTreeClassifier
from sklearn.model_selection import GridSearchCV
from sklearn.metrics import classification_report, accuracy_score
import numpy as np

# Define the hyperparameter grid for the decision tree ❶
param_grid = {
    'max_depth': [3, 5, 10, None],
# Try different depths of the tree
    'min_samples_split': [2, 10, 20],
# Minimum number of samples required to split an internal node
    'min_samples_leaf': [1, 5, 10],
# Minimum number of samples required to be at a leaf node
    'criterion': ['gini', 'entropy']
# Criterion for splitting ❷
}
```

```
# Initialize the decision tree model
dt = DecisionTreeClassifier(random_state=42)

# Set up the GridSearchCV with 5-fold cross-validation
grid_search = GridSearchCV(estimator=dt, param_grid=param_grid,
                           cv=5, scoring='accuracy', n_jobs=-1, verbose=1) ❸

# Train the model using the SMOTE-resampled training data with GridSearchCV
grid_search.fit(X_train_smote, y_train_smote)

# Get the best parameters from GridSearchCV
best_params = grid_search.best_params_
print(f"Best Parameters from GridSearchCV: {best_params}") ❹

# Use the best model found by GridSearchCV to make predictions on the
validation set
best_dt_model = grid_search.best_estimator_
y_pred_dt = best_dt_model.predict(X_val) ❺

# Evaluate the best decision tree model
print("\nBest Decision Tree Model Evaluation:")
print(classification_report(y_val, y_pred_dt))
print("Accuracy:", accuracy_score(y_val, y_pred_dt)) ❻

# Print the feature importance values
importances = best_dt_model.feature_importances_
indices = np.argsort(importances)[::-1]  # Sort feature importances in
descending order

print("\nFeature Importance (in descending order):")
for idx in indices:
    print(f"{X.columns[idx]}: {importances[idx]:.4f}")

Fitting 5 folds for each of 72 candidates, totalling 360 fits
Best Parameters from GridSearchCV: {'criterion': 'entropy', 'max_depth': 10,
'min_samples_leaf': 1, 'min_samples_split': 2} ❼

Best Decision Tree Model Evaluation:
              precision    recall  f1-score   support

           0       0.73      1.00      0.84     14599
           1       0.31      0.00      0.01      5401

    accuracy                           0.73     20000
   macro avg       0.52      0.50      0.43     20000
weighted avg       0.62      0.73      0.62     20000

Accuracy: 0.72835
Feature Importance (in descending order):
PaymentMethod: 0.6032
StreamingMovies: 0.1393
```

```
TechSupport: 0.1055
Contract: 0.0421
InternetService: 0.0209
StreamingTV: 0.0203
MultipleLines: 0.0129
OnlineSecurity: 0.0116
TotalCharges: 0.0115
tenure: 0.0077
MonthlyCharges: 0.0069
DeviceProtection: 0.0068
Partner: 0.0033
OnlineBackup: 0.0031
Dependents: 0.0026
SeniorCitizen: 0.0011
gender: 0.0006
PaperlessBilling: 0.0004
PhoneService: 0.0002
```

❶ GridSearchCV is started by defining the hyperparameter grid. The `param_grid` parameter specifies the hyperparameters to tune. The parameter `max_depth` controls the maximum depth of the tree. Other parameters are `min_samples_split`, which defines the minimum number of samples required to split an internal node, and `min_samples_leaf`, which defines the minimum number of samples required to be at a leaf node. Last, we determine the criterion, which is the function to measure the quality of a split (`gini` or `entropy`).

❷ Gini and Entropy are metrics used in decision trees to measure the impurity of a node, guiding the process of splitting data. *Gini Index (Gini Impurity)* quantifies the likelihood of incorrectly classifying a randomly chosen element from a dataset if it were randomly labeled based on the distribution of labels in the dataset. A lower Gini value indicates better splits. Gini ranges from 0 (perfectly pure node) to 0.5 (maximum impurity). The Entropy metric comes from information theory and measures the amount of disorder or randomness in the dataset. It helps in identifying the best split by minimizing uncertainty. Entropy ranges from 0 (pure node) to 1 (maximum impurity for binary classification).

❸ GridSearchCV is used to perform an exhaustive search over the specified hyperparameter grid. Then it performs 5-fold cross-validation (`cv=5`) to find the optimal combination of hyperparameters. The parameter of `n_jobs=-1` ensures that all available processors are used for parallel computation, speeding up the process.

❹ The best parameters are then identified after training. `grid_search.best_params_` gives the best combination of hyperparameters found during the search.

❺ The best model found by GridSearchCV is stored in `best_dt_model`, which is then used to make predictions on the validation set. The model is evaluated using `classification_report` and `accuracy_score` to see how well the model performs.

❻ The feature importance values are printed, showing which features contributed most to the model's predictions.

❼ The output from GridSearchCV identifies the best parameters for the decision tree. This line indicates that GridSearchCV is performing 5-fold cross-validation for each hyperparameter combination (`candidate`) in the grid. There are 72 different combinations of hyperparameters to try based on the grid you provided (`max_depth`, `min_samples_split`, `min_samples_leaf`, and `criterion`). Since each combination is evaluated using 5-fold cross-validation, the model is trained and evaluated 5 times for each combination, leading to a total of 360 model fits (72 candidates * 5 folds = 360 fits).

After completing the grid search, GridSearchCV identifies the best combination of hyperparameters that resulted in the highest cross-validated performance (in this case, accuracy). These were the best hyperparameters found:

`criterion: 'entropy'`
: The model used the `'entropy'` criterion to measure the quality of splits (which uses information gain).

`max_depth: 10`
: The maximum depth of the tree is 10, meaning the tree can have up to 10 levels of splits.

`min_samples_leaf: 1`
: A minimum of 1 sample is required to be in a leaf node.

`min_samples_split: 2`
: A minimum of 2 samples are required to split an internal node.

This combination of parameters was found to produce the best performance during cross-validation across the 5 folds, and this configuration was then used to create the final decision tree model.

By carefully evaluating the characteristics of the algorithm, the hyperparameter space, and the tools available for tuning, business analysts can optimize their models more effectively, leading to better performance and more efficient use of resources.

Working with a Small Dataset

When working with small datasets, selecting the appropriate machine learning algorithm requires careful consideration of model complexity, overfitting risks, and the inherent limitations of the data. Small datasets pose unique challenges, including a limited representation of the broader population and increased sensitivity to model selection and preprocessing. To navigate these challenges, it's crucial to prioritize techniques that balance simplicity with generalizability.

For small datasets, simpler models often outperform more complex ones due to their lower likelihood of overfitting. Overfitting happens when a model learns the noise in the training data rather than the underlying patterns, leading to poor performance on new, unseen data. Algorithms such as linear regression, logistic regression, and decision trees are typically better suited for small datasets because of their simplicity and interpretability. To further mitigate overfitting, regularization techniques such as ridge regression and LASSO regression can be highly effective. These methods penalize large coefficients or select a subset of features, helping to reduce model complexity and enhance generalization.

While individual complex models may be prone to overfitting, ensemble methods like bagging and boosting can still be adapted for small datasets by carefully tuning their parameters. Cross-validation techniques, such as k-fold cross-validation, are essential when working with small datasets, as they help maximize the efficient use of limited data and provide a more reliable assessment of model performance. Bayesian models, which incorporate prior knowledge into the modeling process, offer another advantage for small datasets. By integrating prior beliefs about the data or model parameters, Bayesian methods like Bayesian regression or Gaussian processes can make well-informed predictions and account for uncertainty, even with limited data.

In addition to these traditional methods, more advanced techniques like semisupervised learning and transfer learning can also enhance model performance on small datasets. Semisupervised learning leverages both labeled and unlabeled data to improve predictions, while transfer learning applies knowledge gained from a model trained on a larger, related dataset to the smaller dataset at hand, allowing the model to benefit from previously learned patterns.

Ultimately, working with small datasets requires an emphasis on simplicity, regularization, and the efficient use of data through techniques like cross-validation. By also exploring advanced methodologies such as Bayesian approaches or transfer learning, it is possible to unlock valuable insights and achieve robust performance even from limited data.

Working with a Large Dataset

Large datasets provide a wealth of information, enabling the development of complex models that can capture subtle patterns and relationships. However, working with large amounts of data requires careful consideration of computational efficiency, scalability, and strategies to improve model performance. The scalability of algorithms is critical when dealing with large datasets, as they must efficiently handle high volumes of data without requiring excessive computational resources. Techniques like stochastic gradient descent (SGD) are particularly effective for large-scale learning because they offer computational efficiency and support incremental, online learning approaches that are useful in deep learning and other complex models.

To further optimize the use of large datasets, algorithms that support parallelization and distributed computing are essential. These techniques reduce training time by distributing the computational workload across multiple processors or machines. For example, ensemble methods such as random forests and gradient boosting frameworks (e.g., XGBoost, LightGBM) are inherently scalable and include built-in support for parallel processing. Similarly, deep learning frameworks like TensorFlow and PyTorch allow for efficient handling of large-scale data by leveraging GPUs and distributed training environments.

When algorithms struggle with large data volumes, techniques like batch processing and minibatch learning can improve computational efficiency. Minibatch learning updates the model in small subsets of the data, balancing the computational speed of stochastic methods with the stability of batch methods. This approach is particularly useful for algorithms that are not natively optimized for large datasets, ensuring faster and more stable convergence during training.

Large datasets can also introduce challenges related to imbalanced class distributions or biased data representations. To address this, algorithms that incorporate methods for handling imbalanced data, such as weighted loss functions, oversampling, or undersampling, can ensure that the model performs well across all classes. This is especially important for maintaining the fairness and accuracy of the model in real-world applications.

By leveraging advanced algorithms, efficient processing techniques, and appropriate computational infrastructure, large datasets can be transformed into high-performing models that capture both complex relationships and nuanced insights.

Feature Interaction

Feature interactions happen when the effect of two or more features on the predicted variable is not additive, meaning the combined effect differs from the sum of their individual effects. Properly modeling these interactions can lead to more accurate and insightful predictions, especially in datasets where interactions play a critical role.

When feature interaction is an important criteria—often important in supporting model interpretability—then it is important to consider how algorithms incorporate or compensate for feature interactions. For instance, not all algorithms consider feature interactions. Examples of algorithms that do consider feature interactions include trees, neural networks, kernel methods, and regularization techniques.

Decision trees

Decision trees naturally model interactions between variables by splitting the data along feature values in a hierarchical manner. Each decision node in a tree represents a feature interaction, making trees naturally capable of capturing complex relationships. Ensemble tree methods like random forests and GBM (e.g., XGBoost, LightGBM) leverage multiple trees to model interactions more robustly, resulting in improved performance and reduced risk of overfitting compared to individual trees.

Using the telecom churn example, the features are plotted based on importance, as shown in Figure 7-3.

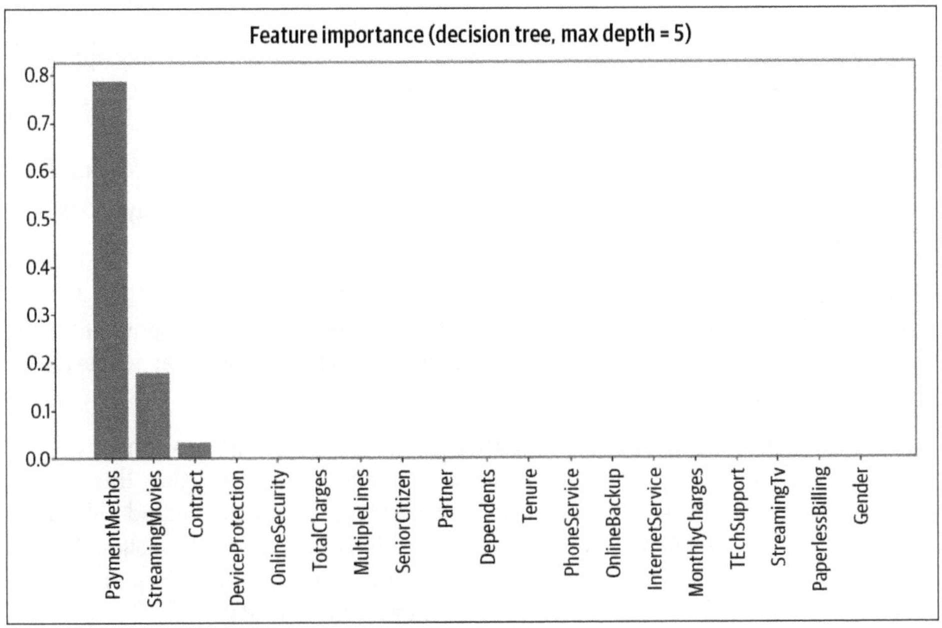

Figure 7-3. Decision tree feature importance for telecom churn

The feature importance values quantify how much each feature contributes to the model's decision-making process.

Here's the feature importance list, with the most important features at the top:

PaymentMethod (0.7864)
: This feature dominates the decision tree, with a feature importance of 78.64%. This indicates that how a customer pays is by far the most influential factor in predicting churn. Payment methods like electronic checks are often associated with higher churn rates, which may explain its importance.

StreamingMovies (0.1788)
: This feature is the second-most important, with an importance of 17.88%. Whether or not a customer streams movies may correlate with their engagement level or the value they place on the service, influencing their likelihood to churn.

Contract (0.0320)
: The type of contract (month-to-month, yearly, etc.) has some importance, contributing 3.2%. Short-term contracts are generally associated with higher churn risk, which is why this feature plays a role in the model.

DeviceProtection (0.0012) and OnlineSecurity (0.0011)
: These two features have very small importance values. Although they appear in the tree, they don't significantly influence the model's predictions.

TotalCharges (0.0005)
: While this feature might be expected to play a larger role, it has very little influence in this model. This might be because other features like MonthlyCharges or contract type are capturing similar information.

Other Features (MultipleLines, SeniorCitizen, Partner, etc.)
: These features all have an importance of 0.0000, meaning they did not contribute to the model's decision making at all. This suggests they are either irrelevant or redundant in this specific model.

So what can we conclude about the features? The payment method dominates. The overwhelming importance of PaymentMethod (78.64%) suggests that this feature plays a critical role in determining churn. If a certain payment method (like electronic check) is correlated with higher churn, this is an actionable insight for the business. The company could consider offering promotions or incentives to customers who use high-risk payment methods to encourage them to stay.

It is also clear that other features are less influential. While StreamingMovies and Contract are somewhat important, their influence is much smaller compared to PaymentMethod. This means that, while these features help fine-tune the model's predictions, they don't drive the decision-making process as much. It could be worth investigating whether additional features could help better capture customer behavior.

Several features (e.g., MultipleLines, PhoneService, InternetService) have zero importance. This might indicate that these features are not useful in this particular dataset for predicting churn. You could consider removing these features to simplify the model and improve interpretability.

Deep learning models

Neural networks, especially deep learning models (neural networks with many payers), excel at learning feature interactions through their layered architecture. Each layer can learn increasingly complex representations of the data, with deeper layers capturing higher-level interactions. Convolutional neural networks (CNNs) and recurrent neural networks (RNNs) are particularly effective for spatial and temporal data interactions, respectively. CNNs focus on learning visual data, and RNNs focus on learning text data. The flexibility and capacity of deep learning models make them suitable for datasets with intricate patterns and interactions, but they come at the cost of increased computational complexity and possible interpretability challenges.

Kernel methods

SVMs with nonlinear kernels, such as the radial basis function (RBF) kernel, model complex interactions between features by mapping the input data into a higher-dimensional space where the data is more easily separable. Kernel methods are tools that can capture nonlinear interactions without expanding the feature space, offering a balance between model complexity and interpretability.

Regularization techniques

When incorporating interaction terms in a model, regularization techniques like LASSO (L1 regularization) are highly effective for feature selection. LASSO works by adding a penalty term to the loss function, which is proportional to the absolute value of the model's coefficients. This penalty forces the coefficients of less important features to shrink to zero, effectively removing them from the model. As a result, LASSO can simplify models with many interaction terms by eliminating those that do not significantly contribute to the predictive power, preventing overfitting and enhancing generalizability.

For example, in cases where interaction terms between features might introduce unnecessary complexity, LASSO prioritizes the most impactful interactions and excludes weaker ones, which can lead to a more interpretable model. Ridge regression (L2 regularization), on the other hand, reduces the magnitude of coefficients but does not perform feature selection, as it tends to shrink coefficients toward zero without eliminating them completely. Elastic net, a combination of L1 and L2 regularization, balances the benefits of both LASSO and ridge, allowing for more flexibility in managing interaction terms by both penalizing large coefficients and performing feature selection.

In feature selection, these regularization techniques help in reducing model complexity, improving prediction accuracy, and preventing overfitting, especially when the dataset includes numerous features or when interaction terms are added. Regularization techniques, therefore, allow the model to focus only on the most relevant predictors and their interactions, ensuring a more robust and efficient model.

A range of algorithms are available to tackle the challenge of feature interactions. The ones listed in this chapter are some of the primary algorithms used when feature interactions are important.

Data Characteristics

Data demographics refer to the properties of the dataset, including its size, dimensionality, distribution, and the nature of the features (e.g., categorical versus numerical, balanced versus imbalanced classes). Each of these characteristics can influence the choice of the algorithm, because different algorithms have different strengths and weaknesses that make them suitable for specific types of data.

Dimensionality

High-dimensional datasets, where the number of features significantly exceeds the number of observations, may lead to the curse of dimensionality. The curse of dimensionality simply means too many dimensions exist and dimensionality reduction is needed. Dimensionality reduction techniques or models with built-in feature selection capabilities, such as LASSO regression or tree-based ensemble methods, can be particularly effective. For datasets with a manageable number of features, models that can capture complex interactions without the need for manual feature engineering, like kernel SVMs or neural networks, might be preferred.

Feature type

The nature of the features in the dataset (categorical versus numerical) can influence algorithm selection. For instance, naive Bayes and decision trees naturally handle categorical data well, while SVMs and neural networks are better suited for numerical data unless preprocessing steps are applied. Models like GBMs (e.g., XGBoost, LightGBM) are versatile in handling both types of features effectively.

Class distribution (balanced versus imbalanced)

For datasets with imbalanced class distributions, algorithms that have mechanisms to handle imbalance or that are less sensitive to it are preferable. Techniques such as class weighting in logistic regression, the use of appropriate evaluation metrics (like the F1 score or AUC-ROC), and ensemble methods designed to improve balance, like synthetic minority oversampling technique (SMOTE) with random forests, can be effective strategies.

Data quality and missing values

Datasets with a large amount of missing values or noise require algorithms resilient to such imperfections or require preprocessing steps like imputation. Tree-based models, including random forests and certain implementations of gradient boosting, can handle missing values intrinsically. For other models, data cleaning and preprocessing become critical steps before model training.

Imputation is a common technique used to handle missing values in datasets before applying machine learning algorithms that cannot inherently deal with such data gaps. While some models, like tree-based algorithms (e.g., random forests and certain gradient boosting implementations), can handle missing values during training, others—such as linear regression and SVMs—require explicit data cleaning and preprocessing steps. Imputation methods replace missing values with substituted values based on various strategies, such as using the mean, median, or mode of a feature, or employing more complex methods like KNN or multiple imputation.

For example, suppose you are working with a dataset of customer purchase behavior, and several entries for the customer age field are missing. Using mean imputation, you can replace the missing ages with the average age of the available data. This ensures that no data points are dropped due to missing values while maintaining the overall distribution of the feature. However, it's important to note that the choice of imputation strategy can affect model performance, so selecting the most appropriate method depends on the data structure and the nature of the missing values.

Imputation, when paired with appropriate algorithms, can enhance the robustness of the model by ensuring that incomplete datasets still contribute to the learning process rather than being discarded entirely.

Underlying data distribution

The assumptions about the data distribution are crucial in guiding the selection of appropriate machine learning algorithms. For instance, linear regression assumes that there is a linear relationship between the independent variables (features) and the dependent variable (target), meaning it works best when the data follows this linear structure. In contrast, logistic regression, often used for binary classification tasks, assumes a sigmoid (logistic) relationship, where probabilities of the outcomes follow an S-shaped curve. On the other hand, nonparametric models like KNN and kernel-based SVMs are more flexible as they do not make strict assumptions about the underlying data distribution. These models can adapt to complex, nonlinear patterns, making them suitable for more intricate datasets where the relationships between variables do not follow predefined mathematical functions.

The selection of an algorithm based on data demographics and characteristics involves a thorough analysis of the dataset at hand. Understanding these properties helps narrow down the choice of algorithms given the specific challenges and opportunities presented by the data.

Example: Selecting the Right Algorithm

Let's consider a scenario where a retail company, "FashionForward," aims to develop a machine learning model to predict monthly sales based on various factors such as product features, pricing, store location, and promotional activities. The primary goal is to use these predictions to optimize inventory levels, adjust marketing strategies, and improve sales forecasts. We will explore the steps a business analyst might follow to choose the right algorithm for this project, considering the problem type, model interpretability, accuracy, training and prediction speed, parameter tuning, dataset size, and feature interaction.

Choosing the Right Algorithm to Predict Sales

Selecting an appropriate algorithm for predicting monthly sales is a critical decision that impacts the accuracy, interpretability, and efficiency of the forecasting process. FashionForward aims to leverage data-driven insights to make informed business decisions, making it essential to choose an algorithm that balances predictive power with the ability to understand the factors driving those predictions.

Step 1: Understanding the problem type

The first step is recognizing that the task at hand is to predict a continuous variable—monthly sales—making this a regression problem. This understanding narrows down the algorithm choices to those suitable for regression tasks, such as linear regression, decision trees, GBMs, and neural networks. Each algorithm comes with strengths and trade-offs, particularly regarding complexity, performance, and interpretability, which need to be balanced according to FashionForward's business needs.

Step 2: Model interpretation

FashionForward places significant value on the ability to understand how different factors—like pricing and promotions—affect sales predictions. This insight is crucial for making data-driven business decisions. Therefore, model interpretability is a key criterion. Algorithms like linear regression and decision trees are highly interpretable, allowing stakeholders to visualize the relationships between input features and their impact on sales. However, more complex algorithms like GBMs or neural networks may offer higher accuracy but lower transparency. To bridge this gap, techniques like SHAP can be applied to more complex models, like GBMs, to make their predictions more interpretable by quantifying the contribution of each feature to individual predictions. SHAP provides a breakdown of how much each factor contributes to a particular prediction, offering a way to extract interpretable insights even from black-box models.

Step 3: Model accuracy

To ensure reliable sales forecasts, prediction accuracy is paramount. Simple models like linear regression are easier to interpret but might not capture the complex, non-linear relationships between features and sales. In contrast, models like GBMs and neural networks are more adept at learning these interactions and can potentially offer superior accuracy. However, this increased accuracy often comes at the expense of interpretability. Fortunately, SHAP can step in here to provide clear explanations for predictions from these more accurate models, making them more palatable for business decision making.

Step 4: Training speed

Given that FashionForward plans to refine their model with new data each month, training speed is another important consideration. Algorithms such as linear regression and decision trees generally require less computational time to train, making them ideal for quick iterations. In contrast, more complex models like GBMs and neural networks tend to be more computationally intensive, especially with large datasets, which could slow down the iterative refinement process. However, these more complex models might still be feasible if the training infrastructure supports faster computation (e.g., GPUs for neural networks or distributed processing for GBMs).

Step 5: Prediction speed

For real-time applications—such as dynamically adjusting marketing strategies based on current sales predictions—prediction speed is crucial. Both linear regression and decision trees offer rapid prediction times, making them suitable for real-time systems. GBMs and neural networks, while slightly slower, can still be optimized for faster predictions with techniques like model compression or serving optimized versions of the model.

Step 6: Parameter tuning

Achieving optimal model performance often requires hyperparameter tuning. Algorithms like GBMs and neural networks have numerous parameters that can be fine-tuned to enhance their accuracy. However, this tuning process can be resource-intensive and time-consuming. Tools like GridSearchCV or automated hyperparameter optimization frameworks such as Hyperopt can help streamline this process. Decision trees and linear regression models require less tuning, but their performance improvements are typically less dramatic compared to the gains possible with fine-tuned GBMs or neural networks.

Step 7: Size of dataset

FashionForward has accumulated several years of sales data across multiple store locations, resulting in a large dataset. This size can support more complex models like GBMs and neural networks, which generally perform better as the volume of data increases. These models can effectively capture intricate patterns in the data, such as interactions between store location, product type, and promotional activity, leading to improved performance over simpler models.

Step 8: Feature interaction

Sales predictions are influenced by complex interactions between features (e.g., the effect of promotions might vary depending on store location and product type). GBMs and neural networks are particularly well-suited to learning these feature interactions automatically, often resulting in more accurate models. Simpler models like linear regression may not capture these interactions without manual feature engineering, limiting their potential accuracy. Importantly, even though GBMs can capture complex interactions, SHAP can be used to interpret these interactions by showing the individual contributions of each feature in a prediction, helping stakeholders understand how different factors work together to drive sales.

Evaluating the Criteria

Interpretability is essential for business insights, making models like linear regression and decision trees favorable due to their transparency. However, with the use of SHAP, even more complex models like GBMs can become interpretable, offering valuable insights into how individual features contribute to predictions. Accuracy is critical for reliable forecasts, favoring more powerful models like GBMs and neural networks, which excel at capturing complex interactions in the data. Training and prediction speed are also important, particularly for iterative model refinement and real-time applications, which makes linear regression and decision trees more attractive. Still, GBMs and neural networks can remain feasible if the infrastructure is optimized for speed. Parameter tuning is another consideration, as GBMs and neural networks require more effort to fine-tune but can deliver higher performance when optimized, making them suitable if adequate resources are available. Additionally, large datasets support the use of complex models like GBMs and neural networks, which are well equipped to handle feature interactions, automatically improving performance in more intricate scenarios.

Decision and Implementation

Given the need for both interpretability and accuracy, a GBM offers the best balance for FashionForward's sales prediction model. GBMs are capable of learning complex, nonlinear relationships and feature interactions, which is crucial for accurate predictions. Though GBMs are not inherently interpretable, tools like SHAP can be employed to provide actionable insights into how various factors influence sales predictions, making the model more transparent and easier to trust for decision makers. The large dataset available further supports the choice of a GBM, as these models thrive on more data. Additionally, while GBMs require hyperparameter tuning, tools like GridSearchCV can automate this process, optimizing the model for maximum performance with relatively minimal manual effort.

To move forward, FashionForward would start with a baseline GBM model using default parameters and a subset of the most relevant features identified through domain knowledge and preliminary analysis. The model would then be iteratively refined through hyperparameter tuning and feature engineering, including testing different ways to model feature interactions explicitly. Cross-validation would be used throughout to ensure the model generalizes well to unseen data, and the model's predictions would be evaluated against a holdout test set to estimate real-world performance.

This example demonstrates how a business analyst at FashionForward could navigate the selection of a machine learning algorithm for sales prediction by carefully weighing various criteria against the company's goals and resources.

Summary

The journey of selecting the optimal machine learning algorithm for a modeling project is a multifaceted endeavor that requires a deep understanding of both the problem at hand and the characteristics of the available algorithms. As we have explored throughout this chapter, the decision-making process is influenced by many factors, including the nature of the problem, the need for model interpretability, the desired level of accuracy, the computational resources available for training and prediction, the complexity of parameter tuning, the volume of the data, and the importance of capturing feature interactions.

Understanding the type of problem you are solving is the first step in narrowing down the algorithm choices, whether it's a classification, regression, clustering, or dimensionality reduction task. From there, the trade-offs between interpretability and accuracy must be carefully balanced, with simpler models offering greater transparency and complex models often providing higher predictive performance. The practical considerations of training and prediction speed play a critical role in environments where computational resources are limited or real-time predictions are required. The art of parameter tuning can significantly enhance model performance, albeit at the cost of additional computational effort and expertise.

The size of the dataset and the nature of the features within it further complicate the selection process. Large datasets may benefit from the sophistication of algorithms like deep learning, while smaller datasets might require the simplicity and efficiency of algorithms like decision trees or linear models. The ability of an algorithm to effectively capture feature interactions can also be a deciding factor, particularly in complex domains where such interactions significantly impact the outcome.

The choice of algorithm is as much an art as it is a science, requiring a blend of technical knowledge, practical experience, and even intuition. There is no one-size-fits-all solution. Experimentation, coupled with a thorough evaluation process, remains the best approach to identifying the most suitable algorithm for a specific project. The goal is not just to select an algorithm but to solve a business problem or answer a research question effectively and efficiently.

CHAPTER 8

Model Operations

Model operations, or ModelOps, play a crucial role in the deployment, management, and maintenance of analytical models in real-world applications. While the development of accurate and robust analytical models is essential, the successful deployment into production environments requires careful consideration of various operational aspects. The bottom line is this: if the analytical model is not deployed or not being used, the value of the model is not being achieved by the organization.

Overview of Model Operations

ModelOps ensure that models perform reliably in technical production environments by monitoring their performance, detecting anomalies, and triggering retraining when necessary. Accordingly, having strong ModelOps enhances model reliability.

A technical production environment refers to the operational infrastructure where software applications or analytical models are deployed and executed for real-world use. It encompasses the hardware, software, networks, and configurations necessary to support the continuous operation and performance of these applications. Business analysts may need to partner with technical teams to work on model deployment into the technical production environment. The components of a technical production environment are as follows:

Computational resources
: Sufficient computing power and storage capacity to execute the model efficiently, handle large datasets, and support concurrent user requests.

Software environment
: The software stack necessary to run the model, including programming languages, libraries, frameworks, and runtime environments like Python, R, TensorFlow, or PyTorch.

Data pipelines
: Robust data pipelines for ingesting, preprocessing, and transforming input data required by the model. This includes data extraction, cleaning, normalization, and feature engineering processes.

Model serving infrastructure
: A system for hosting and serving the trained model to end users or applications, often implemented using technologies like REST APIs, microservices, or serverless computing platforms.

Monitoring and logging
: Mechanisms to monitor the performance and behavior of the model in real time, including metrics such as accuracy, latency, throughput, and resource utilization. Logging is essential for recording events, errors, and debugging information.

Security and compliance
: Measures to safeguard sensitive data, protect against unauthorized access, and comply with relevant regulations and industry standards. This includes encryption, access controls, audit trails, and privacy-preserving techniques.

Deployment and versioning
: Processes and tools for deploying new versions of the model into production, managing dependencies, and ensuring backward compatibility. Version control is crucial for tracking changes and facilitating rollback if needed.

ModelOps helps maintain model accuracy and effectiveness over time. In addition, it enables the deployment of machine models at scale, allowing organizations to leverage predictive analytics across large datasets and complex systems. Effective ModelOps data and development pipelines enable agile development and deployment of models, allowing teams to iterate quickly on model improvements and deploy updated versions seamlessly. Lastly, ModelOps also assists with continuous improvement, mitigates risks, and optimizes resource allocation. Overall, ModelOps is essential for maximizing the value of machine learning investments, ensuring the reliability and scalability of deployed models, and driving continuous innovation and improvement in data-driven decision-making processes.

Model Operations Processes

ModelOps encompasses a series of interconnected processes aimed at managing analytical models throughout their life cycle, from development to deployment and maintenance. The key processes are outlined here:

Model scoring
: This process involves applying trained machine learning models to new data to generate predictions or classifications. Model scoring can be performed in real

time or batch mode and is a critical step in operationalizing machine learning models for decision making.

Model monitoring
: Model monitoring involves continuously tracking the performance and behavior of deployed models in production environments. This includes monitoring key performance metrics, detecting anomalies or drifts in model behavior, and triggering alerts or interventions when issues are detected.

Model retraining
: Model retraining is the process of updating machine learning models with new data to ensure their continued relevance and accuracy over time. This may involve periodic retraining based on predefined schedules, triggered events, or changes in data distributions.

Final report generation
: Final report generation involves creating comprehensive reports summarizing the performance, insights, and outcomes of machine learning models. These reports provide stakeholders with valuable insights into model effectiveness, business impact, and areas for improvement.

Let's explore each of these processes in more detail, starting with model scoring.

Model Scoring

Model scoring, also known as model evaluation or model inference, is the process of using a trained analytical model to make predictions or generate insights from new, unseen data. It occurs after a model has been trained on historical or labeled data and is deployed into a production environment for real-world use. Model scoring begins with a model that has been trained and accepted by business stakeholders. A model is developed using historical data to learn patterns, relationships, or classifications relevant to the problem domain. This training phase involves techniques like supervised learning, unsupervised learning, or reinforcement learning.

Once accepted by business stakeholders via evaluation (model evaluation was covered in Chapter 6), the model is deployed into a production environment where it can receive new input data and generate predictions or insights. This deployment may involve setting up APIs, microservices, or batch processing pipelines to serve the model's predictions to end users or downstream applications.

During model scoring, the deployed model receives input data from various sources, such as user interactions, sensor readings, or database queries. This input data typically represents the features or variables that the model was trained to analyze or predict. Using the received input data, the deployed model applies the learned patterns or rules to generate predictions, classifications, or other outputs. For example, it might predict customer churn, classify images, recommend products, or detect anomalies in sensor data.

The model's predictions or insights are delivered to the appropriate destination, such as a UI, database, dashboard, or downstream application. This output may inform decision-making processes, trigger automated actions, or provide valuable insights for stakeholders. After generating predictions, the model's performance may be evaluated based on how well its predictions align with ground truth or expected outcomes. This evaluation can help identify areas for model improvement or refinement and inform future iterations of the model training process.

Overall, model scoring is a critical step in the life cycle of an analytical model, enabling organizations to leverage the model's predictive capabilities to drive business decisions, optimize processes, and deliver value to stakeholders.

There are different requirements that can determine the model scoring technique. One of the requirements is timing, which can be addressed through batch or real-time scoring. In batch scoring, predictions are generated for a large batch of data all at once. This approach is suitable when dealing with offline or non-real-time data processing scenarios, such as batch processing jobs or periodic model updates. Real-time scoring involves generating predictions or insights in real time as new data becomes available. This approach is suitable for applications requiring immediate responses or low-latency processing, such as online recommendation systems or fraud detection.

Model Scoring in R: Using Shiny Apps for Real-Time Scoring

There can be multiple approaches to implementing model scoring in R. One popular approach is to use Shiny applications, which can be adapted to do both real-time and batch scoring. Shiny is an R package that enables the creation of interactive web applications directly from R scripts. It allows R users to build web-based dashboards, data visualizations, and interactive tools without needing to know HTML, CSS, or JavaScript. With Shiny, R users can leverage their existing knowledge of R programming to create dynamic and interactive web applications that can be easily shared and accessed through a web browser.

Shiny works by combining the computational power of R with the interactivity of web applications. Users can define UI elements (such as sliders, buttons, and text inputs) and connect them to R code to generate dynamic outputs. Shiny applications can be deployed locally or on a web server, making them accessible to a wide audience.

Using Shiny Apps for real-time scoring can include developing a Shiny web application in R that takes input data from users, processes it in batches, and generates predictions using the trained model. Shiny allows for interactive data visualization and real-time updates, making it suitable for deploying batch scoring applications. For batches, one could develop a Shiny web application with reactive elements that continuously update predictions as users input new data. Shiny's reactivity model allows for real-time updates and interactivity, making it well suited for building interactive dashboards or applications with real-time scoring capabilities.

Let's consider an example. For simplicity, let's assume we're building a binary classification model using logistic regression. The following is an example of R code that can be used to develop a Shiny app that takes input data from users, processes it in batches, and generates predictions using a pretrained model:

```
library(shiny)
library(dplyr)

# Define UI ❶
ui <- fluidPage(
  titlePanel("Batch Prediction App"),

  sidebarLayout(
    sidebarPanel(
      fileInput("file", "Upload CSV file",
                accept = c(".csv")),
      br(),
      actionButton("process", "Process Data"),
      br(),
      textOutput("status")
    ), ❷

    mainPanel(
      tableOutput("data_preview"),
      tableOutput("predictions")
    ) ❸
  )
)

# Define server logic
server <- function(input, output) {

  # Reactive values to store data and predictions ❹
  data <- reactiveVal(NULL)
  predictions <- reactiveVal(NULL)

  # Read uploaded CSV file ❺
  observeEvent(input$file, {
    data(read.csv(input$file$datapath))
    output$data_preview <- renderTable({
      head(data(), 5)
    })
  })

  # Process data and generate predictions ❻
  observeEvent(input$process, {
```

```
      if (!is.null(data())) {
        # Dummy prediction (replace with actual model prediction)
        predictions(data() %>% mutate(prediction = sample(c(0, 1), nrow(.),
            replace = TRUE))) ❼
        output$predictions <- renderTable({
          head(predictions(), 5)
        })
        output$status <- renderText({
          "Data processed and predictions generated."
        })
      } else {
        output$status <- renderText({
          "Please upload a CSV file first."  ❽
        })
      }
    })
  }

  # Run the application
  shinyApp(ui = ui, server = server)
```

This Shiny app is designed for batch prediction using a CSV file as input. Here's a more detailed explanation of what the code does:

❶ The interface consists of a title panel (`"Batch Prediction App"`); a file input to upload a CSV file; a button to process the data; and output areas to display status messages, a data preview, and predictions.

❷ The `sidebarPanel` contains the file upload feature (`fileInput`), a button (`actionButton`) to trigger processing, and a status message area (`textOutput`).

❸ The `mainPanel` shows two tables: one for previewing the uploaded data (`tableOutput("data_preview")`) and another for displaying predictions (`tableOutput("predictions")`).

❹ `reactiveVal()` is used to store and update the uploaded data (data) and predictions (predictions).

❺ When a CSV file is uploaded, the app reads the file and previews the first five rows in the Data Preview section.

❻ When the Process Data button is clicked, the app checks if the data is uploaded. If data is available, it performs a dummy prediction by adding a prediction column to the dataset. This column is randomly generated with the value 0 or 1 for each row.

❼ The `mutate()` function is used to add a prediction column to the dataset with random 0 or 1 values (as placeholders). In a real-world application, this part would be replaced with actual predictions from a machine learning model.

❽ The first five rows of the predictions are then displayed, and a status message indicates that the predictions were generated. If no file is uploaded, the app displays a prompt asking the user to upload a file first.

The application flow is as follows:

1. The user uploads a CSV file.
2. The app previews the first five rows of the file.
3. After you click Process Data, the app adds dummy predictions and displays them alongside the original data.

The CSV file structure required for the upload should follow these guidelines:

- The file should have relevant features (columns) needed for the prediction model. Each column represents a variable, such as numerical or categorical features that the prediction model expects. If this is for a classification model, for example, these columns would include the features that the model uses to make its predictions.
- The CSV should include a header row with clear column names that match the data expected by the app or the machine learning model. For example, if your model expects features like Age, Income, and Education, the first row should contain these headers.
- Each subsequent row should contain values corresponding to the features in the header. Each row represents an observation or data instance.
- Since this app is for batch prediction, the CSV should not contain the target variable (e.g., labels) that you want to predict. The app will append the predicted values after processing.

Figures 8-1, 8-2, and 8-3 depict the Shiny application UI that shows the screens to upload the file, review the uploaded file, and then review the predictions.

In summary, this Shiny app allows users to upload a CSV file, process the data (in this case, generate dummy predictions), and display both the data and predictions in a tabular format. This code creates a Shiny app with a sidebar that allows users to upload a CSV file containing the input data. After uploading the file, users can click the "Process Data" button to generate predictions using a pretrained model. In this example, the predictions are dummy values generated randomly and are displayed in the main panel. To use this code for practical purposes, replace the dummy prediction code with your actual model prediction logic.

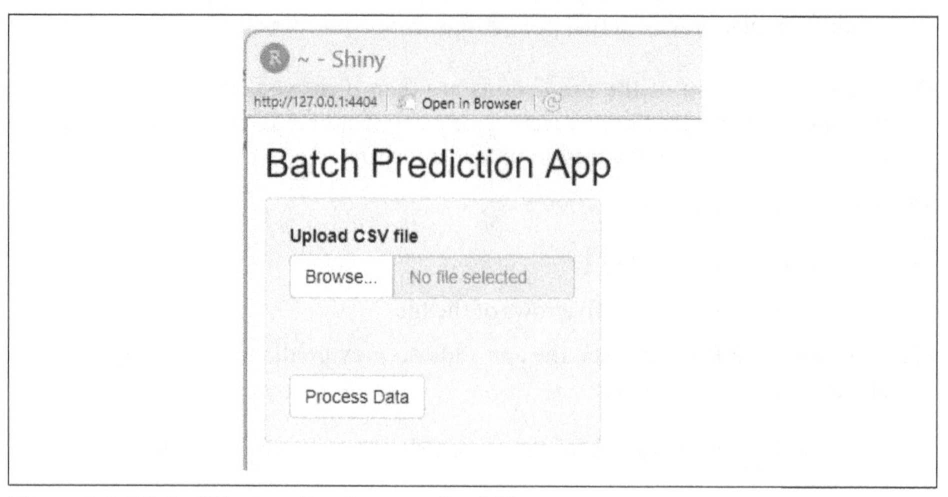

Figure 8-1. UI for Shiny application to upload file

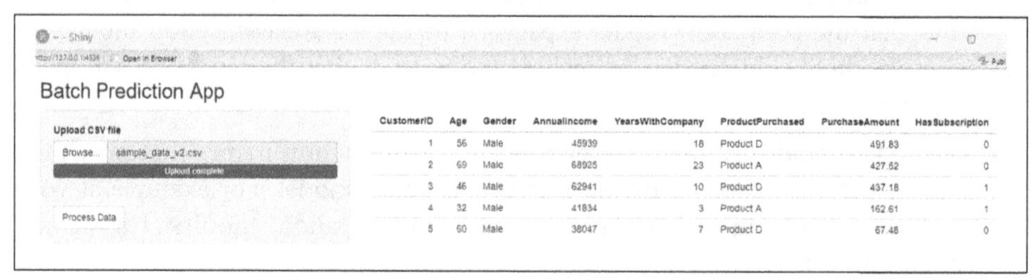

Figure 8-2. Application showing uploaded file

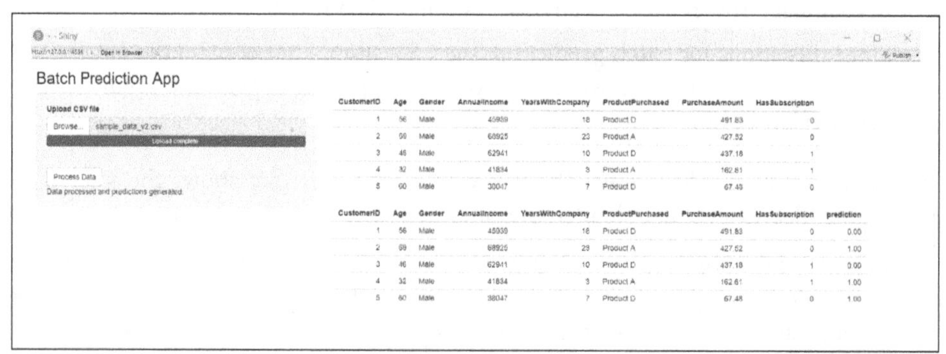

Figure 8-3. Application showing predicted values

Model Scoring in Python: Deploying Models with Streamlit

As with R, Python could be used in multiple ways to implement model scoring. Python provides multiple ways to implement model scoring, and one popular approach is using Streamlit. Streamlit is a powerful and user-friendly framework designed specifically for building interactive data science and machine learning applications. It simplifies the process of creating intuitive UIs, enabling both batch and real-time scoring of machine learning models with minimal effort.

With Streamlit, you can quickly create a frontend interface that allows users to upload data, interact with the model, and view predictions in real time. One of the key advantages of Streamlit is its simplicity; you don't need advanced web development knowledge to create fully functional apps. Instead, you can focus on integrating your machine learning models and generating insights, while Streamlit takes care of the UI and presentation:

```
import streamlit as st
import pandas as pd
import numpy as np

# Example of loading a pretrained model (replace with your model)
# from joblib import load
# model = load('path_to_your_model.joblib')

# Function for prediction (replace with your model's prediction logic)
def predict(data):
    # Example: Replace with model prediction logic
    # predictions = model.predict(data)
    return {"result": "Prediction result based on input data"}

# Streamlit app UI
st.title('Real-time Model Scoring with Streamlit')

st.header('Input Data')
uploaded_file = st.file_uploader("Upload your CSV file", type=['csv']) ❶

if uploaded_file is not None:
    # Reading the uploaded CSV file
    input_data = pd.read_csv(uploaded_file)
    st.write("Input data preview:")
    st.dataframe(input_data.head())

    # Process the input data and make predictions
    if st.button("Make Predictions"):
        predictions = predict(input_data)
        st.write("Prediction Results:")
        st.json(predictions)

# For real-time single input ❷
st.header('Real-Time Prediction')
```

```
user_input = st.text_input("Enter input data for real-time prediction")

if st.button("Predict"):
    # Convert user input into the appropriate format
    # Example assumes single input, modify as per your model's requirement
    data_for_prediction = np.array([user_input])
    prediction_result = predict(data_for_prediction)
    st.write("Prediction Result:")
    st.json(prediction_result)
```

❶ Users can upload a CSV file, and the app will read and display the data. Upon clicking the Make Predictions button, the app processes the input data and displays the model's predictions.

❷ Users can input data directly into a text box for real-time predictions. After clicking the Predict button, the app returns the result from the model.

Streamlit's simplicity, along with its ability to seamlessly integrate with machine learning models, makes it a practical solution for business analysts or data scientists looking to deploy their models interactively. Whether you're scoring batches of data or generating real-time predictions, Streamlit provides an easy-to-use platform for building effective machine learning applications.

Model Monitoring

Model monitoring is a critical component of ModelOps that involves continuously assessing the performance and behavior of deployed machine learning models in production environments. As a point to remember, models are built on static training data that represents a point in time. Over time, data changes as an organization changes. For example, new products and services offered to customers will change patterns in data resulting in models that may no longer predict or perform as accurately as they once did.

Model monitoring helps ensure model reliability, provides early detection of issues, helps mitigate risk, and optimizes model performance. It also helps detect deviations in model performance, ensuring that models continue to provide accurate and reliable predictions over time. Monitoring includes analyzing metrics to identify issues such as data drift, concept drift, or model degradation early on, allowing for timely intervention and maintenance. Effective monitoring reduces the risk of making erroneous decisions or acting on inaccurate predictions, thereby safeguarding against potential financial, reputational, or regulatory risks. Continuous monitoring provides insights into areas for model improvement or optimization, enabling organizations to enhance the overall performance and effectiveness of deployed models.

Key Metrics and Indicators for Model Performance Monitoring

Effective model monitoring relies on the tracking and analysis of various metrics and indicators that reflect the performance, behavior, and health of deployed models. While key metrics may vary depending on the model, common metrics indicators for model performance monitoring include:

- Accuracy and error metrics: Measures such as accuracy, precision, recall, F1 score, and mean squared error provide insights into the predictive accuracy and error rates of the model.
- Data drift: Monitoring changes in the distribution or characteristics of input data compared to the training data indicates potential shifts in the underlying data generating process.
- Concept drift: Assessing changes in the relationship between input features and target variables over time may indicate shifts in the relevance or applicability of the model.
- Model degradation: Detecting declines in model performance or predictive power over time (possibly due to changes in the data environment or model decay).

Techniques for Automated Model Monitoring

When models are operationalized, it is important to automate the monitoring. Automating model monitoring processes streamlines the detection and analysis of performance anomalies, ensuring timely responses and interventions. Multiple techniques are leveraged to alert stakeholders to performance that may need attention. These techniques include threshold-based alerts, statistical analysis, using other models, and development approaches.

Threshold-based alerts include setting predefined thresholds for key metrics and triggering alerts or notifications when thresholds are exceeded, indicating potential issues or deviations.

Statistical analysis employs different methods to detect significant changes or anomalies in model performance metrics, such as control charts or time-series analysis. It is possible to also train additional machine learning models to detect anomalies or deviations in model predictions, leveraging techniques such as anomaly detection or outlier detection algorithms. Additionally, integrating model monitoring into continuous integration and deployment (CI/CD) pipelines to automate the monitoring process as part of the model deployment life cycle is a common best practice.

All of these techniques include using a visual dashboard to track and visualize model performance. Let's explore how this can be accomplished in R and Python.

Implementation in R: Building dashboards with shinydashboard

R provides the Shiny Dashboard framework, allowing users to create interactive web-based dashboards for visualizing and monitoring model performance metrics. To implement model monitoring dashboards the key steps of data collection, dashboard design, metric identification, and alerting mechanisms are followed. Here are the steps to build a model monitoring dashboard:

Data collection
: Collecting model performance metrics and relevant data sources, such as prediction results, input data distributions, and drift detection outputs

Dashboard design
: Designing the dashboard interface using Shiny Dashboard components, including widgets, plots, and tables, to visualize key metrics and indicators

Metric tracking
: Implementing functionality to dynamically update dashboard components with real-time or periodic updates of model performance metrics

Alerting mechanisms
: Integrating alerting mechanisms into the dashboard to notify stakeholders of critical events or deviations in model performance

The following code example demonstrates a simple setup for a Shiny Dashboard, including data collection, dynamic updates, visualization, and alerting mechanisms. This example assumes that you have some model performance metrics stored, perhaps updated regularly through scheduled scripts or real-time data feeds.

Here's a basic setup in R:

```
# Install Shiny and Shinydashboard if not already installed
if (!require("shiny")) install.packages("shiny")
if (!require("shinydashboard")) install.packages("shinydashboard")
if (!require("dplyr")) install.packages("dplyr")
```

Server and UI code: Develop the server logic and UI using shinydashboard:

```
library(shiny)
library(shinydashboard)
library(dplyr)

# Sample data simulation ❶
generate_data <- function() {
  data.frame(
    Time = Sys.time(),
    Accuracy = runif(1, 0.8, 1),
    Loss = runif(1, 0, 0.2)
  )
}
```

```
# UI
ui <- dashboardPage(
  dashboardHeader(title = "Model Performance Dashboard"),
  dashboardSidebar(disable = TRUE),
  dashboardBody(
    fluidRow(
      box(title = "Accuracy", status = "primary", solidHeader = TRUE,
          plotOutput("accuracyPlot")),
      box(title = "Loss", status = "warning", solidHeader = TRUE,
          plotOutput("lossPlot")) ❷
    )
  )
)

# Server logic
server <- function(input, output) {
  data <- reactive({
    invalidateLater(5000, session)  # Update every 5 seconds
    generate_data()
  }) ❸

  output$accuracyPlot <- renderPlot({
    data <- data()
    plot(data$Time, data$Accuracy, type = 'o', col = 'blue', ylim = c(0.8, 1),
         xlab = "Time", ylab = "Accuracy", main = "Model Accuracy Over Time")
  })

  output$lossPlot <- renderPlot({
    data <- data()
    plot(data$Time, data$Loss, type = 'o', col = 'red', ylim = c(0, 0.2),
         xlab = "Time", ylab = "Loss", main = "Model Loss Over Time")
  })
}

# Run the application
shinyApp(ui, server)
```

❶ This script simulates data collection by generating random values for `Accuracy` and `Loss`. In a real scenario, you would replace `generate_data()` with a function that pulls updated performance metrics from your model or a database.

❷ The dashboard includes two boxes showing plots for `Accuracy` and `Loss`. You can extend this by adding more metrics or visual components.

❸ The `reactive()` function with `invalidateLater()` is used to update the dashboard every five seconds, simulating real-time data monitoring.

While the basic script does not include explicit alerts, you can integrate conditional logic within the server function to send notifications (e.g., via email or another communication channel) when metrics fall outside acceptable thresholds.

Figure 8-4 shows and example Shiny Dashboard for model monitoring.

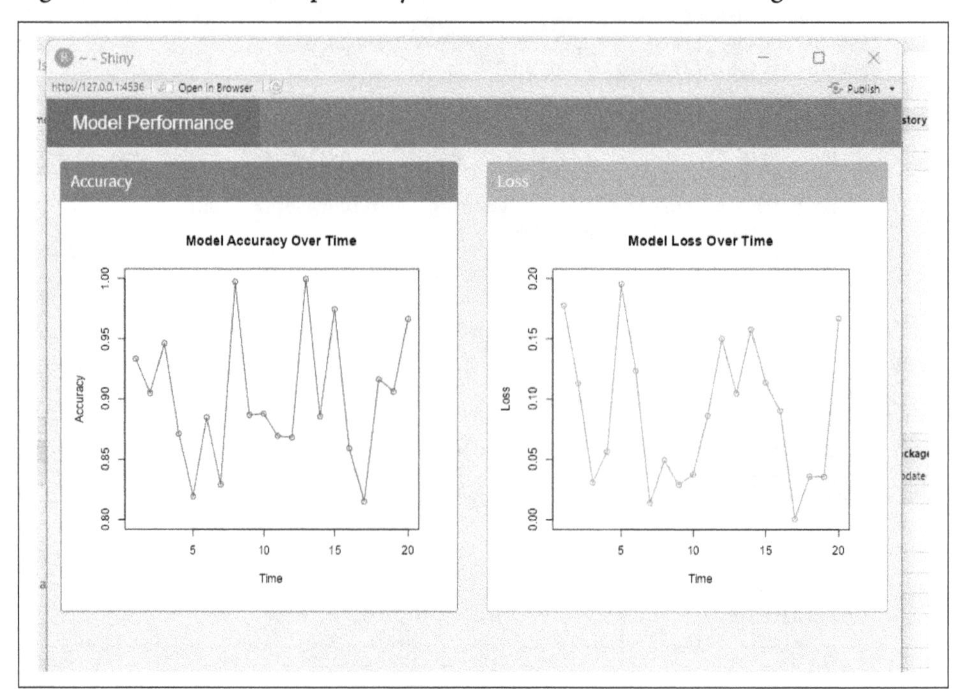

Figure 8-4. Example Shiny Dashboard for model monitoring

This code provides a fundamental framework. Depending on your specific needs and the complexity of your data, you may need to expand or modify it extensively. For sophisticated setups, consider incorporating more advanced reactive programming concepts and possibly connecting the dashboard to live data streams or databases.

Implementation in Python: Utilizing visualization libraries like Matplotlib and Seaborn

Python offers powerful visualization libraries such as Matplotlib and Seaborn, which can be leveraged to create customized visualizations for model monitoring. The steps in building a dashboard can be repeated in Python. Python has several frameworks similar to R's Shiny that allows you to build interactive applications. One of the more popular ones is Dash, which is developed by Plotly.

Dash is an open source framework for building analytical web applications. All coding is done with Python (although it supports other languages like R and Julia, too). It's especially suited for creating complex, interactive visualizations and connecting them to Python code, much like how Shiny operates with R.

Dash applications are composed of two parts. The first part is the `layout` of the app, and it describes what the application looks like. The layout is composed of a hierarchy of components such as HTML tags (via dash_html_components) and graphs (via dash_core_components). The second part describes the interactivity of the application. Dash apps are made interactive through `callbacks`, which are Python functions that are automatically called by Dash whenever an input component's property changes. `Callback` functions can read and modify the properties of the app components. Dash apps are web servers running Flask and communicating JSON packets over HTTP requests. They can be deployed on servers or platforms that support Python applications, such as Heroku, AWS (Amazon Web Services), or your own Linux server.

To create a model monitoring dashboard using Python, we can use libraries such as Matplotlib or Seaborn for creating static visualizations, and Plotly for dynamic, interactive dashboards. This code example assumes you have some data to monitor model performance such as accuracy, error rates, and potentially data drift statistics.

Step 1: Install necessary packages. First, make sure you have the necessary Python packages installed. You can install these using pip if they're not already installed:

```
pip install pandas numpy matplotlib seaborn plotly dash
```

Step 2: Sample Python code. Here's a Python script that demonstrates the preparation of data, creation of visualizations, and building an interactive dashboard using Plotly and Dash:

```
import pandas as pd
import numpy as np
import plotly.graph_objs as go
from dash import Dash, dcc, html, Input, Output

# Sample data generation ❶
np.random.seed(0)
dates = pd.date_range(start='2023-01-01', periods=100, freq='D')
accuracy = np.random.uniform(low=0.8, high=1.0, size=(100,))
error_rates = np.random.uniform(low=0.0, high=0.2, size=(100,))

data = pd.DataFrame({
    'Date': dates,
    'Accuracy': accuracy,
    'ErrorRate': error_rates
})

# Create a Dash application  ❸
app = Dash(__name__)

# App layout
app.layout = html.Div([
    html.H1("Model Performance Monitoring Dashboard"),
    dcc.Graph(id='accuracy-graph'),
    dcc.Graph(id='error-rate-graph'),
```

```
        dcc.Interval(
            id='interval-component',
            interval=1*1000,  # in milliseconds
            n_intervals=0
        )
    ])

# Callbacks to update graphs ❹
@app.callback(
    Output('accuracy-graph', 'figure'),
    Output('error-rate-graph', 'figure'),
    Input('interval-component', 'n_intervals')
)
def update_graph_live(n):
    # This function could be modified to pull data from a live data source
    fig1 = go.Figure(data=[ ❷
        go.Scatter(x=data['Date'], y=data['Accuracy'], mode='lines+markers')
    ])
    fig1.update_layout(title='Accuracy Over Time', xaxis_title='Date',
        yaxis_title='Accuracy')

    fig2 = go.Figure(data=[
        go.Scatter(x=data['Date'], y=data['ErrorRate'], mode='lines+markers',
                marker=dict(color='red'))
    ])
    fig2.update_layout(title='Error Rate Over Time', xaxis_title='Date',
        yaxis_title='Error Rate')

    return fig1, fig2

# Run the app
if __name__ == '__main__':
    app.run_server(debug=True)
```

❶ This script generates sample data for accuracy and error rates over time. You would replace this with data loading and preparation from your model's output.

❷ Using Plotly, the script creates line graphs for accuracy and error rates. Plotly's interactive capabilities allow zooming and hovering for more detailed inspection.

❸ The dashboard is built using Dash, which is a web application framework for Python built on top of Plotly.

❹ The Dash app is set up to refresh data dynamically, which could be linked to live data for real-time monitoring. Integration for alerting can be added by checking thresholds and triggering alerts within the `callback` function.

Dash can be used to display multiple chats that can track model accuracy and error rates, as depicted in Figures 8-5 and 8-6.

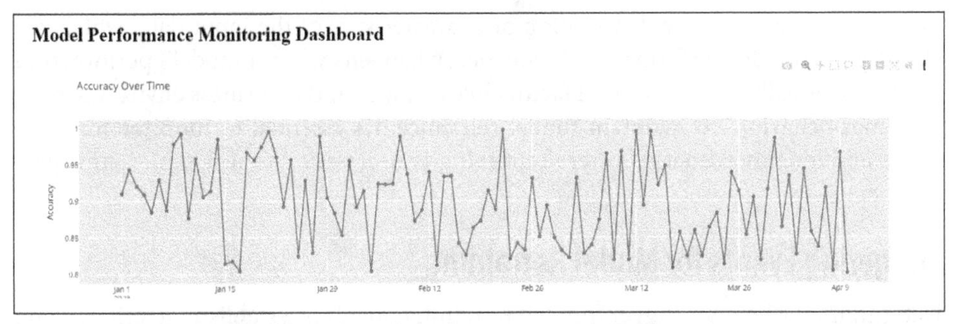

Figure 8-5. Example dashboard tracking accuracy using Dash

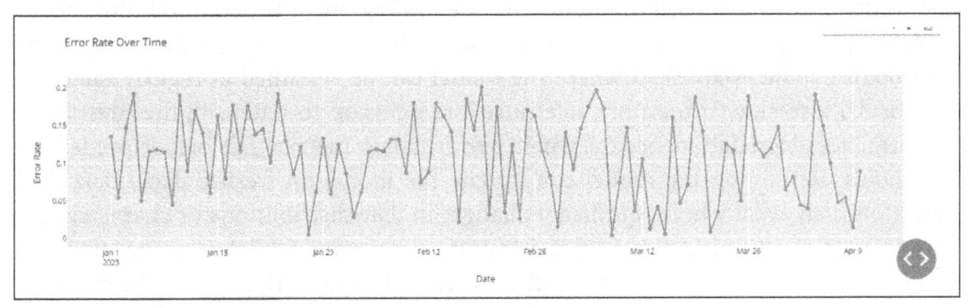

Figure 8-6. Example dashboard tracking error rate using Dash

This example is used for simplicity and demonstration. For a production environment, consider robust data handling, error checking, and possibly more complex interaction with backend systems for real-time monitoring and alerting.

Automating model monitoring processes using these techniques in R or Python ensures proactive identification and resolution of issues, enhances model reliability, and facilitates continuous improvement of deployed machine learning models in production environments. Model monitoring identifies opportunities for model retraining, which is addressed next.

Model Retraining

Model retraining occurs when data changes. This occurs when organizations, doing daily business, have a dynamic environment resulting in data changes. Model retraining is essential in dynamic environments where data distributions, patterns, and relationships change over time. There are three primary areas where model retraining needs to be considered. These are concept drift, data drift, and overall model decay. Drift refers to the change in the data that has occurred since the current model was created with training data.

Concept drift occurs when the relationships between input features and target variables change over time, reducing a model's effectiveness and accuracy. Similarly, data drift refers to shifts in the distribution or characteristics of the input data, which can also degrade model performance. Model decay happens when a model's performance declines gradually due to external factors like changes in the business environment or customer behavior. To maintain model relevance, it's essential to monitor for these issues and identify scenarios where retraining is necessary. We will explore this process next.

Triggering Events for Model Retraining

Determining when to trigger model retraining involves establishing criteria and events that signal the need for updates or adjustments to the model. Common events and criteria that are used to determine retraining include temporal thresholds, thresholds on performance degradation, data drift, and concept drift. Temporal thresholds can be established where the model can be retrained at regular intervals, such as daily, weekly, or monthly, to ensure it remains up-to-date with the latest data. Performance degradation occurs when performance metrics fall below predefined thresholds. Retraining the model can update the model on fresher data. Data drift detection is an event where significant changes in data distributions or characteristics compared to the training data trigger retraining. Concept drift detection occurs when shifts in the relationship between input features and target variables have been identified and retraining the model supports adapting to new patterns or trends.

Techniques for Automated Model Retraining

Automating model retraining simplifies the process and helps ensure that models remain up-to-date and effective. Retraining can be performed using various approaches, such as incremental learning, reactive retraining, integration into an automated pipeline, and utilizing version control.

Incremental learning occurs when the mode is updated continuously or incrementally with new data, leveraging techniques like online learning or minibatch updates. This avoids large changes to the model, allowing organizations to manage the impacts of retraining. Reactive retraining is triggered by specific events or conditions, such as detecting data drift during model monitoring or when performance metrics reach certain thresholds. Retraining as part of an automated pipeline would typically occur when a model is mature and well understood. This occurs when retraining is part of end-to-end automated pipelines that encompass data ingestion, preprocessing, model training, evaluation, and deployment, facilitating seamless retraining workflows. Because retraining effectively changes the model, it is necessary to manage the model as code where version control is put into place. This allows for management of model versions and tracking changes over time, enabling rollback to previous versions if necessary.

Implementation in R: Using cron Jobs for Scheduled Retraining

In R, scheduled model retraining can be implemented using cron jobs, which allow users to schedule and automate repetitive tasks at specified intervals. cron is a time-based job scheduler in Unix-like computer operating systems including Linux and macOS. It enables users to automate the execution of scripts, commands, or software programs at specified times, dates, or intervals. cron is particularly useful for scheduling repetitive tasks such as backups, system updates, or, in this case, running R scripts.

Steps for implementing scheduled retraining using cron jobs include script development, cron configuration, and establishing logging and monitoring. Script development includes developing R scripts that encompass data retrieval, preprocessing, model training, evaluation, and deployment. cron configuration focuses on configuring cron jobs to execute the R scripts at predefined intervals, specifying the frequency and timing of retraining tasks. Logging and monitoring establishes mechanisms to track the execution of retraining tasks, capture errors or anomalies, and ensure reliability. Coding examples for cron go beyond the scope of what a business analyst typically is responsible for, and often business analysts will partner with technical teams to schedule R retraining jobs.

Implementation in Python: Leveraging Tools Like Airflow for Workflow Management

Python provides workflow management tools like Apache Airflow, which facilitate the automation and orchestration of complex data workflows, including model retraining. Apache Airflow is an open source platform designed to programmatically author, schedule, and monitor workflows. Developed initially by Airbnb and later contributing to the Apache Software Foundation, Airflow has become widely adopted for managing complex computational workflows and data processing pipelines. It is particularly valued for its ability to manage tasks that are dependent on each other in a robust, scalable way.

Steps for implementing model retraining using Airflow include workflow definition, task configuration, scheduler configuration, and setting up monitoring and alerting. Airflow, like many pipeline tools, uses the concept of a workflow. Creating a workflow definition involves defining a directed acyclic graph (DAG) in Airflow to represent the retraining workflow, specifying tasks, dependencies, and execution logic. A scheduler is then used to execute the workflow, telling the DAG to run at specified intervals, ensuring timely and automated retraining.

As with any automated process, implementing monitoring and alerting mechanisms is needed within Airflow to track task execution, detect failures or issues, and notify stakeholders as needed. Setting up workflows in Airflow is usually beyond the responsibilities of a business analyst. Business analysts will partner with technical teams to complete the implementation.

Automating model retraining using these techniques in R and Python ensures that machine learning models remain adaptive, accurate, and effective in dynamic environments, thereby enhancing their utility and value in real-world applications.

Generating Reports

Final reports play a crucial role in ModelOps by documenting the outcomes, insights, and performance of machine learning models. Reports can come in many formats, but the primary purpose of the reports are to ensure transparency and accountability, support decisions, help with regulatory compliance, and enable knowledge sharing. They can be generated as part of the model evaluation process and then created as part of ModelOps.

Reports serve many purposes, including:

- Providing stakeholders with transparency into the model development process, enabling them to understand the methodologies, assumptions, and limitations underlying the models
- Serving as decision-making aids for stakeholders, providing them with actionable insights, recommendations, and conclusions based on the analysis of model outcomes and performance
- Helping organizations comply with regulatory requirements by documenting model development, validation, and deployment processes, ensuring adherence to regulatory standards and guidelines
- Facilitating knowledge sharing and collaboration among team members, allowing for the dissemination of best practices, lessons learned, and model insights across the organization

Let's consider how to create and structure reports.

Content and Structure of Final Reports

Effective final reports should encompass comprehensive and well-structured content that communicates key findings, insights, and recommendations to stakeholders. The structure of final reports typically includes:

Executive summary
: A concise overview of the report's objectives, methodology, findings, and recommendations, designed to provide senior stakeholders with a high-level understanding

Introduction
: Background information on the problem statement, objectives, scope, and methodology of the analysis or modeling project

Data description and preprocessing
: Description of the data sources, data collection process, data preprocessing steps, feature engineering, and data cleaning techniques applied

Model development
: Details of the machine learning models developed, including algorithms used, model architecture, hyperparameters, and model evaluation metrics

Results and analysis
: Presentation and analysis of model performance, including evaluation metrics, validation results, insights into model behavior, strengths, weaknesses, and areas for improvement

Recommendations
: Actionable recommendations based on the analysis, highlighting opportunities for optimization, further research, or model refinement

Conclusion
: Summary of key findings, implications, and next steps for stakeholders, emphasizing the value and impact of the model outcomes

Here is an example of what a final report might include. It is common to leverage tools like Jupyter to prepare the final report:

```
# Final Report: Machine Learning Model Evaluation

## 1. Executive Summary

Provide a high-level overview of the project, objectives, and outcomes.

- **Objective**: Predict [Target Variable] using [Algorithm/Model].
- **Key Results**: Summarize the performance of the model and its business impact.
- **Recommendations**: State recommendations for model deployment or improvement.

---

## 2. Introduction

### 2.1 Background

Describe the problem, its significance, and any context that led to this project.

### 2.2 Objectives

State the goals of the analysis and the questions you aim to answer using the
model.
```

````
---

## 3. Data Description and Preprocessing

### 3.1 Data Sources

```python
# Load dataset and describe data
import pandas as pd

# Example loading dataset
data = pd.read_csv('your_dataset.csv')
data.head()

# Handle missing values
data.fillna(data.mean(), inplace=True)

# Convert categorical variables
data = pd.get_dummies(data)

# Feature engineering example
data['new_feature'] = data['feature1'] / data['feature2']

from sklearn.model_selection import GridSearchCV

# Example grid search
param_grid = {'n_estimators': [100, 200, 300]}
grid_search = GridSearchCV(estimator=rf_model, param_grid=param_grid, cv=5)
grid_search.fit(X_train, y_train)
grid_search.best_params_

from sklearn.metrics import classification_report, accuracy_score

# Predict on validation set
y_pred = rf_model.predict(X_val)

# Model performance
print(classification_report(y_val, y_pred))

import matplotlib.pyplot as plt
from sklearn.metrics import roc_curve, auc

# ROC curve example
fpr, tpr, _ = roc_curve(y_val, rf_model.predict_proba(X_val)[:, 1])
roc_auc = auc(fpr, tpr)

plt.figure()
plt.plot(fpr, tpr, color='blue', label=f'ROC curve (area = {roc_auc:.2f})')
plt.plot([0, 1], [0, 1], color='navy', linestyle='--')
plt.xlabel('False Positive Rate')
plt.ylabel('True Positive Rate')
````
````

```
plt.title('ROC Curve')
plt.legend(loc="lower right")
plt.show()
```

The recommendations section of the report highlights key insights gained from the analysis, specifically identifying the most important features driving the model and assessing the model's confidence in making predictions. These insights help prioritize the variables that significantly influence the outcome and provide an understanding of the model's reliability. In terms of next steps, suggestions include deploying the model into a production environment, improving it through further refinement, or gathering additional data to enhance model accuracy and robustness.

The conclusion summarizes the findings, overall model performance, and implications for stakeholders, providing an overview of how the model can impact decision making. Furthermore, any limitations in the model, data quality, or analysis should be addressed, identifying areas where improvements can be made. Lastly, a references section is included, citing relevant sources, frameworks, and tools used throughout the project to ensure transparency and provide context for further research or model enhancement.

Another final report is a model card, which is a structured report that provides essential information about a machine learning model, including its purpose, performance, limitations, and intended use. The concept was introduced by Google to promote transparency and accountability in the deployment of machine learning models. Model cards are particularly useful for ensuring that stakeholders, including nontechnical users, understand the model's capabilities and constraints, leading to more responsible and informed usage. Here are the components of a model card:

Model overview
: A brief description of the model's purpose, what it predicts or classifies, and the context in which it should be used.

Intended use
: Information about the expected use cases of the model, including the target audience and scenarios where the model is appropriate.

Model performance
: A summary of the model's performance metrics (e.g., accuracy, precision, recall) across different evaluation datasets. It may include performance on specific subgroups (such as age, gender, etc.) to reveal potential biases.

Limitations
: Any known limitations of the model, such as areas where it may underperform or risks of bias.

Training data
: Information on the dataset used to train the model, including its size, source, and any preprocessing applied.

Evaluation data
: A description of the data used to evaluate the model's performance, including how it differs from the training data.

Ethical considerations
: Ethical issues, such as potential biases in the model, fairness considerations, and any risks associated with using the model.

Caveats and recommendations
: Suggestions for how the model should (or should not) be used, along with guidance for ensuring that the model is applied appropriately and responsibly.

A model card could accompany a model used to predict loan approvals, outlining performance metrics for different demographic groups, ensuring fairness, and helping stakeholders assess if the model is suitable for their intended application. By providing this transparency, model cards help mitigate risks such as unintended bias and misuse of models in production.

Techniques for Automated Report Generation

Automating report generation streamlines the process of creating consistent, standardized, and reproducible reports, saving time and ensuring accuracy. Techniques for automated report generation include template-based reporting, leveraging report generating libraries, integrating as part of the workflow, and scheduling. Template-based reporting uses tools like LaTeX, Markdown, or HTML, with placeholders for dynamic content such as model results, graphs, and tables. Report-generating libraries and frameworks such as R Markdown, Jupyter notebooks, or ReportLab in Python can be used to automate the generation of reports from code and data. Once reports are created, the report generation can be integrated into existing workflow management systems or pipelines, ensuring reports are automatically generated as part of the model deployment process. Reports should also be set up as scheduled tasks or jobs to generate reports at predefined intervals, ensuring regular updates and distribution to stakeholders.

Implementation in R: Generating Reports with R Markdown and knitr

R provides powerful tools like R Markdown and knitr for generating dynamic, reproducible reports from R code and data. These are tools within the R ecosystem for dynamic report generation, combining comprehensive data analysis and the ability to produce high-quality reports directly from R code. They are essential for creating reproducible research and for sharing analysis results with collaborators, stakeholders, or a broader audience through various formats like HTML, PDF, or Word documents.

Steps for implementing report generation with R Markdown and knitr include:

Document setup
: Creating an R Markdown document (*.Rmd*) specifying the report structure, content, and formatting using Markdown syntax.

Code integration
: Embedding R code chunks within the Markdown document to execute data analysis, model training, and visualization tasks.

Knitting process
: Knitting the R Markdown document to produce a final report in various formats such as PDF, HTML, or Word, automatically executing code chunks and rendering output.

Automation
: Automating the knitting process using scripts or workflows, enabling scheduled or event-triggered report generation.

Next, let's review the steps how you can implement each step and provide an example of an R Markdown script.

Step 1: Document setup

You start by creating an R Markdown file, which typically has the extension *.Rmd*. This file includes both metadata (in YAML format) that defines the output format and the mix of text and R code chunks.

Here is an example R Markdown document (`my_report.Rmd`):

```
---
title: "Dynamic Report"
author: "Your Name"
date: "`r Sys.Date()`"
output:
  html_document: default
  pdf_document: default
  word_document: default
--
```

This is an automated report generated by R Markdown. The report provides insights into data analysis, model training, and visualizations:

````
# Data Analysis

```{r load-data}
# Load data
````
````

```
data <- read.csv("path/to/your/data.csv")
summary(data)

# Fit a linear model
model <- lm(SalePrice ~ ., data = data)
summary(model)

# Plotting
library(ggplot2)
ggplot(data, aes(x=variableX, y=variableY)) +
  geom_point() +
  geom_smooth(method="lm")
```

Step 2: Code integration

In the Step 1 code example, R code chunks are integrated directly within the document. These chunks are enclosed in triple backticks and prefixed with `{r}`. You can run data analysis and model training and even create visualizations directly within these chunks.

Step 3: Knitting process

To knit the document into a final report in formats such as HTML, PDF, or Word, you can use the knit button in RStudio, which typically looks like a ball of yarn. Alternatively, you can use the command line:

```
rmarkdown::render("my_report.Rmd")
```

This command will generate the report in all specified formats (as per the YAML header in the *.Rmd* file).

Step 4: Automation

To automate the knitting process, especially for generating reports on a schedule or event-trigger, you can use R scripts that call `rmarkdown::render()` and schedule them to run using cron jobs on Linux/macOS or Task Scheduler on Windows.

Here is an example R script for automation (*run_report.R*):

```
library(rmarkdown)

# Set the working directory to the directory containing the Rmd file
setwd("/path/to/directory/")

# Render the report
render("my_report.Rmd")
```

This setup ensures your reports are dynamically generated based on the latest data and R analysis, providing up-to-date insights in an automated, reproducible fashion.

Implementation in Python: Creating Reports with Jupyter Notebooks and nbconvert

Python users can utilize Jupyter notebooks and nbconvert to create and automate report generation workflows. Jupyter notebooks are an open source web application that allows you to create and share documents that contain live code, equations, visualizations, and narrative text. Originally part of the IPython project, Jupyter notebooks support over 40 programming languages, including Python, R, Julia, and Scala. They are widely used in data science, scientific computing, and academic research for interactive data visualization, machine learning, and much more. nbconvert is a command-line tool that comes with Jupyter that allows you to convert Jupyter notebooks (*.ipynb* files) into various static formats such as HTML, PDF, LaTeX, Markdown, reStructuredText, and more. It is a flexible tool that can be used for automatic report generation, presentations, or even preparing notebooks for publication.

Steps for implementing report generation with Jupyter notebooks and nbconvert include:

Notebook development
: Developing Jupyter notebooks (*.ipynb*) containing code, visualizations, and narrative text to document the analysis or modeling process

Markdown cells
: Adding Markdown cells within the notebook to provide context, explanations, and interpretations of the analysis results

nbconvert conversion
: Using nbconvert, a command-line tool for converting Jupyter notebooks, to export notebooks to various formats such as HTML, PDF, or LaTeX

Automation
: Incorporating nbconvert commands into scripts or workflow systems to automate the conversion process, enabling scheduled or on-demand report generation

Automating final report generation using these techniques ensures consistency, reproducibility, and efficiency in communicating model outcomes and insights to stakeholders, facilitating informed decision making and knowledge sharing within organizations.

To automate report generation with Jupyter notebooks and nbconvert, you will need to set up a notebook with comprehensive documentation and visualization, and then use nbconvert to convert the notebook to various formats. Here's how to implement each of these steps, including automating the conversion process.

Step 1: Notebook development

Create a Jupyter notebook (*analysis.ipynb*) that contains your code, visualizations, and text explaining your methodology, analysis, and results. For instance, your notebook might include sections structured as follows:

```
# Cell with Markdown
# ## Data Loading and Cleaning

# Cell with code
import pandas as pd

data = pd.read_csv('data.csv')
data_clean = data.dropna()

# Cell with Markdown
# ## Data Analysis

# Cell with code
import seaborn as sns
import matplotlib.pyplot as plt

plt.figure(figsize=(10,6))
sns.histplot(data_clean['variable_of_interest'], kde=True)
plt.title('Distribution of Variable of Interest')
plt.show()
```

Step 2: Markdown cells

In your Jupyter notebook, intersperse Markdown cells to add narrative text. Markdown cells allow you to write formatted text and insert images, links, and even HTML code to enhance the interpretability of your notebook.

Step 3: nbconvert conversion

To convert the notebook to different formats using nbconvert, you can use the following command-line commands. First, ensure nbconvert is installed:

```
pip install nbconvert

To convert the notebook to HTML, PDF, or LaTeX, use:

# Convert to HTML
jupyter nbconvert --to html analysis.ipynb

# Convert to PDF (requires TeX to be installed)
jupyter nbconvert --to pdf analysis.ipynb

# Convert to LaTeX
jupyter nbconvert --to latex analysis.ipynb
```

Step 4: Automation

To automate the conversion process, you can use a simple script that invokes nbconvert and then schedule this script to run at specific times or events using a scheduler like cron on Linux or Task Scheduler on Windows.

Here's a simple Python script called *automate_report.py*, which uses nbconvert:

```
import os

# Function to convert notebook
def convert_notebook(notebook_name, output_format):
    cmd = f"jupyter nbconvert --to {output_format} {notebook_name}"
    os.system(cmd)

# Convert to HTML
convert_notebook('analysis.ipynb', 'html')

# Convert to PDF
convert_notebook('analysis.ipynb', 'pdf')
```

To run this script daily at a specified time, you can add a cron job (on Linux/macOS):

```
# Open crontab editor
crontab -e

# Add a cron job to run the script every day at 7 AM
0 7 * * * /usr/bin/python /path/to/automate_report.py
```

This approach ensures your Jupyter notebook is automatically converted and updated reports are generated regularly, maintaining consistency, reproducibility, and efficiency in communicating important insights to stakeholders.

Version Control and Model Reproducibility

Version control and model reproducibility are essential for ensuring transparency, consistency, and traceability in ModelOps. Best practices in this area include managing source code, model versioning, containerization, and reproducibility of model outputs. Typically, technical teams responsible for ModelOps handle these tasks. Source code management involves using version control systems like Git to track changes in code, scripts, and configuration files, which enables collaboration, version tracking, and the ability to roll back to previous versions.

Model versioning is a related process that tracks machine learning models and associated artifacts, such as datasets, preprocessing steps, and hyperparameters, to allow for reproducibility and comparison of model versions. Containerization, which uses tools like Docker or Singularity, packages models and their dependencies to ensure consistent execution across different platforms and deployment environments.

Model reproducibility is critical for auditing and troubleshooting models when changes are made. Technical teams often implement automated pipelines to re-create model training and evaluation processes, ensuring the validation and verification of model performance across various environments. These practices ultimately contribute to robust and reliable ModelOps workflows.

Collaboration and Documentation Practices

All organizations typically have a standard when it comes to required documentation and knowledge sharing. In ModelOps, this is no different. Effective collaboration and documentation practices are important for facilitating knowledge sharing, communication, and collaboration among the stakeholders involved in ModelOps. Considerations include documentation standards, what repositories are used, collaboration platforms, and training new team members. Establishing documentation standards includes creating templates for documenting model development, deployment, and operational procedures to ensure consistency and clarity. Knowledge repositories are centralized repositories or knowledge bases to store documentation, code, and best practices related to model development, deployment, and maintenance, enabling easy access and sharing of information. Many technical teams leverage collaboration platforms and tools such as Confluence, SharePoint, or Slack to facilitate communication, collaboration, and knowledge sharing among team members, stakeholders, and subject matter experts. Additionally, collaboration and documentation practices provide training and onboarding programs to educate team members and stakeholders on ModelOps practices, tools, and processes.

ModelOps Use Cases

To see how companies in different industries automate scoring, monitoring, and retraining processes to enhance accuracy, adaptability, and responsiveness, let's consider a few ModelOps examples. Each scenario illustrates the integration of machine learning models to not only predict outcomes but also to dynamically refine strategies and provide valuable insights, ensuring business agility and competitive advantage in rapidly changing environments. These examples have been generalized from several industry use cases.

Retail Sales Forecasting: Automation of Scoring and Monitoring

A retail company leverages machine learning models for sales forecasting. By automating model scoring and monitoring processes, the company can automatically generate sales forecasts based on historical data, product attributes, and market trends using machine learning models. Also, this company can monitor model performance metrics such as accuracy, error rates, and drift over time to ensure the reliability and effectiveness of sales forecasts. ModelOps also assists in alerting stakeholders of significant deviations or anomalies in sales predictions, enabling timely intervention and decision making.

Fraud Detection: Dynamic Model Retraining and Reporting

A financial institution utilizes machine learning models for fraud detection in transactions. By employing dynamic model retraining and reporting, the institution can continuously update fraud detection models based on evolving fraud patterns, new data sources, and regulatory changes to enhance detection accuracy and adaptability. ModelOps supports creating real-time reports and dashboards summarizing fraud detection outcomes, performance metrics, and actionable insights for stakeholders, enabling informed decision making and risk management. Last, the benefits also include incorporating feedback loops and human-in-the-loop processes to validate model predictions, investigate false positives, and refine fraud detection strategies iteratively.

Customer Churn Prediction: Scheduled Model Retraining and Final Report Generation

In this use case, a subscription-based service provider utilizes machine learning models to predict customer churn. By implementing scheduled model retraining and final report generation, the provider is able to retrain churn prediction models using updated customer data, behavioral features, and engagement metrics. This allows them to improve predictive accuracy and adaptability to changing customer dynamics. ModelOps supports automatically generating final reports summarizing churn prediction outcomes, model performance metrics, and actionable insights for decision makers, facilitating strategic planning, customer retention efforts, and revenue optimization. They have also integrated churn prediction insights into customer relationship management (CRM) systems and marketing campaigns to proactively identify at-risk customers, tailor retention strategies, and minimize churn rates.

Integration with Existing Systems and Infrastructure

It is important to note that ModelOps is primarily the responsibility of technical teams who manage a large technical ecosystem of existing systems. Throughout this chapter, I've highlighted how models need to integrate into this larger ecosystem, which presents challenges in ensuring compatibility, data consistency, and seamless interoperability. In addition to completing the R or Python scripting for ModelOps, additional work is often needed to operationalize models. The scope of work will vary depending on the ecosystem, but it is possible that APIs, data pipelines, ETL (extract, transform, load) processes, and implementing change management and version control will be necessary for ModelOps.

For instance, developing robust APIs and microservices architectures can enable seamless integration of machine learning models with production systems, databases, and business applications. Establishing data pipelines and ETL processes may be required to ensure data consistency, quality, and lineage across heterogeneous data sources and systems. Change management and version control practices to track and manage changes to model configurations, dependencies, and deployment environments may be needed if they do not already exist.

Future Direction of MLOps

The future of ModelOps may see increased adoption of cloud-based platforms and services that offer scalable, cost-effective solutions for model development, deployment, and management. Key trends include a migration toward serverless computing and managed services for model hosting, monitoring, and orchestration, enabling organizations to focus on model development while offloading infrastructure management to cloud providers. Cloud-based platforms are integrating AI/ML technologies such as Kubernetes, Docker, and serverless frameworks to streamline deployment workflows, improve resource utilization, and enhance scalability and reliability.

Another key trend is the integration of AI/ML Ops and DevOps practices. There is a growing convergence between AI/ML Ops and DevOps practices, driven by the need for end-to-end automation, collaboration, and continuous delivery in model development and deployment pipelines. Technology teams are adopting DevOps principles and tools such as CI/CD, version control, and infrastructure as code (IaC) in AI/ML Ops workflows to accelerate model deployment cycles, increase agility, and improve quality and reliability. Teams are also developing DevOps toolchains and platforms tailored for machine learning and data science workflows, incorporating features such as model versioning, experiment tracking, and automated testing to support end-to-end DevOps for AI/ML.

The maturity of analytics has also created a renewed focus in automated model governance and compliance. The future of ModelOps will witness advancements in automated model governance and compliance solutions to address regulatory requirements, ethical considerations, and risk management concerns. As analytical capabilities increase, organizations are building automated model governance frameworks and tools that provide capabilities for model cataloging, lineage tracking, and policy enforcement to ensure compliance.

The future of ModelOps is increasingly moving toward cloud-based platforms and services, which offer scalable and cost-effective solutions for model development, deployment, and management. Cloud platforms like AWS, Microsoft Azure, and Google Cloud provide businesses with managed services that eliminate the need for on-premises infrastructure. By leveraging cloud services, organizations can quickly deploy models at scale without worrying about infrastructure management. For example, Netflix uses AWS for managing and deploying machine learning models that personalize user recommendations, allowing them to handle millions of requests per day while maintaining optimal performance. Additionally, cloud-based platforms enable serverless computing, where companies can run machine learning models without managing servers. This is beneficial for businesses as it reduces operational costs while providing flexibility and scalability for handling fluctuating workloads. For instance, companies like Zillow use serverless computing to scale real estate price predictions based on demand, processing vast amounts of data in real time.

One of the main technological trends driving the future of ModelOps is the integration of AI/ML technologies like Kubernetes, Docker, and serverless frameworks. These technologies streamline deployment workflows and improve resource utilization. Kubernetes, an open source container orchestration platform, helps businesses manage complex machine learning models by automating deployment, scaling, and operations. Companies such as Airbnb use Kubernetes to handle large-scale machine learning workflows, ensuring their models can scale as demand fluctuates. Docker, a containerization technology, allows machine learning models to be packaged with all necessary dependencies, ensuring consistent execution across environments. This has become crucial in environments where multiple teams work on different models, ensuring consistency and reducing the chances of errors when models move from development to production. The integration of serverless frameworks allows businesses to focus on model development without worrying about the underlying infrastructure. Uber, for example, uses serverless frameworks to build scalable machine learning models for tasks like dynamic pricing and fraud detection, ensuring that their operations remain smooth as the business scales globally.

Another key trend is the convergence of AI/ML Ops and DevOps practices. Business analysts may be familiar with DevOps, the integration of software development and IT operations to shorten the system development life cycle. In the context of AI/ML, the convergence of DevOps and AI/ML Ops introduces CI/CD pipelines, IaC, and version control to the machine learning life cycle. This allows companies to accelerate their model deployment cycles while maintaining high standards of quality and reliability. For instance, JPMorgan Chase integrates AI/ML Ops with DevOps practices by utilizing automated pipelines to deploy models that detect fraud and ensure compliance. Their machine learning models undergo continuous testing and monitoring, ensuring the models are always up to date and performing optimally. The use of CI/CD pipelines enables quicker deployment of machine learning models, reducing the time from model development to production from months to days.

As machine learning and data science mature, organizations are placing greater emphasis on automated model governance and compliance. In industries like healthcare and finance, compliance with regulations such as GDPR (General Data Protection Regulation) and HIPAA (Health Insurance Portability and Accountability Act) is critical. ModelOps frameworks are evolving to integrate automated governance, allowing organizations to track model lineage, enforce policies, and ensure ethical considerations are met. For example, Pfizer uses automated model governance to ensure their machine learning models for drug discovery comply with regulatory standards and are ethically sound. This is achieved through tools that provide model cataloging, lineage tracking, and policy enforcement, ensuring that every decision made by the models is traceable and compliant with regulatory guidelines. This automated governance ensures that organizations mitigate risks, adhere to regulatory standards, and maintain the trust of stakeholders, especially when handling sensitive data.

In summary, the future of ModelOps will rely heavily on cloud adoption, the integration of AI/ML Ops with DevOps, and automated governance frameworks. These advancements will streamline model deployment, improve scalability, and ensure compliance, allowing businesses to stay competitive while leveraging the full power of machine learning.

Summary

This chapter provided an in-depth exploration of ModelOps and its critical role in achieving business value. ModelOps ensures not only the scalability and reliability of models but also facilitates their continuous enhancement and efficient deployment, which are vital for maintaining competitive advantage in business. Business analysts can leverage ModelOps tools and techniques in R and Python, which are invaluable for those looking to leverage predictive analytics without deep technical expertise. Tools like Shiny Apps for R and Flask for Python facilitate real-time model scoring, which is essential for timely business insights. Overall, this chapter was designed to provide comprehensive guidance for business analysts on maximizing the value of machine learning through ModelOps, ensuring models are not only effective but also aligned with strategic business objectives.

CHAPTER 9

Advanced Visualization

In the realm of business analytics, the ability to visualize data effectively is not just a supplementary skill but a central component of data analysis. Visualization transforms raw data into a form that is easier to understand, revealing trends, uncovering insights, and supporting decision-making processes. Research shows that humans process visuals 60,000 times faster than text and that 90% of the information transmitted to the brain is visual. This highlights how critical visualizations are in helping people grasp complex information quickly. For example, in 1854, Dr. John Snow famously used a simple map (*https://oreil.ly/Db3Sq*) to plot cholera cases in London, visually demonstrating that cases clustered around a contaminated water pump. This visualization not only helped explain the cause of the outbreak but also spurred immediate action, leading to the removal of the pump and the prevention of further deaths.

This chapter focuses on the critical role of advanced visualization techniques in interpreting complex datasets, enabling analysts to communicate findings clearly and persuasively. As we delve deeper into the world of data visualization, two main tools emerge as frontrunners: R Shiny and Python visualization packages. R Shiny offers a framework for building interactive web applications directly from R, allowing users to interact with their data and visualizations dynamically. On the other hand, Python, with its robust libraries such as Matplotlib, Seaborn, and Plotly, provides versatile options for creating static, animated, and interactive visualizations. Each tool has its strengths and caters to different aspects of data visualization needs, ensuring that insights are not only understood but also acted upon.

It's also important to acknowledge the leading roles of Tableau and Power BI in the visualization space. Both tools are highly favored in enterprise environments for their ease of use and powerful drag-and-drop interfaces. Tableau excels in creating sophisticated, interactive dashboards that can handle large datasets with ease, while Power

BI integrates seamlessly with Microsoft's ecosystem, making it ideal for organizations already using Office 365. These platforms offer rich visualizations, automated reporting, and advanced analytics capabilities, making them front-runners for business intelligence.

However, as this book is primarily focused on R and Python, we won't be diving into the specifics of Tableau and Power BI. Instead, we will concentrate on how R Shiny and Python's visualization libraries provide the flexibility and customization required for more tailored and complex data science applications.

Advanced Visualization with R Shiny

R Shiny is a tool from RStudio (now called Posit) that transforms R statistical software into an interactive web application platform. Cleveland Clinic, a renowned healthcare provider, used R Shiny to develop an interactive application that improved patient care. The Shiny app allowed healthcare professionals to visualize patient data in real time, helping with faster and more accurate clinical decision making. This adoption reduced hospital stays and improved the quality of patient care by enabling personalized treatment plans. Posit reported that R Shiny is used by over two million users worldwide, and it's rapidly becoming the go-to tool for interactive data visualization in sectors like finance, healthcare, and academia.

What Is R Shiny?

R Shiny is an open source R package that provides an elegant and comprehensive framework for building interactive web applications using R. Unlike traditional R scripts that run independently and produce static outputs, Shiny applications are web apps that can interact dynamically with user inputs. The outputs change instantaneously based on user interactions, which can range from simple data inputs to complex algorithm adjustments.

Key Features and Capabilities of R Shiny

R Shiny transforms static data analysis into dynamic, interactive web applications without the need for HTML, CSS (cascading style sheets), or JavaScript knowledge. Its interface features drag-and-drop components like sliders and buttons, making it accessible for data scientists to create professional-looking applications with minimal effort. Shiny's core strength lies in its reactivity system, which updates outputs instantly as users modify inputs, enhancing the user experience by providing immediate feedback. This makes it ideal for presenting complex data interactively.

Shiny apps are also highly accessible; they can be easily hosted online or embedded in various formats like R Markdown documents, reaching a broad audience beyond those familiar with R. For those needing more customized solutions, Shiny supports integration with standard web technologies, allowing advanced users to tailor applications extensively. Supported by a vibrant community and continuous updates from RStudio, Shiny offers a comprehensive platform for developing sophisticated web applications in the data science realm. The key features are:

Reactive programming
: Shiny automatically updates outputs when inputs change, offering real-time interaction without refreshing the page.

No web development skills required
: Shiny allows the creation of interactive web applications using only R, without the need for HTML, CSS, or JavaScript.

Drag-and-drop UI components
: Simplified UI creation through input widgets like sliders, drop-downs, and buttons.

Cross-platform accessibility
: Shiny apps can be hosted online via platforms like shinyapps.io, or embedded into R Markdown documents and dashboards.

Extensibility with web technologies
: While no web knowledge is required, advanced users can incorporate custom HTML, CSS, and JavaScript for further customization.

Interactive visualizations
: Shiny seamlessly integrates with popular R packages like ggplot2, Plotly, and leaflet to deliver powerful, interactive visualizations.

Scalability and deployment
: Apps can scale from simple prototypes to enterprise-level applications and can be deployed on various cloud platforms.

Community support and documentation
: A broad ecosystem backed by Posit ensures continuous updates, tutorials, and community contributions.

Figure 9-1 shows a simple Shiny application that we will use as a basis for examples. The application provides a choice of datasets and enables interaction to see data distribution via a histogram. Data distribution can be reexamined by altering the number of bins.

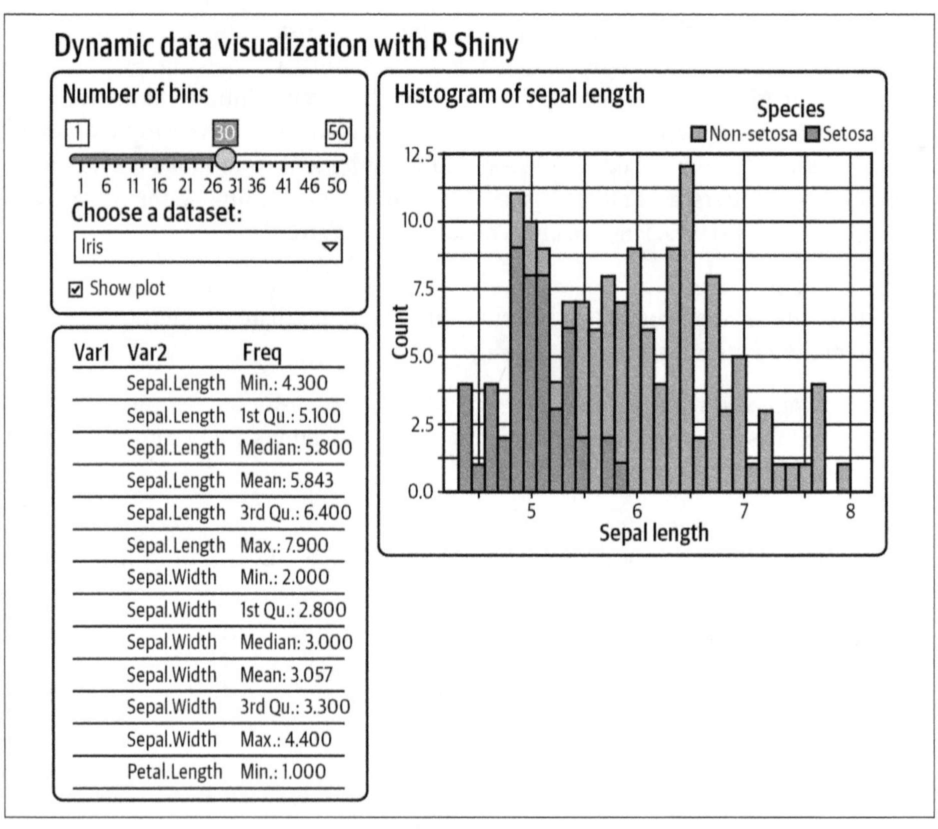

Figure 9-1. Example of a Shiny application

Interactive web applications

R Shiny enables users to create fully interactive web applications directly from R. This means you can transform analyses that would typically be static into engaging, dynamic tools that users can interact with in real time. Examples of Shiny's drag-and-drop UI components—like sliders, buttons, and text inputs—means that no prior knowledge of web development languages like HTML, CSS, or JavaScript is required to get started. For analysts, this reduces the barrier to creating polished, professional-looking web applications that are both functional and accessible.

Reactivity

One of the core strengths of R Shiny is its built-in reactivity system. This feature automatically updates the output in your app whenever user inputs change, without the need for reloading or navigating to a different page. This means your app can respond to user interactions immediately. For example, changing a parameter in a drop-down menu can instantaneously update a graph or a statistical analysis on the same page, facilitating a truly interactive user experience that is ideal for exploring and presenting complex datasets.

Reactivity in Shiny works through on a dependency tracking mechanism that automatically updates outputs when inputs or other reactive expressions change. This is accomplished using Shiny's reactive graph system, which tracks relationships between reactive inputs, expressions, and outputs.

Under the hood, when a user interacts with the app by adjusting inputs (like sliders or checkboxes), Shiny reevaluates any reactive expressions that depend on those inputs. These reactive expressions (created with functions like `reactive()`, `observe()`, or `reactiveVal()`) respond to changes by recalculating and updating the relevant outputs. Shiny distinguishes between:

- Reactive inputs like `input$slider`, which trigger updates
- Reactive expressions that perform calculations when dependencies change
- Reactive outputs such as `renderPlot()` or `renderTable()`, which are updated in response to changes in reactive expressions

For more advanced users interested in how Shiny's reactivity system functions, including details about dependency graphs, lazy evaluation, and observers, the official RStudio documentation provides an excellent resource: Posit Shiny Reactivity (*https://oreil.ly/UjfdF*). Additionally, the book *Mastering Shiny* by Hadley Wickham contains in-depth explanations of Shiny's reactivity system.

Accessibility

Shiny apps are highly accessible. They can be easily hosted and shared online as standalone applications, or embedded within R Markdown documents or dashboards. This flexibility allows Shiny developers to disseminate their work widely, reaching an audience beyond those familiar with R. Whether you need to integrate a Shiny app into a larger report, showcase it during presentations, or share it with decision makers, the platform supports various formats that help make your insights more accessible.

Extensibility

While Shiny does not require web development skills for basic usage, it supports extensive customization through integration with standard web technologies. Advanced users can enhance their applications by incorporating HTML, CSS, and JavaScript, allowing for custom UIs and interactions that go beyond the default Shiny framework. This extensibility makes Shiny a versatile tool that can cater to sophisticated web application needs, aligning both functionality and aesthetic aspects as required.

Community and support

R Shiny is supported by a robust community of developers and users who contribute to an extensive ecosystem of resources, tutorials, and example projects. Whether you are facing a technical challenge, looking for inspiration, or trying to learn new techniques, the community platforms, including forums, blogs (*https://shiny.posit.co/blog*), and user groups, offer invaluable support. The continuous development of the package supported by RStudio ensures that new features and improvements are regularly available, keeping the technology up-to-date with the latest trends in web development and data science.

These features collectively make R Shiny a powerful tool for anyone looking to leverage the power of R in a web-based interactive format, expanding the reach and impact of their data-driven projects.

Setting Up Your Environment

Once R and RStudio are installed (this was covered in Chapter 1), the next step is to install the Shiny package and any required dependencies. This can be done directly within RStudio.

Open RStudio and then open the console. Type the following command to install the Shiny package (see Figure 9-2):

```
install.packages("shiny")
```

This command will download and install Shiny along with any dependencies it requires.

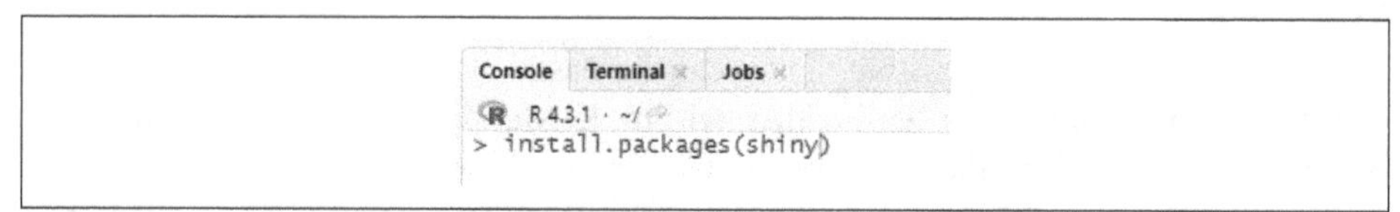

Figure 9-2. R console showing Shiny installation

By following these steps, you will have a fully equipped environment to start building your own interactive web applications using R Shiny. The simplicity of getting started is one of the many advantages that make R Shiny an invaluable tool for data scientists looking to bring their data analysis to life.

Building Your First Shiny App

Creating your first Shiny app is a journey into interactive web application development using R. This section guides you through the essential components and steps to create a simple yet functional Shiny app.

Structure of a Shiny app: UI and server components

A Shiny app is fundamentally structured into two key components:

UI
: This is where you define the layout and appearance of your app. The UI is created using R code that specifies the various elements like plots, tables, and input controls (like sliders or drop-down menus).

Server
: The server component contains the instructions that your computer needs to build and manage the app's reactive components. It is a script that takes the inputs from the UI, processes them, and returns the results back to the UI.

Both components are usually defined in a single R script called `app.R`, which is placed in a directory that will represent your Shiny app.

A simple example app explained

Here's a basic Shiny app that creates a histogram:

```
library(shiny) ❶

# Define the UI
ui <- fluidPage( ❷
  titlePanel("My First Shiny App"), ❸
  sidebarLayout( ❹
    sidebarPanel( ❺
      sliderInput("bins", ❻
                  "Number of bins:",
                  min = 1,
                  max = 50,
                  value = 30)
    ),
    mainPanel( ❼
      plotOutput("distPlot")
    )
  )
)

# Define server logic to draw a histogram
server <- function(input, output) {
  output$distPlot <- renderPlot({ ❽
    x    <- rnorm(500)
    bins <- seq(min(x), max(x), length.out = input$bins + 1) ❾
    hist(x, breaks = bins, col = '#75AADB', border = 'white') ❿
  })
}

# Run the application
shinyApp(ui = ui, server = server) ⓫
```

❶ This command loads the Shiny package into R, enabling you to use its functions to build interactive web applications.

❷ `fluidPage()` creates a page with a fluid layout that adjusts to the browser's width.

❸ `titlePanel("My First Shiny App")` adds a title to the application.

❹ `sidebarLayout()` organizes the layout into a sidebar for inputs and a main panel for outputs.

❺ `sidebarPanel()` contains UI elements that allow user interaction—in this case, a slider.

❻ `sliderInput()` creates a slider for input labeled `Number of bins`. It allows users to choose a value between 1 and 50, starting at 30. The input value from this slider is accessible via the input ID bins.

❼ `mainPanel()` displays outputs, here expecting a plot with the output ID `distPlot`.

❽ The `server` function takes input and output as its arguments. `output$distPlot` uses `renderPlot()`, a reactive expression that automatically updates the plot when the input value (`input$bins`) changes. Inside `renderPlot()`, `rnorm(500)` generates 500 random values from a standard normal distribution.

❾ `bins` calculates the sequence of bin edges for the histogram, adjusted to include the number of bins specified by the user. The `length.out` argument ensures the right number of breaks based on the slider's value.

❿ `hist()` generates a histogram of the data `x`, with the color (col) of the bars set to a blue shade and the borders set to white.

⓫ The function call `shinyApp(ui = ui, server = server)` starts the Shiny app, linking the defined UI and server logic so that the application runs as specified.

This R Shiny script is a simple example of an interactive web application that creates a histogram with a user-defined number of bins.

The UI of the Shiny app is shown in Figure 9-3. Again, the application allows a user to manipulate the number of bins for the histogram with slider functionality on the left, and the histogram adjusts to the number bins chosen by the user interactively.

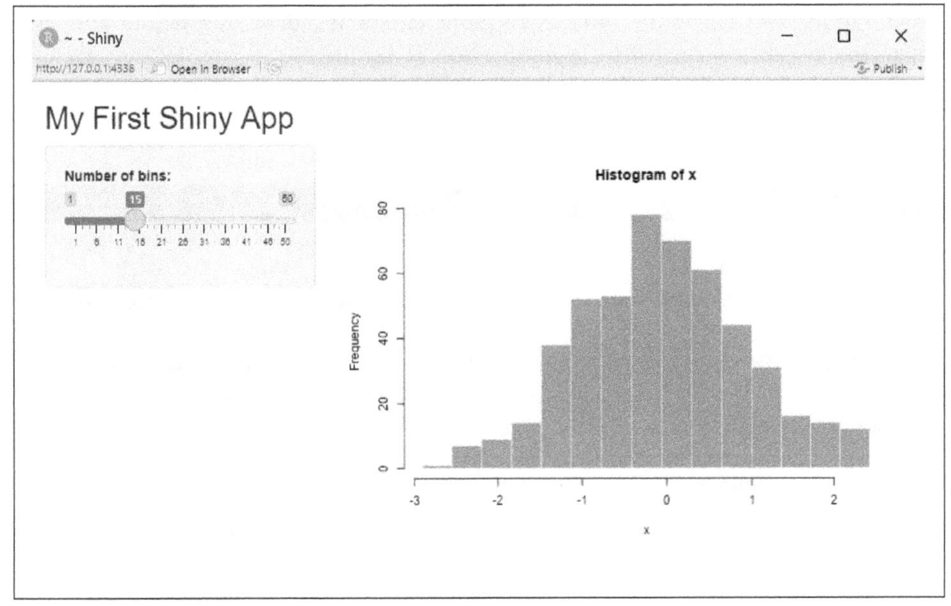

Figure 9-3. UI example for Shiny application

Interactive graphics in Shiny

Shiny's reactive expressions are one of its most powerful features, enabling the automatic updating of outputs when user inputs change. This reactivity is what makes the app truly interactive. Here's an example:

```
# Server logic
server <- function(input, output) {
  output$distPlot <- renderPlot({
    # Access the value of the input$bins, which is reactive
    bins <- seq(min(x), max(x), length.out = input$bins + 1)
    hist(x, breaks = bins)
  })
}
```

This example of server code will produce the same results as the previous example. Here, the code is highlighting the reactive nature of the Shiny app.

Using Plotly and ggplot2 for dynamic plots

Plotly and ggplot2 are two powerful tools for data visualization that cater to different needs. ggplot2, part of the R ecosystem, is widely used for creating static, high-quality plots with a grammar-of-graphics approach, making it ideal for exploratory data analysis and presentation. Plotly, on the other hand, extends the functionality of ggplot2 by enabling dynamic, interactive plots that can be used in dashboards and web applications. By combining these tools, users can create both detailed static visualizations and engaging, interactive plots, enhancing data storytelling and user engagement. Table 9-1 provides a comparison of features across ggplot2, Plotly, and Shiny.

Table 9-1. Comparison of ggplot2, Plotly, and Shiny

Feature/Aspect	ggplot2	Plotly	Shiny
Primary use	Static and aesthetically pleasing plots	Interactive and dynamic plots	Building interactive web applications with R
Interactivity	Limited (mainly static)	Highly interactive (e.g., zoom, hover, click)	High interactivity through reactive elements
Customization	Extensive customization for static plots	Good customization with interactive controls	Customization possible through reactive UIs
Learning curve	Moderate, requires understanding of grammar of graphics	Moderate, simpler for interactive plotting but can become complex for advanced features	Moderate to high, especially for advanced UI and reactivity
Typical output	Static images (e.g., PNG, PDF)	Interactive web-based visualizations (e.g., HTML)	Interactive web apps deployed on browsers
Integration with web	Limited (requires external tools like Shiny or R Markdown)	Directly integrates with web and dashboards	Fully web-based, integrates seamlessly into websites and dashboards
Ease of use	Straightforward for static plots	Relatively easy to use for interactivity	More complex due to the app-building framework
Supported file formats	Exports to image formats (PNG, PDF, etc.)	Interactive web formats (HTML, JavaScript)	Full web applications; not primarily for static outputs
3D plot support	No	Yes	No (depends on integration with Plotly or other libraries)
Animations	Limited, basic	Excellent support for animations	Limited, but possible through custom interactions
Best for	Publication-quality static plots	Interactive dashboards and web visualizations	Full-fledged, dynamic web applications with data interactivity
Community support	Large and active community with extensive documentation	Strong, with lots of online resources and examples	Large community, with active support from RStudio and forums
Requires knowledge of	R	R, Python, or JavaScript (depending on use case)	R and basic web development concepts (HTML, CSS, JS for advanced users)
Deployment	Output to documents (e.g., R Markdown)	HTML and web-based platforms	Web deployment using Shiny Server, shinyapps.io, etc.
Dependency	Part of the tidyverse	Standalone or can integrate with ggplot2	Dependent on R, integrates with ggplot2, Plotly, and others for visuals

ggplot2

ggplot2 is a data visualization package for the R programming language, based on the principles of the Grammar of Graphics. The Grammar of Graphics is a framework or theoretical model for describing and building a wide range of statistical graphics systematically. It was originally conceptualized by Leland Wilkinson in his book *The Grammar of Graphics*, published in 1999. This framework forms the foundation of several data visualization tools, most notably ggplot2 in R, which was directly inspired by Wilkinson's ideas.

It provides a coherent system for describing and building graphs, using a layered approach. With ggplot2, users define a graphic using a logical sequence where data is mapped to aesthetic attributes (like color, size, and shape) of geometric objects (like points, lines, and bars). ggplot2 handles many of the complexities of plotting, such as drawing legends and axes, automatically. Its versatility and expressive power make it one of the most popular tools for data visualization in R, ideal for creating complex multiplot layouts and customizing plots with themes and faceting.

Plotly

Plotly is a versatile graphing library that enables users to create interactive plots using a variety of programming languages, including R, Python, and JavaScript. In R, the Plotly package can be used to convert static ggplot2 figures into interactive visualizations or to build complex, dynamic plots directly. It enhances visualizations with features such as tooltips, zooming, panning, and clickable legends, which are particularly useful for web-based data exploration. Plotly plots are web-friendly and can be easily integrated into web applications, especially Shiny apps for R, making them accessible to a broad audience. This interactivity makes Plotly an invaluable tool for data analysis and presentation, particularly when engaging and real-time user feedback is essential.

To make the histogram more interactive using ggplot2 for plotting and Plotly for interactivity, let's look at another example. This application will include the UI setup to incorporate a slider for adjusting the number of bins dynamically, along with the server logic that uses ggplot2 and Plotly (see Figure 9-4):

```
library(shiny)
library(ggplot2)
library(plotly) ❶ # This line must be executed without error

# Define the UI
ui <- fluidPage( ❷
  titlePanel("Interactive Histogram with ggplot2 and plotly"), ❸
  sidebarLayout( ❹
    sidebarPanel( ❺
      sliderInput("bins", ❻
                  "Number of bins:",
```

```
                  min = 1,
                  max = 50,
                  value = 30)
    ),
    mainPanel( ❼
      plotlyOutput("distPlot")  # Ensure this matches the output type
    )
  )
)

# Define server logic
server <- function(input, output) { ❽
  output$distPlot <- renderPlot({ ❾
    # Generate a data frame of random normal data
    data <- data.frame(x = rnorm(500)) ❿

    # Create a ggplot histogram with dynamic bins
    p <- ggplot(data, aes(x)) + ⓫
      geom_histogram(bins = input$bins) +
      labs(title = "Interactive Histogram",  ⓬
           subtitle = "500 random samples from a normal distribution",
           x = "Value",
           y = "Frequency") +
      theme_minimal()

    # Convert ggplot object to an interactive plotly object
    ggplotly(p)  # This function must be recognized ⓭
  })
}

# Run the application
shinyApp(ui = ui, server = server) ⓮
```

❶ Load the libraries. Shiny is essential for creating interactive web applications in R. ggplot2 is used for creating complex and aesthetically pleasing statistical graphics. Plotly converts static charts into interactive plots and integrates with Shiny for interactive web applications.

❷ `fluidPage()` creates a page with a fluid layout that adjusts to the size of the user's browser window.

❸ `titlePanel()` places a title at the top of the UI.

❹ `sidebarLayout()` organizes the layout into a sidebar and a main panel.

❺ `sidebarPanel()` contains interactive elements. Here, it has a `sliderInput()`, which lets users control the number of bins in the histogram.

❻ `sliderInput()` creates a slider for the user to select a value. It's labeled "Number of bins" and allows values between 1 and 50, defaulting to 30.

❼ `mainPanel()` displays outputs, here specifying `plotlyOutput()` to show the interactive plot created by Plotly.

❽ `server` function defines how the app should respond when users interact with UI elements (like the slider).

❾ `renderPlot()` is a reactive function that automatically updates the output when inputs change.

❿ `data.frame(x = rnorm(500))` generates a dataframe with 500 normally distributed random values.

⓫ `ggplot()` with `geom_histogram()` creates a histogram of the data `x` with a number of bins specified by the user via the slider (`input$bins`).

⓬ `labs()` and `theme_minimal()` add labels and apply a minimal theme to the plot for better aesthetics.

⓭ `ggplotly()` converts the ggplot object into an interactive Plotly object, enhancing the plot with interactive features such as tooltips and zooming capabilities.

⓮ `shinyApp()` takes the UI and server definitions to run the Shiny application, linking the UI and server logic together to make the app operational.

The Shiny app depicted in Figure 9-4 creates an interactive web application that displays a histogram, which can be dynamically adjusted by the user through a slider input to change the number of bins. This is the same functionality as the prior Shiny app, except the visuals are depicted now in the Viewer pane in R using ggplot2 and Plotly.

This code provides an example of integrating ggplot2 and Plotly within a Shiny application to create interactive and dynamic web applications, which is very useful for statistical analysis and data visualization.

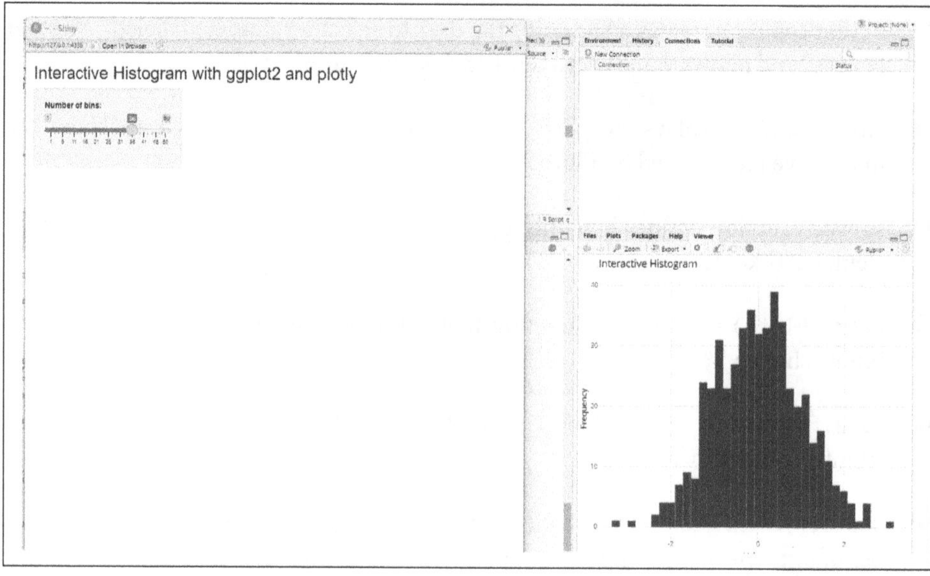

Figure 9-4. Shiny app with Plotly and ggplot2

Advanced UI Development

While many frameworks and libraries offer built-in styling components, customizing the appearance with HTML and CSS allows you to have full control over the look and feel of your applications. Advanced UI development involves going beyond basic layouts, using HTML to structure content and CSS to implement custom designs, animations, and responsive features. By mastering these techniques, you can create unique, branded interfaces that offer both functionality and aesthetic appeal, providing users with a seamless and engaging experience.

Customizing appearance with HTML and CSS

HTML and CSS are foundational technologies for building web pages and can be incorporated with Shiny to customize the appearance of the application.

HTML (HyperText markup language)

HTML is the standard markup language used to create and structure sections on the web. It forms the skeletal framework of all websites and is primarily responsible for the layout and structure of the web content. HTML uses tags (elements enclosed within angle brackets like `<div>`, `<p>`, `<header>`, etc.) to mark different parts of the content such as headings, paragraphs, links, and other items. Each tag serves a specific purpose and helps browsers understand how to display the content to users. For instance, `<p>` is used for paragraphs, `<a>` for hyperlinks, `<img>` for images, and so on.

Here's a short and simple example of HTML that includes a paragraph and a hyperlink for a reference:

```
<!DOCTYPE html>
<html>
<head>
    <title>Sample HTML Page</title>
</head>
<body>
    <h1>Welcome to My Web Page</h1>

    <p>This is a simple web page with an external reference. For more
    information on web development, you can visit
    <a href="https://www.w3schools.com"
    target="_blank">W3Schools</a>.</p>

</body>
</html>
```

The `<a>` tag is used to create a hyperlink, and the href attribute specifies the URL of the page the link goes to. The `target="_blank"` attribute makes the link open in a new tab. In this case, the hyperlink points to the W3Schools website (*https://w3schools.com*) as a reference for more information on web development.

CSS

CSS (*https://oreil.ly/GvwPX*) is used to control and enhance the visual appearance of web pages. It allows you to specify the style of your web content, including colors, fonts, and layout options, making your pages aesthetically pleasing. CSS works by selecting elements based on their HTML tags, classes, IDs, or more complex criteria, and applying various stylistic properties to them, such as color, margin, padding, and border. CSS is powerful for formatting a web page, without changing its structure. With CSS, you can also create responsive designs, which means your web pages will look good on all devices, from desktops to smartphones.

Relationship and usage

HTML and CSS go hand in hand: HTML structures the web content, while CSS styles it. Although they are separate technologies, they interact closely in web development. Typically, HTML files include references to CSS files that apply styles globally across multiple pages of a website. This separation of structure and style allows for more flexibility, easier maintenance, and faster page loading.

Integrating HTML and CSS can tweak the appearance of a Shiny app far beyond what is possible with standard Shiny functions. This integration allows you to precisely control the layout, style, and responsiveness of your application. By using HTML, you can directly manipulate the structure of the app's components, adding custom classes, IDs, and other HTML attributes that are not natively supported by Shiny's built-in functions. CSS can then be used to apply detailed styling rules to these elements. This includes altering colors, fonts, margins, paddings, and more, as well as implementing complex layouts and animations that enhance the UI.

Furthermore, CSS provides the tools to make your application visually consistent with broader corporate branding or to adhere to specific design guidelines. It also allows for responsive design practices, making sure your app looks good on various devices, from desktops to mobile phones. You can add CSS rules directly within the app's UI definition or link to external stylesheets for easier maintenance and reusability. These are the advantages of integrating HTML and CSS into a Shiny App:

Precise control over layout, style, and responsiveness
: Go beyond what standard Shiny functions offer by manipulating the structure and appearance of your application.

Direct manipulation of app structure using HTML
: Add custom classes, IDs, and other HTML attributes that are not natively supported by Shiny's built-in functions.

Detailed styling with CSS
: Alter colors, fonts, margins, paddings, and more. Implement complex layouts and animations to enhance the UI.

Consistency with branding and design guidelines
: Ensure that the application visually aligns with corporate branding or specific design standards.

Responsive design
: Make sure the app looks good on a variety of devices, from desktops to mobile phones.

Flexibility in adding CSS
: Apply CSS rules directly within the app's UI definition or link to external stylesheets for easier maintenance and reusability.

Here is a code example incorporating the use of HTML (Figure 9-5):

```
ui <- fluidPage(
  tags$head(tags$style(HTML("
    .myClass {
      font-weight: bold;
      color: #4CAF50;
    }
  "))),
  titlePanel("My Styled Shiny App"),
  div(class = "myClass", "This text will be styled")
)
```

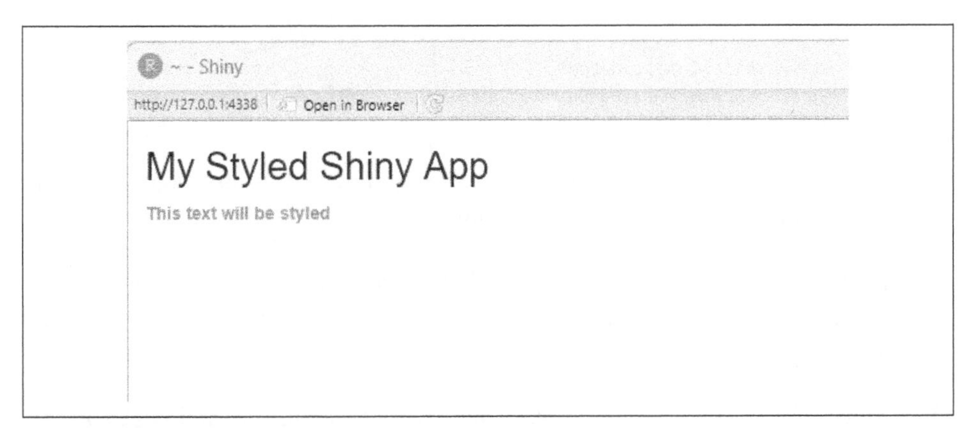

Figure 9-5. Example of Shiny UI with HTML

The following is an example of how you can extend your Shiny UI using CSS to style multiple elements. This example will incorporate more CSS styling options to demonstrate how CSS can modify the layout and appearance of various components within a Shiny application (Figure 9-6):

```
library(shiny)

ui <- fluidPage(
  tags$head(
    tags$style(HTML(" ❶
      .myClass {
        font-weight: bold;
        color: #4CAF50; /* Sets the color */
        font-size: 20px; /* Sets the font size */
        text-align: center; /* Centers the text */
      }
      .customPanel {
        background-color: #f0f0f0; /* Light grey background */
        border: 1px solid #ccc; /* Adds a border */
        padding: 15px; /* Adds space inside the borders */
        border-radius: 8px; /* Rounded corners */
        margin-top: 20px; /* Adds margin on the top */
      }
      .myHeader {
        color: #333; /* Sets the color */
        text-transform: uppercase; /* Capitalizes all letters */
      }
    "))
  ),
  titlePanel(div(class = "myHeader", "My Styled Shiny App")), ❷
  div(class = "myClass", "This text will be styled"),
  div(class = "customPanel", ❸
    "This is some content in a styled panel."
  )
)
```

```
server <- function(input, output) { }

shinyApp(ui = ui, server = server)
```

❶ The CSS rules are defined within a <style> tag inserted in the document's head section. This is done using `tags$head` and `tags$style`.

❷ Instead of passing an ID, we wrapped the title text inside a `div` with a class `myHeader` for CSS targeting. This allows for CSS customizations using the class selector `.myHeader`.

❸ Additional `div` elements are styled using the classes `myClass` and `customPanel`, which, respectively, apply specific styles to text and create a styled panel area.

This solution adheres to Shiny's syntax requirements and provides a way to customize the appearance of the title panel and other UI elements effectively using CSS.

Figure 9-6. Shiny UI using CSS

Together, HTML and CSS can enhance the UI portion of the Shiny app, and both enable business analysts to create structured pages that are also styled according to modern aesthetics and usability standards.

Using shinydashboard for creating dashboards

The shinydashboard package is an extension of Shiny that provides functions and themes to create dashboards in R. Dashboards can be interactive web applications (dashboards can be implemented in different ways) that present data and analytics in a visually appealing and concise format, making them ideal for business intelligence tasks, administrative interfaces, or any scenario where data needs to be monitored and accessed frequently.

These are the key features of shinydashboard:

Dashboard layouts (see Figure 9-7 for an example)

The *dashboard header* is typically used for placing the title of the dashboard, logo, and other global controls like logout buttons or links to different sections.

The *dashboard sidebar* is a vertical column on the left side that contains navigation elements—usually in the form of a menu with items and sub-items. These menus facilitate quick navigation across different parts of the dashboard.

The *dashboard body* is the main area where the content (like charts, tables, and text) is displayed. This section is highly customizable and can include multiple tabs and boxes.

Boxes and widgets

Boxes are primary UI components used within the dashboard body for organizing content. Boxes can be used to contain plots, tables, values, or even other UI elements. They can be styled with different colors, can be collapsible, closable, and even draggable depending on user needs.

Value boxes are a special type of box that highlights key metrics or summary data. They are typically colorful and designed to draw attention to important numbers or indicators.

Tab boxes and *info boxes* are for organizing content into tabs or providing brief snippets of information, respectively.

Interactivity and reactivity

shinydashboard uses the reactive programming model of Shiny, meaning that the dashboard can dynamically update and respond as users interact with it. For example, selecting a different time frame in a drop-down can automatically update the charts and tables within the dashboard.

Theming and customization

The package allows for extensive customization of the appearance of the dashboard through themes and CSS. Predefined themes can be used to change the look and feel of the dashboard with minimal effort, or custom CSS can be applied for more specific branding needs.

Applications

Business intelligence is ideal for creating BI tools that allow executives and managers to monitor key performance indicators (KPIs) and other important metrics at a glance.

Data monitoring is useful in scenarios where real-time data monitoring is necessary, such as network operations centers or financial market monitoring.

Administrative panels are often used for building backend administrative panels for managing users, content, or data within enterprises.

Here is an example of using shinydashboard. Additionally, Figure 9-7 depicts a dashboard UI that could be leveraged for applications such as business intelligence or data monitoring:

```
library(shinydashboard) ❶

ui <- dashboardPage( ❷
  dashboardHeader(title = "Simple Dashboard"), ❸
  dashboardSidebar( ❹
    sidebarMenu( ❺
      menuItem("Dashboard", tabName = "dashboard", icon = icon("dashboard")),
      menuItem("Reports", tabName = "reports", icon = icon("file")) ❻
    )
  ),
  dashboardBody( ❼
    tabItems( ❽
      tabItem(tabName = "dashboard", ❾
              h2("Dashboard content")),
      tabItem(tabName = "reports",
              h2("Reports content"))
    )
  )
)

server <- function(input, output) { } ❿

shinyApp(ui, server) ⓫
```

❶ This line loads the shinydashboard package into your R environment, which is necessary to use the functions specific to creating dashboards.

❷ `dashboardPage()` creates the overall page structure and is the container for all other dashboard elements.

❸ `dashboardHeader()` sets up the header of the dashboard. Here, it's configured with the title `"Simple Dashboard"`. The header is typically where you would also manage global settings like user profile links or notifications.

❹ `dashboardSidebar()` defines the sidebar of the dashboard. The sidebar usually contains navigation elements that allow users to switch between different parts of the dashboard.

❺ `sidebarMenu()` creates a menu inside the sidebar, where each item represents a clickable link that users can use to navigate the dashboard.

❻ `menuItem()` corresponds to a different section of the dashboard. Here, there are two items: `"Dashboard"` and `"Reports"`. Each `menuItem` is associated with a `tabName` and can include an icon, which helps users visually identify the sections.

❼ `dashboardBody()` is the main content area of the dashboard where you display the bulk of the interactive elements or results.

❽ `tabItems()` organizes content into tabs according to the navigation defined in the sidebar.

❾ `tabItem()` defines the content for each tab specified in the sidebar menu. Each `tabName` corresponds to a `tabName`. The content within the `tabName` can include any Shiny output or HTML content. Here, simple headers (`h2`) are used to denote placeholder content.

❿ In this basic example, the server function does not contain any operations. This function is where you would typically place the code to reactively update outputs, handle inputs, and perform server-side calculations based on user interactions. Because this example doesn't include interactive elements or data processing, it remains empty.

⓫ This function call initializes the Shiny app, linking the UI and server components. It starts the application, making it live for user interaction.

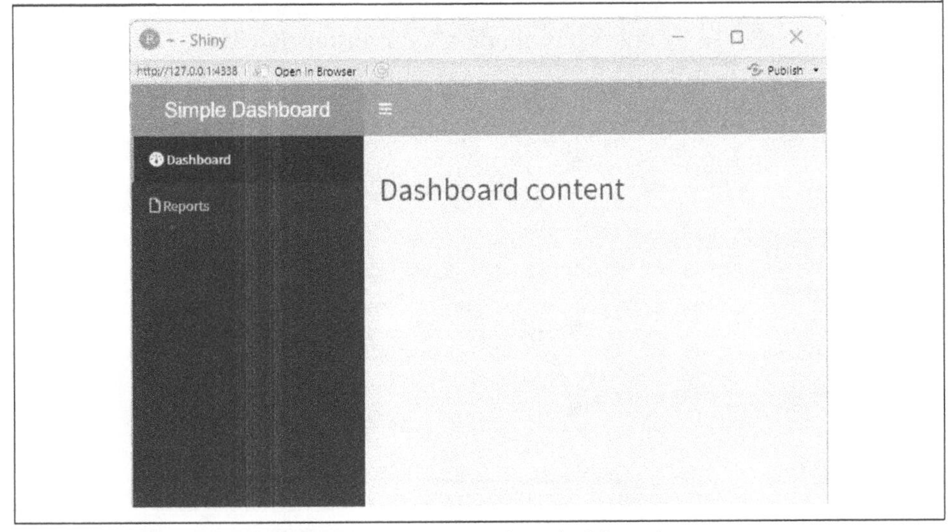

Figure 9-7. Example of a dashboard UI using shinydashboard

Shiny widgets and extensions

In interactive applications, especially those built with frameworks like Shiny, the use of widgets plays a crucial role in enhancing user engagement and functionality. Widgets such as sliders, buttons, and drop-downs are fundamental tools that allow users to interact with the application by inputting preferences, making selections, and initiating actions. These elements not only make the interface more interactive but also empower users to control and manipulate the data visualization or analysis output in real time. I'll provide an overview of these key widgets next, discussing their uses and benefits in creating dynamic and user-friendly web applications.

Widgets are integral components of interactive Shiny applications. They allow users to interact with the app and control its behavior. Here are a few examples:

- Sliders (`sliderInput`) enable users to select a range of values by dragging a bar within a defined interval. They are ideal for adjusting parameters for data visualization, such as setting the number of bins in a histogram.
- Buttons (`actionButton`) can trigger specific actions when clicked, such as recalculating a model or refreshing a dataset. They are crucial for initiating reactive events.
- Drop-downs (`selectInput`) let users choose an item from a list, useful for selecting options or filtering data categories.

Figure 9-8 depicts a filter application that shows the use of sliders, buttons, and drop-downs. This application supports data analysis via filtering where two drop-down menus exist to choose "Number of Cylinders" or Transmission Type. The slider provides Horsepower Range, and Apply Filters is an action button to apply the filtering choices to the dataset.

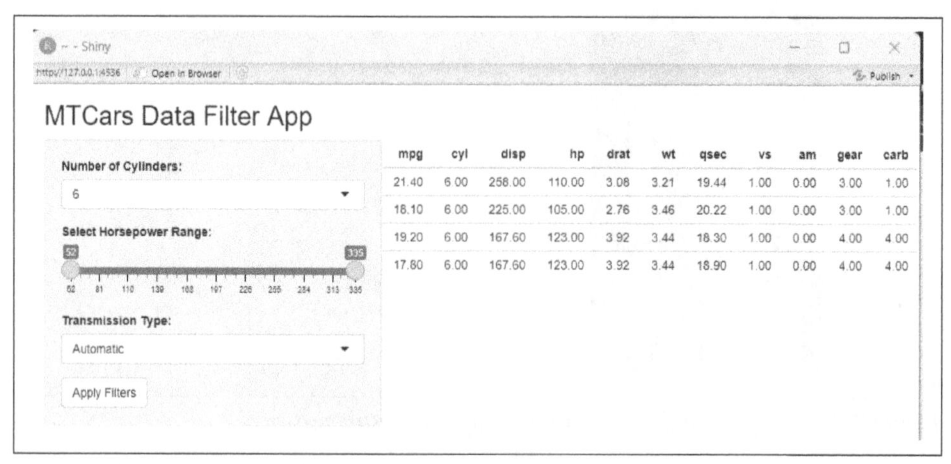

Figure 9-8. Shiny example that uses sliders, buttons, and drop-downs

To further enhance the functionality and user experience of Shiny applications, developers can leverage packages like shinyjs and shinyWidgets.

- The shinyjs package allows the integration of JavaScript functions directly into Shiny applications without requiring users to write any JavaScript code themselves. By utilizing shinyjs, developers can manipulate elements of the UI, hide or show components dynamically, and even run custom JavaScript code, enabling more interactive and responsive applications.
- Offering a range of custom widgets and advanced UI elements, shinyWidgets goes beyond the default Shiny components. It enables developers to create polished and user-friendly applications by incorporating enhanced inputs such as sliders, switches, and pickers. This package allows for more tailored, feature-rich Shiny apps that meet complex user needs with ease.

These tools extend the core capabilities of Shiny, making it easier to build highly interactive and visually appealing applications.

This example shows how to leverage a widget of an action button that provides an alert of `'Hello, this is JavaScript!'` and a drop-down menu with color choices. Figure 9-9 shows the alert after the `"Click Me"` action button is clicked, and Figure 9-10 depicts the color drop-down menu:

```
library(shiny)
library(shinyjs) ❶
library(shinyWidgets) ❷

# Define the UI
ui <- fluidPage( ❸
  useShinyjs(),  # Initialize shinyjs ❹
  actionButton("button", "Click Me"), ❺
  # Proper use of extendShinyjs to include custom JS
  extendShinyjs(text = "shinyjs.myFunction = function() { alert('Hello, this is
  JavaScript!'); }", functions = c("myFunction")), ❻
  pickerInput(inputId = "picker", label = "Pick a color", choices = c('Red',
  'Blue', 'Green'), multiple = FALSE) ❼
)

# Define server logic
server <- function(input, output, session) { ❽
  observeEvent(input$button, { ❾
    # Execute the custom JavaScript function
    js$myFunction() ❿
  })
}

# Run the application
shinyApp(ui = ui, server = server) ⓫
```

❶ shinyjs extends Shiny's capabilities by enabling the use of JavaScript for enhanced interactivity within the Shiny application. It allows you to call JavaScript functions as if they were R functions.

❷ shinyWidgets provides additional custom widgets for Shiny applications, expanding beyond the standard set of UI elements available in Shiny.

❸ `fluidPage()` creates a fluid layout, which adjusts the width of the elements based on the browser window size, making the application responsive.

❹ `useShinyjs()` initializes shinyjs within the application, enabling the use of its functionalities.

❺ `actionButton()` creates a button that users can click. This button is linked to a JavaScript function through server-side reactivity.

❻ `extendShinyjs()` extends the basic functionality of shinyjs by adding custom JavaScript. Here, a new function `myFunction` is defined, which shows a browser alert saying `'Hello, this is JavaScript!'` when called.

❼ `pickerInput()` is a function from shinyWidgets that creates a drop-down menu for picking a color. The choices are defined as `'Red'`, `'Blue'`, and `'Green'`, and `multiple = FALSE` restricts the user to select only one color at a time.

❽ `server()` defines the server-side logic of the Shiny app, where the reactivity is managed.

❾ `observeEvent()` monitors the `"button"` for clicks. When the button is clicked, the `js$myFunction()` is triggered.

❿ `js$myFunction()` calls the custom JavaScript function defined earlier. Since shinyjs treats JavaScript functions as if they were R functions, this line effectively triggers the alert in the user's browser when the button is clicked.

⓫ `shinyApp()` combines the UI and server components to run the Shiny app.

The provided code is a complete example of a Shiny application that integrates both R and JavaScript functionalities using the shiny, shinyjs, and shinyWidgets libraries. This application demonstrates interactive web application capabilities, including a custom JavaScript alert and a color picker.

This code snippet effectively showcases how to integrate JavaScript for added interactivity in a Shiny application, using custom UI elements from shinyWidgets to enhance user interaction. Let's look at a scenario where a Shiny app could be leveraged.

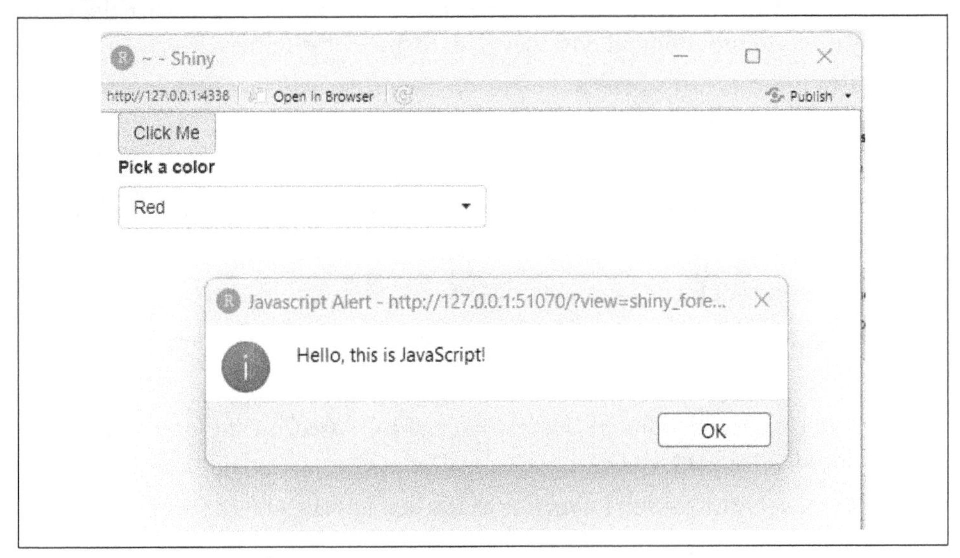

Figure 9-9. JavaScript alert

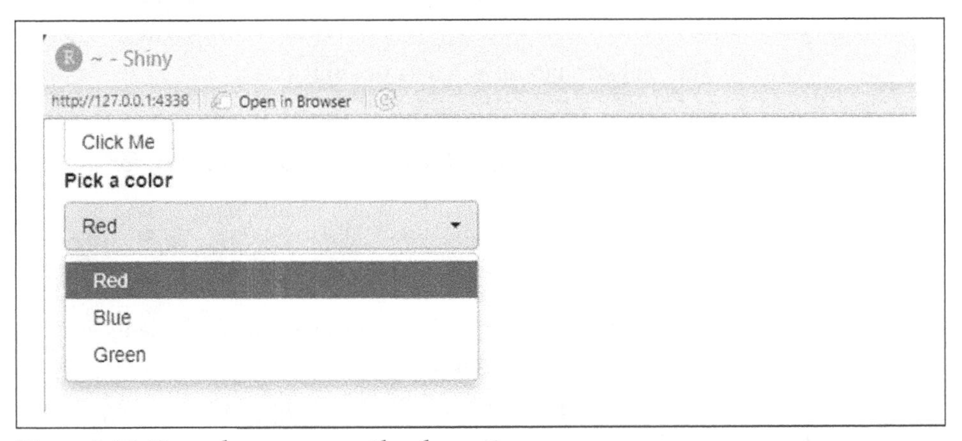

Figure 9-10. Drop-down menu with color options

Example: Creating a Dashboard to Monitor Real-Time Sales

A company needs a dashboard to monitor real-time sales data across different regions and products. The data is stored in various databases and requires interactive visualization to track sales performance and identify trends quickly. Let's look at the steps that you would take and then an example of a dashboard that could be used for this scenario.

Here is a step-by-step guide to building a solution using Shiny:

1. Understand the requirements: gather detailed specifications on what the company wants to monitor and how they wish to interact with the data.
2. Prepare the data: aggregate and clean the data from various databases, ensuring it's in a format suitable for visualization.
3. Build the UI:
   - Use `dashboardPage`, `dashboardHeader`, `dashboardSidebar`, and `dashboardBody` to create the basic structure.
   - Include `selectInput` for users to select regions and products.
   - Implement `dateRangeInput` to allow users to choose time frames.
4. Create the server logic:
   - Write server functions to query the database based on the user's input and return the relevant data.
   - Use `renderPlot` to create dynamic charts that update with the data.
   - Add reactive elements:
     - Utilize reactive expressions to respond to user input and update visualizations accordingly.
     - Implement `actionButton` to refresh data or apply filters.
   - Integrate extensions:
     - Use shinyWidgets to add advanced widgets like `pickerInput` for a better UI.
     - Apply shinyjs for custom JavaScript actions without writing extensive code.
5. Complete your testing:
   - Conduct thorough testing with sample data to ensure that all elements work as intended and that performance is optimized.
   - Deploy the app on a Shiny server or a service like shinyapps.io, making sure it's secure and accessible to the intended users.
   - Create training materials and documentation. Provide training to the company's staff to use the dashboard effectively. Supply documentation for maintenance and updating the dashboard as required.

Here is the code to create the dashboard:

```
# Load required libraries ❶
library(shiny)
library(shinydashboard)
library(shinyWidgets)
```

```
library(shinyjs)
library(ggplot2)
library(dplyr)

# Mock data generation (replace this with actual database query if needed)
set.seed(123)
sales_data <- data.frame( ❷
  Region = rep(c("North", "South", "East", "West"), each = 50),
  Product = rep(c("Product A", "Product B", "Product C", "Product D"),
  times = 50),
  Date = seq(as.Date("2023-01-01"), by = "days", length.out = 200),
  Sales = round(runif(200, 500, 5000), 2)
)

# UI layout
ui <- dashboardPage(
  dashboardHeader(title = "Real-time Sales Dashboard"), ❸

  dashboardSidebar( ❹
    selectInput("region", "Select Region:",
                choices = unique(sales_data$Region),
                selected = unique(sales_data$Region), multiple = TRUE),
    selectInput("product", "Select Product:",
                choices = unique(sales_data$Product),
                selected = unique(sales_data$Product), multiple = TRUE),
    dateRangeInput("dateRange", "Select Date Range:",
                   start = min(sales_data$Date),
                   end = max(sales_data$Date))
  ),

  dashboardBody( ❺
    useShinyjs(),
    fluidRow(
      box(title = "Sales Performance", status = "primary", solidHeader = TRUE,
          plotOutput("sales_plot")),
      box(title = "Sales Summary", status = "warning", solidHeader = TRUE,
          tableOutput("sales_summary"))
    ),
    fluidRow(
      box(title = "Raw Sales Data", status = "info", solidHeader = TRUE,
          tableOutput("raw_data"))
    )
  )
)

# Server logic ❻
server <- function(input, output) {

  # Reactive expression to filter data based on user input
  filtered_data <- reactive({ ❼
    sales_data %>%
      filter(Region %in% input$region,
```

```
               Product %in% input$product,
               Date >= input$dateRange[1],
               Date <= input$dateRange[2])
  })

  # Plot sales data based on selected filters
  output$sales_plot <- renderPlot({ ❽
    data <- filtered_data()
    ggplot(data, aes(x = Date, y = Sales, color = Product)) +
      geom_line(size = 1) +
      labs(title = "Sales Over Time", x = "Date", y = "Sales") +
      theme_minimal()
  })

  # Display summary of sales data
  output$sales_summary <- renderTable({ ❾
    data <- filtered_data()
    summary_data <- data %>%
      group_by(Product) %>%
      summarise(TotalSales = sum(Sales),
                AvgSales = mean(Sales),
                MinSales = min(Sales),
                MaxSales = max(Sales))
    summary_data
  })

  # Display raw sales data for transparency
  output$raw_data <- renderTable({ ❿
    filtered_data()
  })
}

# Run the application
shinyApp(ui = ui, server = server) ⓫
```

❶ Load the necessary libraries: shiny (for creating interactive web applications), shinydashboard (for building a dashboard-style UI), shinyWidgets (for adding more UI elements), shinyjs (for providing JavaScript functionalities in Shiny), ggplot2 (for creating plots), and dplyr (for manipulating data).

❷ A mock sales dataset (`sales_data`) is generated using `data.frame` with random values for `Region` (North, South, East, and West), `Product` (four different products: A, B, C, and D), `Date` (a sequence of dates starting from January 1, 2023), and `Sales` (random sales values between 500 and 5,000). This simulates real-world sales data, but in practice, it could be replaced with actual data from a database query.

❸ `dashboardHeader` defines the dashboard's title ("Real-time Sales Dashboard").

❹ `dashboardSidebar` includes the following input controls for users to filter data: `selectInput`, which allows users to select one or more regions and products, and `dateRangeInput`, which allows users to filter data by selecting a date range.

❺ `dashboardBody` displays the main content and includes `fluidRow` (a layout function to arrange UI elements in rows) and `box` (which encases dashboard components sales performance plot, sales summary table, and raw sales data table).

❻ The server function contains the logic for generating the output based on user input.

❼ `filtered_data` is a reactive expression that filters the sales data based on the user's selected region, product, and date range.

❽ `output$sales_plot` renders a time-series plot using ggplot2. The plot displays sales data over time for each product with lines color-coded by product.

❾ `output$sales_summary` generates a summary table, using dplyr, that groups the data by product and calculates total sales, average sales, and minimum and maximum sales.

❿ `output$raw_data` displays the filtered raw data in a table format for transparency.

⓫ The `shinyApp(ui = ui, server = server)` function launches the Shiny app with the defined UI and server logic.

The key features of the app include dynamic filtering, allowing users to interactively filter the sales data by region, product, and date range. The app provides data visualization through a time-series plot that displays sales trends over time. Additionally, a summary table is available, offering key statistics such as total, average, minimum, and maximum sales by product. For further transparency, the app also includes a raw data table that shows the filtered sales data for users to review and inspect. Figure 9-11 shows what the application would look like.

By following these steps, a robust, user-friendly sales monitoring dashboard can be developed, leveraging Shiny's capabilities to meet real-world business needs.

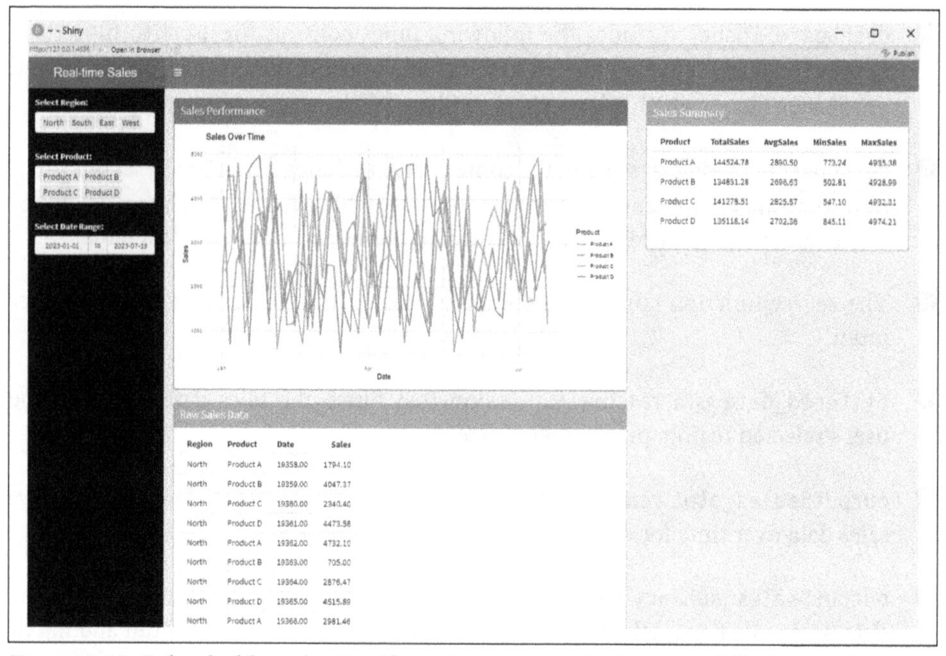

Figure 9-11. Sales dashboard using Shiny (you can see the image at full size in color online (https://oreil.ly/mban_0911png))

Learning Python Visualization

In the vast and versatile ecosystem of Python, data visualization capabilities are used frequently in analytics. The power of visualization in Python lies in its ability to convert data from abstract numbers into graphical representations that can communicate insights clearly and effectively. The ability to present data visually is indispensable for business analysts, and Python offers many options to leverage, each catering to different needs and levels of complexity.

For simple plots and basic charts, Matplotlib is the foundation of Python's visualization toolkit. It provides the flexibility to create static, publication-quality figures with a high level of customization. Analysts can generate bar charts, line graphs, scatter plots, and more with ease, making it ideal for day-to-day analysis where precision and simplicity are key.

Seaborn, built on top of Matplotlib, enhances Python's visualization capabilities by providing a higher-level interface for creating aesthetically pleasing and statistically informative graphics. It is particularly powerful for exploring relationships in datasets with its advanced functionality for visualizing distributions, correlations, and categorical data, such as box plots, violin plots, and heatmaps.

For more dynamic and interactive visualizations, Plotly is a popular choice. It enables users to build interactive dashboards and complex charts, such as 3D visualizations, maps, and even animated plots. Plotly is particularly useful in scenarios where users need to explore data in real time or present interactive results to nontechnical audiences. Additionally, Bokeh offers similar interactivity with the added benefit of being easy to integrate with web applications, making it ideal for live data monitoring or online reporting systems.

Finally, for creating fully interactive data applications, Dash, which is also built on top of Plotly, allows analysts to build dashboards and deploy them as standalone web applications. Dash abstracts away much of the underlying code and complexity involved in creating web apps, making it accessible for Python users who want to transform their visualizations into interactive tools without diving deeply into front-end development.

Overview of Visualization in Python

Python, known for its simplicity and readability, hosts a rich library ecosystem for creating a wide array of visualizations. From simple charts to complex interactive graphics, Python's visualization tools are capable of handling the diverse needs of users. These libraries abstract away the complexities involved in rendering graphics, allowing both novices and experts to create informative and appealing visual representations of data.

Common Libraries: Matplotlib, Seaborn, Plotly, and Dash

In the realm of data visualization, Python boasts a series of libraries designed to cater to various needs, ranging from simple plots to complex interactive visualizations. Among these, Matplotlib, Seaborn, Plotly, and Dash stand out as the most commonly used tools, each with its own unique features and capabilities. Table 9-2 outlines a comparison of these common libraries.

Table 9-2. Comparison of Python visualization libraries

Feature/criteria	Matplotlib	Seaborn	Plotly	Dash
Type of visualizations	Static plots: Line charts, bar plots, scatter plots, histograms, etc.	Static and statistical plots: Heatmaps, box plots, pair plots, violin plots, etc.	Interactive plots: 3D charts, maps, scatter plots, animated plots, etc.	Interactive dashboards and web apps with real-time data
Ease of use	Moderate: Requires more code for customization	High: High-level API, simpler syntax built on top of Matplotlib	Moderate: Intuitive for simple interactive plots, more complex for advanced customization	Moderate: Requires knowledge of Python, but high-level interface simplifies creating dashboards

Feature/criteria	Matplotlib	Seaborn	Plotly	Dash
Customization	Highly customizable, but requires detailed coding	Good default aesthetics, customizable but less flexible than Matplotlib	Highly customizable: Interactive elements like sliders, drop-downs, tooltips	Full control over layout and interactions, as it's a full web app framework
Customization	Simple, precise, and publication-quality plots	Quick exploratory data analysis with beautiful visual defaults	Interactive, dynamic plots with advanced features (3D, maps, etc.)	Building complete, interactive data apps or dashboards
Customization	Static images (no interactivity)	Primarily static, with some limited interactivity (e.g., hover effects with some extensions)	Full interactivity: Zoom, pan, hover, tooltips, animations, 3D rotation, etc.	Fully interactive applications with user inputs and dynamic updates
Customization	Moderate: Extensive options but requires learning detailed syntax	Low: Built on Matplotlib but with easier syntax and built-in styles	Moderate: Easy to get started with, more advanced features require some learning	Moderate: More complex than simple plotting but allows for powerful app-building without needing extensive web development experience
Customization	Works well with Pandas, NumPy, and other scientific libraries	Seamless integration with Pandas, NumPy, and statistical libraries	Integrates well with Pandas and NumPy, and works with web technologies like HTML/CSS	Fully integrates with Plotly for visualizations and supports Pandas, SQL, REST APIs, and more
Customization	PNG, PDF, SVG, etc. (static formats)	PNG, PDF, SVG (static formats)	Interactive HTML, JSON, Jupyter notebook (with inline visualizations)	Full web apps (deployed on a server), or embedded in other platforms
Customization	Simple to very complex (depends on coding skill)	Medium complexity: Great for statistical plots	Very complex: Can handle animations, 3D plots, maps, etc.	Very complex: Full web app functionality with dynamic dashboards
Customization	Good for static, noninteractive plots; can slow down with very large datasets	Good for exploratory analysis on medium-sized datasets	Handles large datasets, but interactivity can slow performance for very large datasets	Performance depends on the complexity of the app and backend infrastructure
Customization	Large, active community with extensive documentation and tutorials	Well documented, especially popular for statistical plots	Large community with strong documentation; used widely for interactive data visualization	Growing community; widely used for building Python-based dashboards and apps
Customization	Part of the Python standard libraries (requires installation)	Part of the Python ecosystem (installed with Matplotlib)	Requires installation (`pip install plotly`)	Requires installation (`pip install dash`)

Together, Matplotlib, Seaborn, Plotly, and Dash provide a valuable tool set for data visualization tasks in Python, catering to a wide spectrum of needs from static plotting to interactive graphics, ensuring that users have the right tools to visualize and interpret their data effectively.

Matplotlib: Foundations of Visualization in Python

Matplotlib is the workhorse of Python visualization libraries, providing a quick way to visualize data stored in Python data structures such as lists or arrays. Basic plotting functions in Matplotlib include:

- `plt.plot()` for creating line plots
- `plt.scatter()` for making scatter plots
- `plt.bar()` for making bar charts
- `plt.hist()` for making histograms

Each of these functions provides a quick and straightforward way to get a visual sense of data. Here is an example of a simple plot in Matplotlib:

```
import matplotlib.pyplot as plt

plt.plot([1, 2, 3, 4]) ❶
plt.ylabel('some numbers') ❷
plt.show() ❸
```

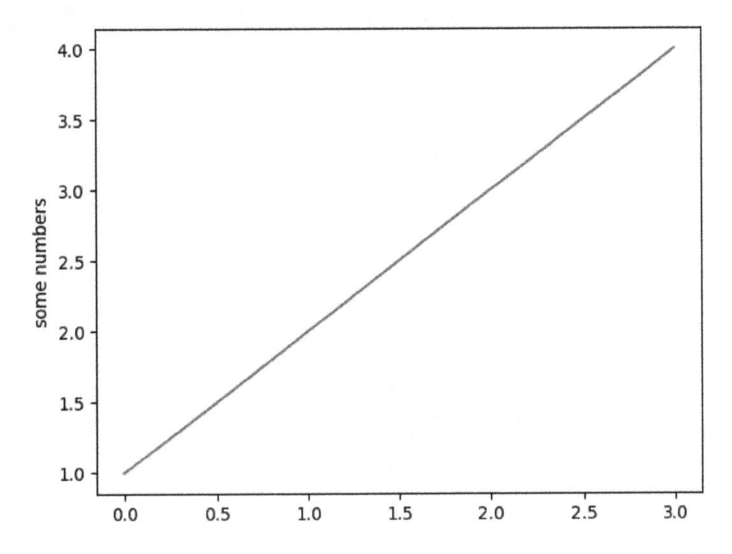

❶ This line is the core of the `plot. plt.plot()` function that generates a line plot. Here, it takes a single list of numbers [1, 2, 3, 4] as an argument. Since only one argument is provided, Matplotlib assumes these numbers to be the y-values of the points to plot. The x-values are automatically assumed to be their indices (starting from 0), making the x-values [0, 1, 2, 3] correspondingly. Therefore, the points plotted are (0,1), (1,2), (2,3), and (3,4).

❷ This line sets the label of the y-axis to `'some numbers'`. `plt.ylabel()`, which is a function that specifies what text to display as the label of the y-axis, which is useful for indicating what kind of data is represented by the y-values.

❸ This line is used to display the `plot`. `plt.show()`, which tells Matplotlib to render the plot and display it in a window. Until you call `plt.show()`, all the commands you give (like plotting data and labeling axes) just set up the plot but do not actually render it visually. This command is crucial for actually viewing the plot, especially when using Matplotlib outside of interactive environments like Jupyter notebooks, where plots may display automatically.

Overall, this simple script demonstrates the creation of a basic line plot with Matplotlib, including setting up axis labels and displaying the plot. It provides a quick way to visualize relationships between sequential data points.

Customizing Plots with Styles and Colors

Matplotlib allows for extensive customization to create a wide variety of plots tailored to the presenter's needs. You can control styles, attributes, and layouts of the plot, making your output publication quality. Customizations can be done through:

- Line styles and colors: `plt.plot(x, y, linestyle='--', color='green')`
- Figure size and DPI: `plt.figure(figsize=(8,6), dpi=80)`
- Setting limits, labels, and titles, and adding a legend

Here's an example that incorporates these customizations:

```
import numpy as np ❶

x = np.linspace(0, 10, 100) ❷

plt.figure(figsize=(8, 6), dpi=80) ❸
plt.plot(x, np.sin(x), color='green', linestyle='--', label='Sine Curve') ❹
plt.plot(x, np.cos(x), color='blue', label='Cosine Curve') ❺
plt.title('Sine and Cosine Curves') ❻
plt.xlabel('x')
plt.ylabel('f(x)') ❼
plt.legend() ❽
plt.show() ❾
```

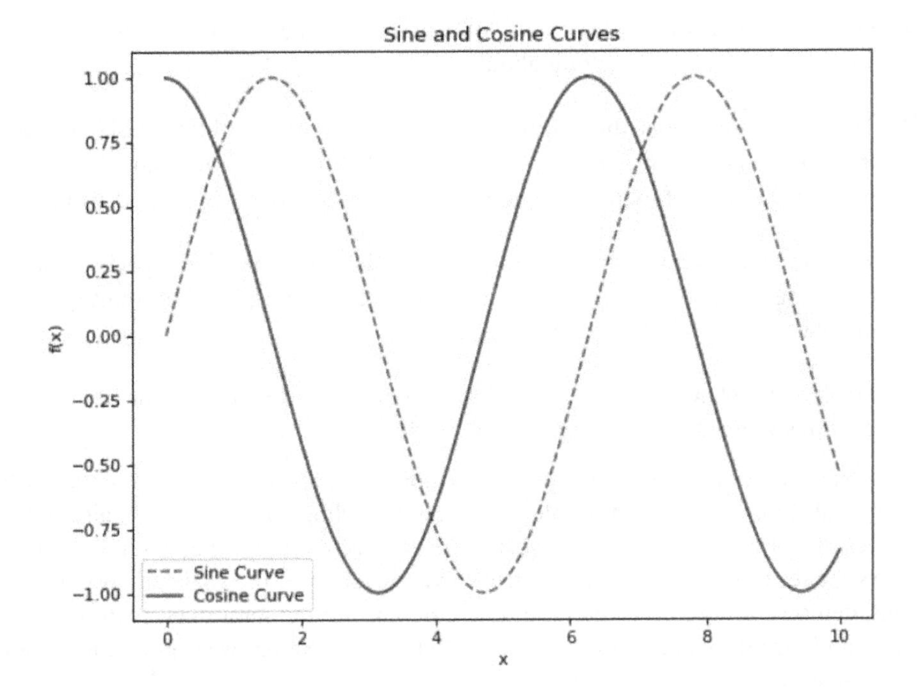

❶ This line imports the NumPy library, which is used for numerical computations in Python. `np` is a common alias used for NumPy, making it shorter to reference. For the visual, it is being used for computations for the curve.

❷ `np.linspace()` generates an array of evenly spaced numbers over a specified interval. Here, it creates an array of 100 points between 0 and 10, inclusive. This array `x` will be used as the domain for the sine and cosine functions plotted.

❸ `plt.figure()` initializes a new figure or plot. The `figsize` parameter sets the dimensions of the figure in inches (width, height), and `dpi` specifies the dots-per-inch (pixel density) of the figure. This line is setting up how large the plot will be and its resolution.

❹ This line plots the sine of `x` against `x`. The `np.sin(x)` function computes the sine of each point in `x`. The plot is styled with a green dashed line (`linestyle='--'`), and it's labeled as `'Sine Curve'` for the legend.

❺ Similarly, this line plots the cosine of `x` against `x`, using the `np.cos(x)` function. This line is styled with a solid blue line and labeled as `'Cosine Curve'` for the legend.

❻ `plt.title()` sets the title of the plot. This provides a heading at the top of the plot indicating what the plot is about.

❼ These functions label the x-axis and y-axis, respectively. `plt.xlabel('x')` sets the x-axis label to `'x'`, and `plt.ylabel('f(x)')` sets the y-axis label to `'f(x)'`, indicating that the function values of sine and cosine are being plotted against `x`.

❽ This command adds a legend to the plot, which uses the labels specified in the `plt.plot()` calls. A legend is essential for distinguishing between multiple lines or datasets shown in the same plot.

❾ Finally, `plt.show()` displays the plot. This function renders all the setup done previously and shows it in a graphical window.

The preceding code snippet demonstrates a more complex plot using Matplotlib, with additional features such as multiple lines, custom styling, and labeling. It also incorporates NumPy, a fundamental package for scientific computing with Python. It shows the features of Matplotlib to create a visualization of the sine and cosine functions, demonstrating how to plot multiple functions on the same axes, customize line styles, and include a legend and axis labels. In the next section, we explore more complex visuals using the Seaborn library.

Statistical Plots: Scatter Plots, Heatmaps, Violin Plots

Seaborn specializes in creating informative and attractive statistical graphics. It operates on top of Matplotlib and integrates closely with Pandas data structures, enhancing the traditional Matplotlib experience with its aesthetic graphics and extended functionality. Seaborn interacts seamlessly with Pandas and data structures, particularly dataframes, which allow for efficient storage and manipulation of tabular data. This integration means you can directly pass Pandas objects to Seaborn functions. Seaborn is ideal for quick, beautiful statistical visualizations with minimal code and is excellent for exploratory data analysis.

Seaborn shines with its variety of plotting functions that cater to different aspects of statistical data visualization, including:

- `sns.scatterplot`, which is ideal for observing relationships between two continuous variables.
- `sns.heatmap`, which is excellent for visualizing matrices of data and highlighting gradients.
- `sns.violinplot`, which combines aspects of boxplots with kernel density estimation (KDE), offering a deeper understanding of the distribution of the data.

The following are code examples for each of the specified Seaborn plots, along with explanations of how they are used and what they are best for. The first example focuses on observing relationships between two continuous variables:

```
import seaborn as sns
import matplotlib.pyplot as plt

# Sample data
tips = sns.load_dataset('tips')

# Create a scatter plot to observe the relationship between total bill and tip
sns.scatterplot(x='total_bill', y='tip', data=tips)
plt.title('Relationship Between Total Bill and Tip')
plt.show()
```

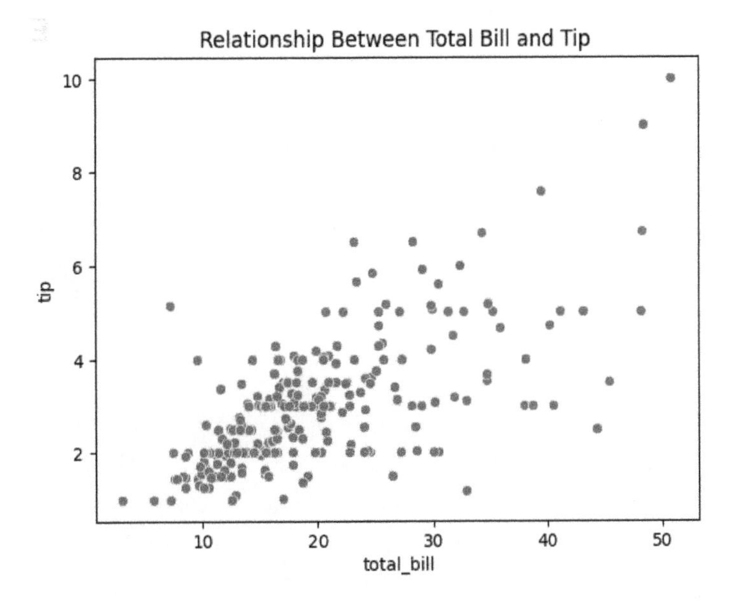

This code uses Seaborn's built-in dataset named `tips` to demonstrate a scatter plot. `sns.scatterplot()` is used to plot `'total_bill'` as the x-axis and `'tip'` as the y-axis. This kind of plot is ideal for visualizing the relationship between two continuous variables, allowing you to see patterns, trends, or correlations:

- `plt.title()` adds a title to the plot.
- `plt.show()` displays the plot.

The next example shows visualizing matrices of data and highlighting gradients. Visualizing matrices of data and highlighting gradients can help quickly identify patterns, trends, or anomalies in the data. For example, in a correlation matrix, a heatmap with color gradients can visually emphasize strong positive or negative correlations between variables, making it easier to spot relationships that might not be obvious from raw numbers alone. This approach is often used in exploratory data analysis to assess variable relationships at a glance:

```
import seaborn as sns
import matplotlib.pyplot as plt

# Load the dataset
flights = sns.load_dataset("flights")

# Pivot the dataset to prepare it for a heatmap
# Use keyword arguments to specify 'month' as the index, 'year' as the columns,
and 'passengers' as the values
flights_pivot = flights.pivot(index='month', columns='year',
values='passengers')

# Create a heatmap using the pivoted data
sns.heatmap(flights_pivot, annot=True, fmt="d", cmap='BuGn')
plt.title('Number of Passengers by Month and Year')
plt.show()
```

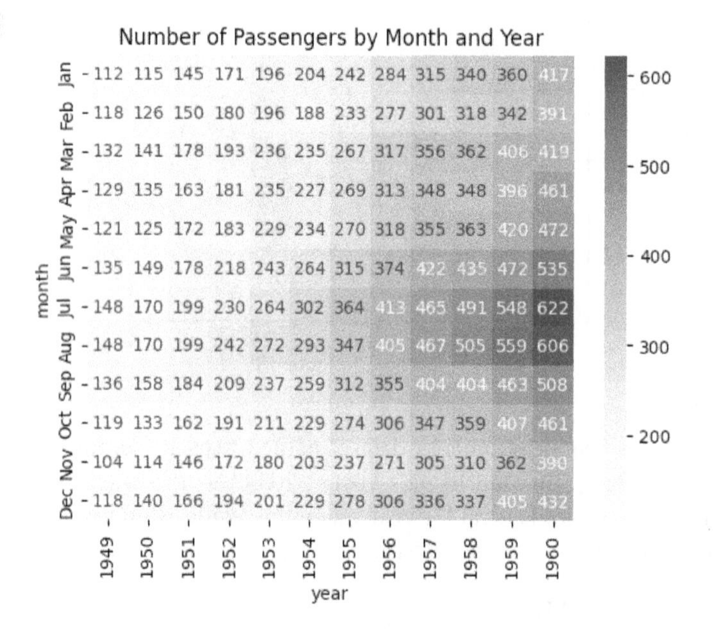

The flight dataset is pivoted to form a matrix where each cell value represents passengers for a particular month and year. `sns.heatmap()` is used to create a heatmap of the pivoted data. `annot=True` adds annotations to each cell showing the passenger numbers, `fmt='d'` formats the numeric values as integers, and `cmap='BuGn'` specifies the color palette. Heatmaps are excellent for displaying matrix-like data with varying shades representing the magnitude of the data values, making it easy to spot patterns and gradients.

Combining boxplots and KDE, like our last example, provides a more comprehensive view of a data distribution. A boxplot shows key statistics such as the median, quartiles, and potential outliers, while the KDE smooths the distribution curve to visualize the overall shape of the data. This combination helps in understanding both the central tendency and the spread, as well as the nuances of the data's distribution (e.g., multimodal distributions), offering deeper insights than either method alone:

```
import seaborn as sns
import matplotlib.pyplot as plt

# Sample data
tips = sns.load_dataset('tips')

# Create a violin plot to visualize the distribution of total bills by day
sns.violinplot(x='day', y='total_bill', data=tips)
plt.title('Distribution of Total Bills by Day')
plt.show()
```

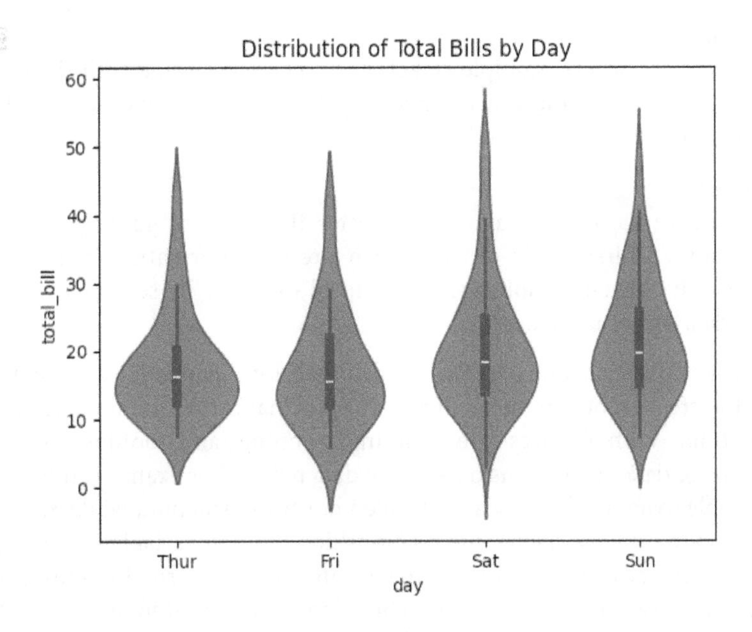

This example uses the tips dataset to create a violin plot. `sns.violinplot()` is used here to show the distribution of `'total_bill'` across different days of the week. The x-axis represents days, and the y-axis represents total bill amounts. Violin plots are useful for comparing the distribution of a variable across different categories. They combine boxplots and KDE to show the distribution's density and range. The wider sections of the violin plot represent a higher density of data points, which means more common values.

These examples illustrate different ways to visualize data using Seaborn, each tailored to highlight specific aspects of the data, whether it's relationships, distributions, or patterns.

Interactive Plots with Plotly

While static images like those created in Matplotlib and ggplot2 are often sufficient for publication or reports, there are situations where they fall short. Interactive visualizations become essential for:

Exploration and drill-downs
: Users need to explore data at different levels, such as zooming in on specific data points, filtering data by category, or viewing subsets interactively.

Real-time data
: When the data being presented is dynamic and changes frequently, such as in dashboards for monitoring business KPIs or real-time financial data.

Collaboration
: In scenarios where multiple stakeholders need to engage with the data, interactive tools offer a more accessible way to present complex data without overwhelming nontechnical audiences.

Storytelling
: Interactive dashboards allow for greater flexibility in guiding users through a narrative. Unlike static charts, which present information in a fixed manner, interactive visuals enable users to control the story by selecting what aspects of the data they want to explore.

Plotly extends visualization to the third dimension: interactivity. Its Python library allows the creation of complex, interactive plots that can be manipulated by end users in real time. With features like zooming, panning, and tooltips, users can easily explore large datasets or focus on specific data points. For example, interactive scatter plots enable dynamic filtering and detailed hover information, while 3D surface plots allow users to rotate and examine relationships between variables in multiple dimensions. Additionally, interactive heatmaps can reveal patterns by letting users zoom into specific areas of the data matrix. These interactive capabilities make Plotly particularly useful for dashboards, data exploration, and presentation, providing an immersive experience that static images cannot offer.

To use Plotly, install the library using pip:

```
pip install plotly
```

Then, import it into your Python script:

```
import plotly.express as px
```

Plotly's syntax is user-friendly and designed to work with a variety of data types. Here's a simple example to create an interactive scatter plot:

```
import plotly.express as px

# Load iris dataset
df = px.data.iris()

# Create an interactive scatter plot with hover data and size scaling
fig = px.scatter(
    df,
    x="sepal_width",
    y="sepal_length",
    color="species",
    size="petal_length",  # Add size based on petal length
    hover_data=["petal_width"],  # Add additional hover information
    title="Interactive Iris Data Scatter Plot"
)

# Show the interactive plot
fig.show()
```

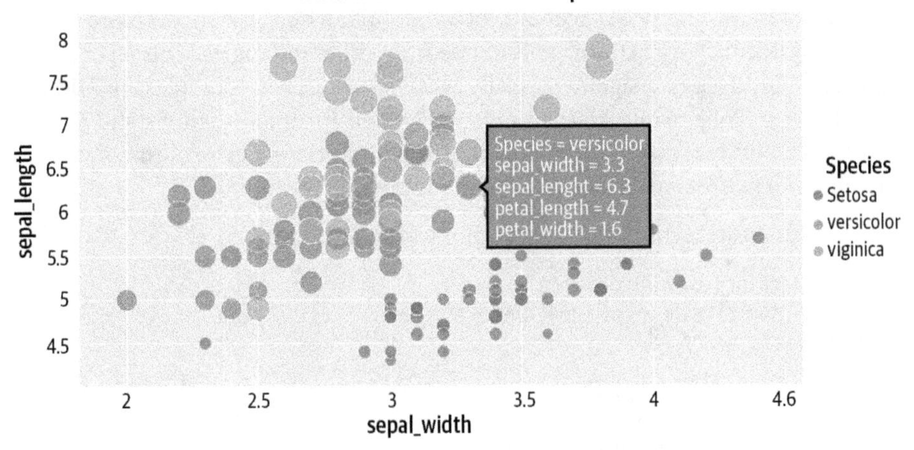

The scatter plot includes the dynamic adjustment of point size based on petal length, using the `size="petal_length"` argument, which makes the visualization more engaging by visually representing an additional dimension of the data. Additionally, the `hover_data=["petal_width"]` argument provides extra information, such as petal width, when hovering over the points, enriching the interactivity. Furthermore, the plot comes with built-in interactive features like zooming, panning, and tooltips, allowing users to explore the data dynamically and gain deeper insights with ease.

3D Plotting with Matplotlib and Plotly

Creating 3D plots can be both informative and visually appealing, particularly when dealing with multidimensional data. For instance, when analyzing a dataset that tracks sales volume, price, and time for multiple products, a 3D plot can reveal relationships between these variables that would be harder to detect in 2D. In Python, both Matplotlib and Plotly offer solutions for visualizing such data, though their capabilities differ significantly.

Matplotlib provides a toolkit called mplot3d for creating 3D plots. It is highly flexible and can generate a wide variety of 3D visualizations, such as wireframes, surface plots, and scatter plots. However, the plots are static, meaning they don't allow for interactive features like zooming, rotating, or dynamically hovering over data points. Once rendered, Matplotlib 3D plots are essentially snapshots of the data. This makes Matplotlib more suitable for static visualizations intended for reports or presentations where interactivity isn't required.

In contrast, Plotly excels at creating interactive 3D plots. With Plotly, users can manipulate the plot directly—rotating it, zooming in and out, or hovering over data points to see additional information. For instance, with the same sales volume, price, and time dataset, a Plotly 3D scatter plot would allow end users to explore the data dynamically, identifying patterns and trends from different angles. This level of interaction provides deeper insight, especially when working with complex, multidimensional datasets. Thus, while Matplotlib is well suited for high-quality static 3D plots, Plotly's interactivity makes it the better choice for real-time data exploration and presentations where users need to engage with the data directly.

Here's how to create a simple 3D scatter plot with Matplotlib:

```
import matplotlib.pyplot as plt
from mpl_toolkits.mplot3d import Axes3D
import numpy as np ❶

# Generating random data ❷
x = np.random.standard_normal(100)
y = np.random.standard_normal(100)
z = np.random.standard_normal(100)

fig = plt.figure()
ax = fig.add_subplot(111, projection='3d') ❸

# Creating the plot
# 'cmap='viridis'' specifies the colormap to use for the color mapping
scatter = ax.scatter(x, y, z, c=z, cmap='viridis') ❹
```

```
# Adding labels and title
ax.set_xlabel('X Coordinate')
ax.set_ylabel('Y Coordinate')
ax.set_zlabel('Z Coordinate')
ax.set_title('3D Scatter Plot Example with Matplotlib') ❺

# Showing color scale
fig.colorbar(scatter) ❻

plt.show() ❼
```

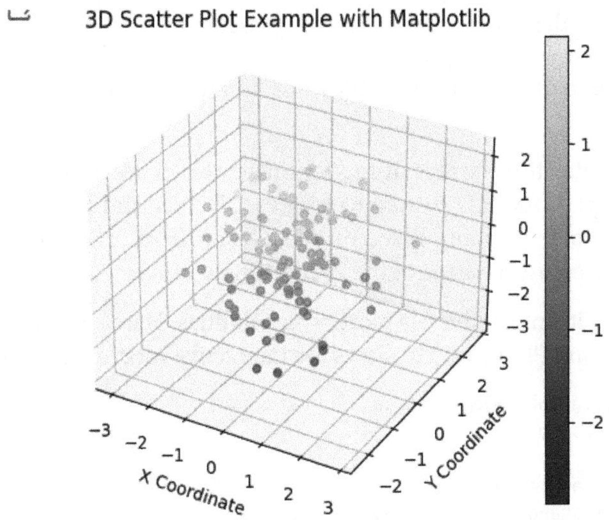

❶ matplotlib.pyplot is a widely used plotting library in Python for creating static visualizations. mpl_toolkits.mplot3d.Axes3D is a toolkit that provides 3D plotting functionality in Matplotlib.

❷ `np.random.standard_normal(100)` generates 100 random numbers for each of x, y, and z from a standard normal distribution (mean = 0, standard deviation = 1). These values are used as coordinates for the 3D scatter plot.

❸ `plt.figure()` creates a new figure window where the plot will be displayed. `fig.add_subplot(111, projection='3d')` adds a 3D subplot to the figure. The `111` argument refers to the grid position (1 row, 1 column, 1 plot), and `projection='3d'` specifies that this will be a 3D plot.

❹ `ax.scatter(x, y, z)` creates a 3D scatter plot where x, y, and z represent the coordinates of the points. `c=z` colors the points based on their z values, so the color will vary with the height (z-axis). `cmap='viridis'` specifies the color map to use, which in this case is `'viridis'` (a color gradient from purple to yellow).

❺ `ax.set_xlabel('X Coordinate')` adds a label to the x-axis, `ax.set_ylabel('Y Coordinate')` adds a label to the y-axis, `ax.set_zlabel('Z Coordinate')` adds a label to the z-axis, and `ax.set_title('3D Scatter Plot Example with Matplotlib')` sets the title of the plot.

❻ Adds a color bar to the figure, which shows the mapping of colors to z values, helping to visualize the range of values represented by the color gradient.

❼ Displays the plot in a new window or inline in environments like Jupyter notebook.

This script generates three arrays of random data, which are used as the coordinates for a 3D scatter plot. Axes3D is used to create a 3D axis, and points are plotted in a 3D space with a color mapping based on the Z-coordinate. The colorbar is added to help interpret the colors in terms of the Z-coordinate values.

Plotly excels with its interactive 3D plotting capabilities. Here's how to create a similar 3D scatter plot using Plotly, which allows for interactive rotation and zoom:

```
import plotly.graph_objects as go
import numpy as np

# Generating random data
x = np.random.standard_normal(100)
y = np.random.standard_normal(100)
z = np.random.standard_normal(100)

# Create a 3D scatter plot
fig = go.Figure(data=[go.Scatter3d(
    x=x,
    y=y,
    z=z,
    mode='markers',
    marker=dict(
        size=5,
        color=z,  # set color to an array/list of desired values
        colorscale='Viridis',  # choose a colorscale
        opacity=0.8
    )
)])

# Adding labels and title
fig.update_layout(
    scene=dict(
        xaxis_title='X Coordinate',
        yaxis_title='Y Coordinate',
```

```
        zaxis_title='Z Coordinate'
    ),
    title='3D Scatter Plot Example with Plotly'
)

fig.show()
```

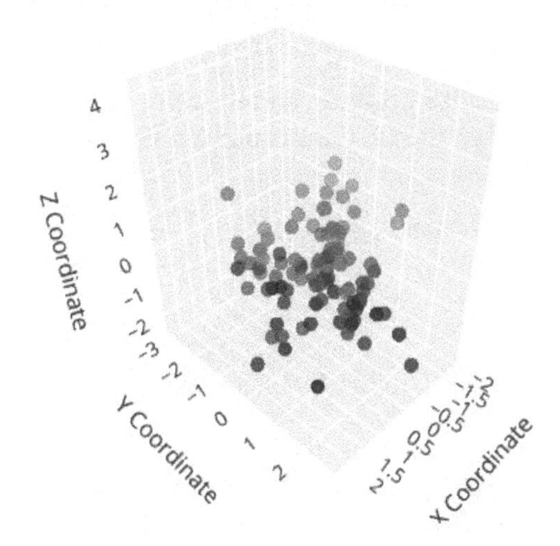

Similar to the Matplotlib example, this script uses Plotly to plot the same set of data points in 3D space, but the resulting plot is interactive. Scatter3d is used to create 3D scatter plots, with colorscale for color mapping based on the Z-coordinate. The plot can be rotated and zoomed interactively, making it easier to explore the data from different angles and depths.

These examples show how both libraries can be used for 3D plotting. While Matplotlib is excellent for quick static views or when embedding plots in static documents, Plotly provides an enhanced experience with interactivity that is particularly useful for presentations or exploratory data analysis where user engagement and detailed examination are required. Seaborn does not natively support 3D plotting.

Geospatial Data Visualization

Geospatial data visualization is a powerful method for spatial analysis and presentation, and the combination of geopandas and folium makes this accessible within the Python ecosystem. Geopandas is a Python library designed to make working with geospatial data easier. It extends the capabilities of the popular data analysis library Pandas by enabling the manipulation of geometries such as points, lines, and polygons. Geopandas allows users to perform operations like spatial joins, buffering, and projections, and can read/write common spatial data formats like Shapefiles and

GeoJSON. It is commonly used for tasks like mapping geographic boundaries, calculating distances, or analyzing spatial relationships in data.

Folium is a Python library used for creating interactive maps. Built on top of the Leaflet.js JavaScript library, Folium makes it easy to visualize geospatial data on a map, with minimal coding. It allows users to overlay features like markers, pop-ups, choropleths (for coloring regions based on data values), and even embed map tiles from online sources like OpenStreetMap. Folium is popular for creating interactive web maps that can be embedded into web pages or displayed in Jupyter notebooks.

In this example, we'll visualize US states and their population estimates using a publicly available shapefile from the US Census Bureau. First, ensure you have the necessary packages installed:

```
pip install geopandas folium
```

```
import geopandas as gpd
import pandas as pd
import folium
```

After the packages are installed, follow this step-by-step example:

1. Load a publicly available dataset of US state shapefiles. You can download US state boundaries shapefile from the US Census Bureau or use any other source. Here, we use an example URL for US state shapefile:

   ```
   url = "https://raw.githubusercontent.com/PublicaMundi/MappingAPI/master/ \
   data/geojson/us-states.json"
   gdf = gpd.read_file(url)
   ```

2. Here is some example population data for US state (replace this with actual data as needed). Here we create a simple dataframe with sample population data:

   ```
   population_data = {
       'State': ['Alabama', 'Alaska', 'Arizona', 'Arkansas', 'California',
           'Colorado'],
       'Population': [4903185, 731545, 7278717, 3017804, 39512223, 5758736]
   }
   population_df = pd.DataFrame(population_data)
   ```

3. Merge the GeoDataFrame with the population data. The shapefile has `'name'` for state names, which we use to merge with our population dataframe:

   ```
   gdf = gdf.merge(population_df, how="left", left_on="name", right_on="State")
   ```

4. Initialize a Folium map:

   ```
   m = folium.Map(location=[37.8, -96], zoom_start=4)
   ```

5. Create a Choropleth map using population data:

   ```
   folium.Choropleth(
       geo_data=gdf.to_json(),  # Convert GeoDataFrame to JSON
       name='choropleth',
   ```

```
        data=gdf,
        columns=['name', 'Population'],  # Use state name and population for
            visualization
        key_on='feature.properties.name',  # Key on state names in GeoJSON
        fill_color='YlGnBu',  # Color gradient from yellow to blue
        fill_opacity=0.7,
        line_opacity=0.2,
        legend_name='Population by State'
    ).add_to(m)
```

6. Add layer control:

```
folium.LayerControl().add_to(m)
```

7. Save the map and display it (if in a Jupyter notebook, just use `m` to display):

```
m.save('us_population_map.html')
m  # In Jupyter, this will display the map inline
```

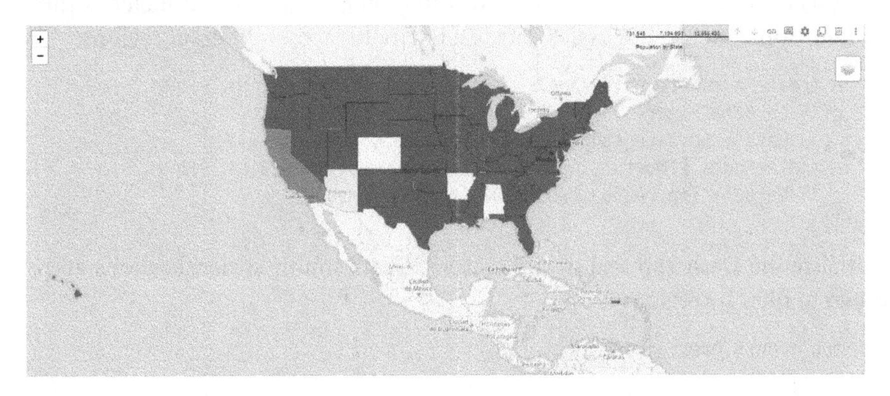

This example uses US state boundaries in GeoJSON format from an external source, though you can easily replace it with any shapefile that fits your needs. The population data provided is a simplified example with fictional values, but you can substitute it with actual population estimates or any other attribute you wish to visualize. The result will be an interactive choropleth map, allowing users to explore population density by state through zooming, panning, and other interactive features, providing a dynamic way to analyze the data. Finally, it demonstrates the synergy between geopandas and folium for effective geospatial data visualization, making complex geographic information accessible and interactive.

Dashboard Creation: Use Plotly Dash

Creating a dashboard combines different visualizations, allowing stakeholders to interact with and explore the outbreak data in real time. Plotly Dash is an excellent choice for building interactive, web-based dashboards using Python. Dash applications are composed of two parts: the layout that defines how the app looks, and the interactions that define the app's functionality. The following example demonstrates how to create a simple dashboard using Dash. This dashboard will include a couple of different visualizations that users can interact with to explore data dynamically.

Ensure you have the necessary packages installed. You can install Dash and its dependencies via pip:

```
pip install dash pandas
```

Next we import the required libraries:

```
import dash
from dash import html, dcc, Input, Output
import plotly.express as px
import pandas as pd
Import numpy as np
```

For this example, let's assume we're working with a dataset that includes time-series data for some metric (e.g., sales data, user activity) across different regions:

```
# Create a sample DataFrame
df = pd.DataFrame({
    "Date": pd.date_range(start='1/1/2020', periods=100),
    "Region": ["North", "South", "East", "West"] * 25,
    "Value": (np.random.randn(100) * 100).cumsum()
})
```

Initialize the Dash app and define its layout with multiple visualizations and a dropdown to filter by region:

```
app = dash.Dash(__name__)

app.layout = html.Div([
    html.H1("Interactive Dashboard"),
    dcc.Dropdown(
        id='region-dropdown',
        options=[{'label': i, 'value': i} for i in df['Region'].unique()],
        value='North',
        clearable=False
    ),
    dcc.Graph(id='time-series-chart'),
    dcc.Graph(id='histogram')
])

# Callbacks to update charts based on dropdown selection
@app.callback(
    Output('time-series-chart', 'figure'),
    Output('histogram', 'figure'),
    Input('region-dropdown', 'value')
)
def update_graph(selected_region):
    filtered_df = df[df['Region'] == selected_region]
```

```
    # Time Series Plot
    fig_time = px.line(filtered_df, x="Date", y="Value", title="Time Series of
        Values")

    # Histogram
    fig_hist = px.histogram(filtered_df, x="Value", nbins=20, title="Value
        Distribution")

    return fig_time, fig_hist

# Run the app
if __name__ == '__main__':
    app.run_server(debug=True)
```

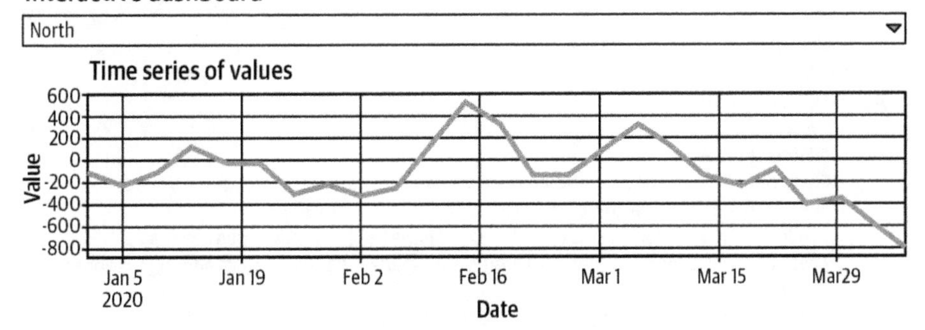

Here is an explanation of the callback section of the code:

- The `update_graph` function is a callback that updates both plots based on the selected region from the drop-down. It filters the dataframe based on the region and then generates a line plot and a histogram.
- The Output objects link the callback function to the Graph components in the layout. The `Input` object captures the selected value from the drop-down.

The `run_server` method starts the Dash app, which will be accessible through a web browser locally.

This dashboard allows stakeholders to interact with and explore data in real time, offering insights into how values change over time and their distribution within any selected region. Dash's interactivity combined with Plotly's powerful plotting capabilities creates a dynamic and user-friendly visualization environment.

Case Study: Using Python for an Advanced Visualization Project

Imagine a retail company wants to better understand customer behavior to optimize their marketing strategies. They've collected a comprehensive dataset over the past year that includes customer demographics, transaction history, website interaction data, and customer feedback. The goal is to create a dashboard that visualizes customer demographics, transaction history, website interactions, and sentiment analysis, combining different libraries such as Pandas, Seaborn, Matplotlib, and Plotly.

Let's walk through each of the necessary steps, data preparation, visualization, and insights. For demonstration purposes, we will create a mock data:

```
import pandas as pd
import numpy as np

# Generate mock customer data
np.random.seed(42)
n_customers = 1000

df = pd.DataFrame({
    'CustomerID': range(1, n_customers + 1),
    'Age': np.random.randint(18, 70, size=n_customers),
    'Gender': np.random.choice(['Male', 'Female'], size=n_customers),
    'TransactionDate': pd.date_range(start='2023-01-01', periods=n_customers,
        freq='D'),
    'PurchaseAmount': np.random.uniform(10, 500, size=n_customers),
    'Clicks': np.random.randint(1, 50, size=n_customers),
    'HourOfDay': np.random.randint(0, 24, size=n_customers),
    'DayOfWeek': np.random.choice(['Monday', 'Tuesday', 'Wednesday', 'Thursday',
        'Friday', 'Saturday', 'Sunday'], size=n_customers),
    'Feedback': np.random.choice(['Great service!', 'Could be better',
        'Loved it!', 'Unsatisfactory', 'Excellent!', 'Not bad'], size=n_customers)
})

# Preview the data
df.head()
```

	CustomerID	Age	Gender	TransactionDate	PurchaseAmount	Clicks	HourOfDay	DayOfWeek	Feedback
0	1	56	Male	2023-01-01	427.091388	5	12	Sunday	Loved it!
1	2	69	Male	2023-01-02	252.621798	35	17	Monday	Great service!
2	3	46	Male	2023-01-03	245.487423	28	10	Friday	Loved it!
3	4	32	Female	2023-01-04	300.279814	29	11	Saturday	Excellent!
4	5	60	Male	2023-01-05	414.093673	9	5	Saturday	Unsatisfactory

Here, data preparation is completed:

```
# Handle missing values if there were any
df.dropna(inplace=True)

# Categorize age into bins
df['AgeGroup'] = pd.cut(df['Age'], bins=[18, 30, 40, 50, 60, 70],
labels=["18-29", "30-39", "40-49", "50-59", "60-69"])

# Convert TransactionDate to datetime
df['TransactionDate'] = pd.to_datetime(df['TransactionDate'])
```

Next we will create distribution charts for customer age and gender:

```
import seaborn as sns
import matplotlib.pyplot as plt

# Set up plot size and style
plt.figure(figsize=(10, 6))

# Age distribution
sns.histplot(data=df, x='AgeGroup', bins=6, kde=True)
plt.title('Customer Age Distribution')
plt.xlabel('Age Group')
plt.ylabel('Frequency')
plt.show()

# Gender distribution
sns.countplot(data=df, x='Gender')
plt.title('Customer Gender Distribution')
plt.show()
```

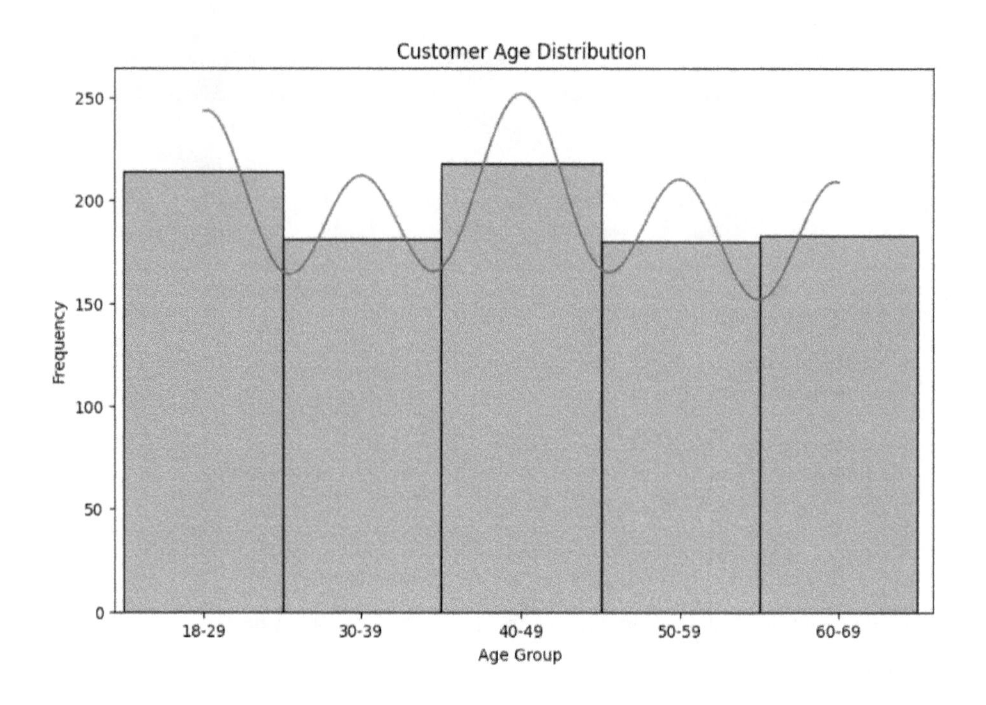

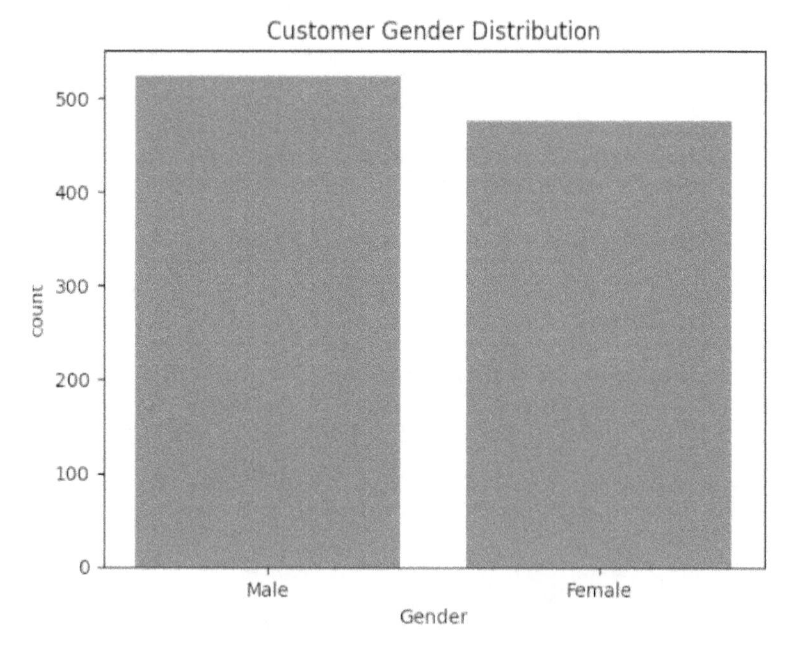

Next we analyze the transaction date:

```
# Set TransactionDate as the index
df.set_index('TransactionDate', inplace=True)

# Resample by month and calculate average purchase amount
monthly_purchase = df['PurchaseAmount'].resample('M').mean()

# Plot the average purchase amount by month
plt.figure(figsize=(10, 6))
monthly_purchase.plot()
plt.title('Average Purchase Amount by Month')
plt.ylabel('Average Purchase Amount')
plt.xlabel('Month')
plt.show()
```

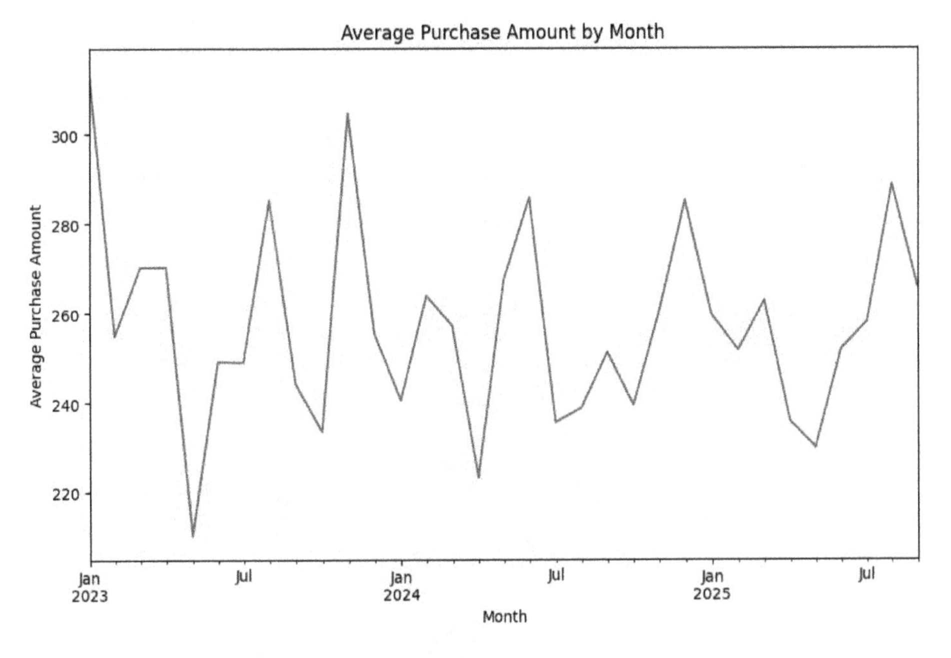

Another chart is created for analyzing website interactions:

```
import plotly.express as px

# Reset the index for further plotting (optional for this step)
df.reset_index(inplace=True)

# Create a heatmap showing website interactions by hour of day and day of week
fig = px.density_heatmap(df, x='HourOfDay', y='DayOfWeek', z='Clicks',
nbinsx=24, nbinsy=7,
                         title="Website Interaction Heatmap (Clicks by Hour
                                                  and Day)")
fig.show()
```

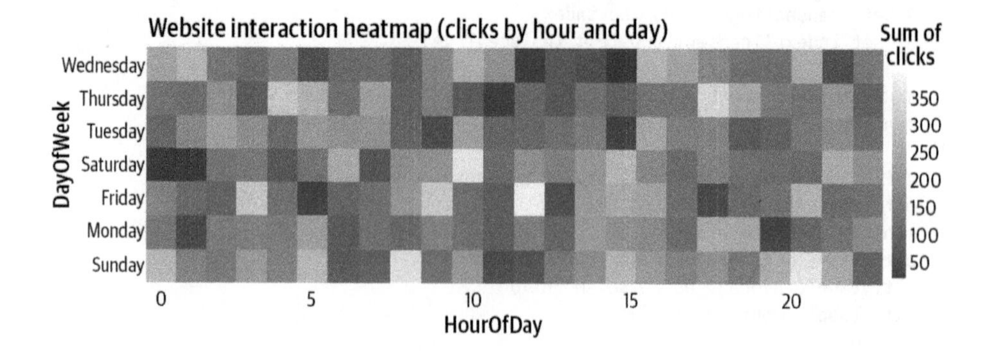

And last we create a chart to analyze sentiment analysis:

```
from textblob import TextBlob

# Perform sentiment analysis to get polarity of feedback
df['Polarity'] = df['Feedback'].apply(lambda x: TextBlob(x).sentiment.polarity)

# Plot sentiment polarity using a violin plot
plt.figure(figsize=(8, 6))
sns.violinplot(data=df, y='Polarity')
plt.title('Sentiment Polarity Distribution from Customer Feedback')
plt.show()
```

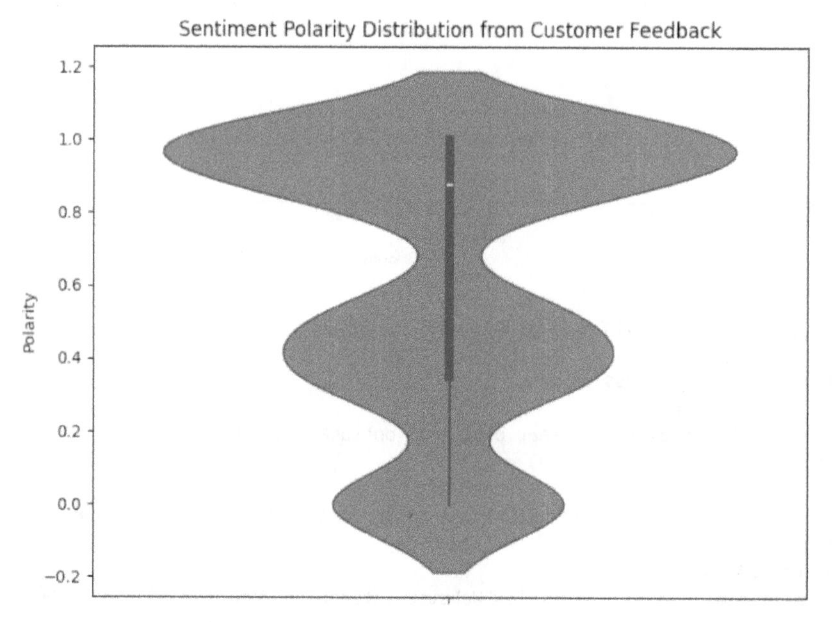

Now that multiple charts have been created, these can be integrated into a dashboard. To combine all of these visualizations into a cohesive dashboard, you can integrate them into a web-based dashboard using Plotly Dash:

```
import dash
from dash import dcc, html
import plotly.graph_objs as go

# Initialize the Dash app
app = dash.Dash(__name__) ❶

# Layout of the dashboard
app.layout = html.Div(children=[ ❷
    html.H1(children='Customer Behavior Dashboard'), ❸

    # Section for customer demographics ❹
    html.Div([
        html.H2('Customer Demographics'),
        dcc.Graph(
            figure=px.histogram(df, x='AgeGroup', title='Customer Age
                        Distribution')
        ),
        dcc.Graph(
            figure=px.histogram(df, x='Gender', title='Customer Gender
                        Distribution')
        ),
    ]),

    # Section for transaction history ❺
    html.Div([
        html.H2('Transaction History'),
        dcc.Graph(
            figure=go.Figure(go.Scatter(x=monthly_purchase.index,
                        y=monthly_purchase.values,
                                mode='lines', name='Average Purchase
                                Amount'))
        )
    ]),

    # Section for website interactions ❻
    html.Div([
        html.H2('Website Interaction Heatmap'),
        dcc.Graph(
            figure=px.density_heatmap(df, x='HourOfDay', y='DayOfWeek',
                        z='Clicks',
                               title="Website Interactions by Hour
                               and Day")
        )
    ]),

    # Section for customer feedback sentiment ❼
    html.Div([
```

```
        html.H2('Customer Feedback Sentiment Analysis'),
        dcc.Graph(
            figure=px.violin(df, y='Polarity', title='Sentiment Polarity from
                        Customer Feedback')
        )
    ])
])

# Run the app ❽
if __name__ == '__main__':
    app.run_server(debug=True)
```

❶ `app = dash.Dash(__name__)` initializes the Dash application, creating a new instance of the app.

❷ The layout is defined using HTML-like components provided by Dash. The layout contains several sections for different visualizations.

❸ `html.H1(children='Customer Behavior Dashboard')` adds a title to the dashboard.

❹ This section includes two histograms created using Plotly's `px.histogram` to show the distribution of customers by age group and gender. `dcc.Graph(figure=px.histogram(df, x='AgeGroup', title='Customer Age Distribution'))` creates a histogram that visualizes the distribution of customer ages. `dcc.Graph(figure=px.histogram(df, x='Gender', title='Customer Gender Distribution'))` visualizes the gender distribution of customers.

❺ In the transaction history section, a line chart is created using Plotly's `go.Figure` and `go.Scatter` objects to visualize average purchase amounts over time. The graph shows how the average purchase amount changes monthly.

❻ A heatmap is generated to show website interactions by time of day and day of the week using `px.density_heatmap`. `dcc.Graph(figure=px.density_heatmap(...))` visualizes user clicks, showing how website interactions vary based on the hour and day of the week.

❼ This section visualizes customer sentiment using a violin plot. `dcc.Graph(figure=px.violin(df, y='Polarity', title='Sentiment Polarity from Customer Feedback'))` displays the distribution of sentiment polarity scores from customer feedback.

❽ `app.run_server(debug=True)` starts the Dash app, making it available at a local URL (e.g., *http://127.0.0.1:8050*). The `debug=True` flag ensures that the app will automatically reload if changes are made to the code.

This code creates a web-based dashboard using the Plotly Dash framework to visualize customer behavior data.

After running the app, all visuals would be shown on a single dashboard where it would be possible to zoom in and out, pan, autoscale, and interact with the charts. What are some of the findings that can be determined with this dashboard?

Customer demographics visualization

The histogram of age groups shows the concentration of customers in specific age ranges. If a particular age group, such as `30-39` or `40-49`, represents a large portion of the customer base, the company can focus marketing efforts on targeting this age demographic. For example, product offerings, advertisements, or promotions can be tailored to suit the interests and needs of the dominant age group.

The bar chart provides insight into the gender breakdown of customers. If one gender significantly outweighs the other, the company may consider balancing its marketing campaigns to attract the underrepresented group or focus on maintaining engagement with the larger group. Additionally, if the genders are evenly split, the company could focus on gender-neutral marketing campaigns.

Transaction history analysis

The line chart illustrating average monthly purchase amounts may reveal seasonal or time-based trends. For example, if there is a spike in purchases during certain months, it could indicate successful seasonal marketing or demand for specific products during certain times of the year (e.g., holiday shopping periods). On the other hand, any declines could signal off-peak seasons where the company might need to implement special promotions or adjust inventory levels to maintain steady sales.

These insights can help the company optimize inventory, prepare for demand surges, and plan promotions accordingly.

Website interaction heatmap

The heatmap allows us to identify periods of peak user activity on the website. If engagement is higher during certain hours (e.g., midday) or days of the week (e.g., weekends), the company can align resources accordingly. For instance, they could allocate more customer support during these peak hours or time promotions and new product launches to coincide with high traffic times.

Understanding peak traffic times helps ensure that marketing and customer support efforts are optimized for maximum user engagement. This can increase conversion rates and improve the overall customer experience.

Customer feedback sentiment analysis

The sentiment analysis through the violin plot offers a breakdown of positive and negative customer feedback. A wider spread of positive sentiment indicates that customers are generally satisfied with the company's services or products. Conversely, a large concentration of negative sentiment could reveal issues with customer service, product quality, or other factors. Understanding the distribution of customer satisfaction can inform improvement initiatives.

If sentiment is mostly positive, the company can leverage this in its branding and marketing, promoting customer testimonials and satisfaction. If there's a notable presence of negative sentiment, the company should investigate recurring issues and take steps to resolve them, such as enhancing product quality, improving customer support, or addressing pain points mentioned in feedback.

Overall findings and actions

Based on the customer age and gender distribution, the company can tailor marketing campaigns, product offerings, and customer engagement strategies to suit the dominant demographics.

The transaction history highlights time-based sales trends, guiding the company to prepare for peak sales periods and manage inventory and promotions more effectively.

Peak engagement times can be used to optimize customer support availability and schedule marketing campaigns during high-traffic periods for better customer outreach.

Customer sentiment analysis offers insights into areas of satisfaction and dissatisfaction, allowing the company to promote strengths and address weaknesses to enhance overall customer experience.

These findings help the retail company make informed, data-driven decisions to optimize marketing strategies, improve customer retention, and maximize operational efficiency.

Choosing Between R Shiny and Python Visualization

Choosing between R Shiny and Python visualization tools often comes down to specific project requirements, user expertise, and the context of the data analysis task at hand. Both platforms offer robust capabilities for data visualization, but they cater to slightly different needs and user bases.

As we have seen, R Shiny is a framework for building interactive web applications directly from R. It is particularly well suited for statisticians and data scientists who are already familiar with R. Shiny excels in creating interactive dashboards that can dynamically update based on user input, making it ideal for delivering complex analytical results directly to end users without requiring them to know R. The integration within the R ecosystem allows Shiny apps to leverage the statistical power of R, making it a go-to choice for applications involving sophisticated statistical models, real-time data exploration, or interactive visualizations of data analysis results.

On the other hand, Python visualization tools like Matplotlib, Seaborn, and Plotly offer powerful options for creating static, interactive, and real-time visualizations. Python's visualization libraries are highly flexible and can be used in a wide range of environments, including web frameworks such as Flask and Django. For projects that require integration with web applications or software requiring heavy use of programming logic alongside data visualization, Python might be the preferred choice. Plotly, for instance, provides tools for creating highly interactive plots that are web-friendly and can easily be embedded into web apps. This makes Python particularly valuable in environments that require seamless integration between the application's backend logic and the frontend visualization.

If your project requires building standalone applications solely focused on data presentation with high interactivity and is developed by users familiar with R, R Shiny is the ideal choice. However, if your project needs to integrate tightly with broader software engineering practices, if it involves a significant amount of system-level programming, or if you are already embedded in the Python ecosystem, then choosing Python visualization tools will likely serve you better. The choice also depends on the scale of the application, the nature of the dataset, and the specific needs of the users or stakeholders involved.

Summary

This chapter covered advanced visualization techniques, particularly through tools like R Shiny and various Python visualization packages, and how they empower business analysts to transform complex datasets into clear, actionable insights. For business analysts, the ability to create interactive, dynamic visualizations is crucial. R Shiny excels in this area by allowing analysts to build web applications directly from R, enabling real-time interaction with data. On the Python side, libraries like Matplotlib, Seaborn, Plotly, and Dash offer versatile solutions for static, animated, and interactive visualizations. The integration of these advanced visualization tools into business analytics workflows allows analysts to effectively communicate complex quantitative insights to nontechnical stakeholders, driving strategic business decisions.

CHAPTER 10

Working with Modern Data Types in Analytics

In today's data-driven landscape, businesses encounter a wide array of data types beyond traditional structured data, such as data in columns and rows. Modern data types include social media data, semistructured formats like JSON, video data, data obtained through web scraping, and data obtained in different formats directly from websites. Effectively handling these diverse data sources is crucial for gaining comprehensive insights and making informed decisions.

This chapter will explore the importance of managing various modern data types in business analytics. In addition to providing an overview of these data types, the chapter will also discuss why their integration is vital for a holistic analytical approach. Furthermore, we will revisit Python and R as well as the functionality used for data handling. Both tools are indispensable for extracting, processing, and analyzing complex datasets, enabling you to unlock the full potential of data.

By mastering the techniques presented in this chapter, you will enhance your ability to work with diverse data sources, thereby improving the depth and accuracy of your business analytics. Whether you are dealing with social media trends, unstructured data formats, or multimedia content, both Python and R offer robust solutions to tackle these data types efficiently. The first data type we'll explore is social media data.

Semistructured Data (JSON)

JavaScript Object Notation (JSON) is a lightweight data interchange format that is easy for humans to read and write, and easy for machines to parse and generate. One of the defining characteristics of JSON is its simplicity and flexibility, making it language-independent and ideal for transmitting data between a server and a web application. JSON's text-based format ensures that it is both human-readable and easily parsed by machines, contributing to its widespread adoption in web services and APIs. Structurally, JSON data is composed of key-value pairs, where keys are strings and values can be strings, numbers, arrays, objects, true, false, or null. JSON is built on two structures:

- A collection of name/value pairs, which is realized as an object, record, structure, dictionary, hash table, keyed list, or associative array
- An ordered list of values, which is realized as an array, vector, list, or sequence

The following is an example of JSON data:

```
{
    "name": "John Doe",
    "age": 30,
    "email": "john.doe@example.com",
    "address": {
      "street": "123 Main St",
      "city": "Anytown",
      "zipcode": "12345"
    },
    "phoneNumbers": [
      {"type": "home", "number": "555-555-5555"},
      {"type": "work", "number": "555-555-1234"}
    ]
  }
```

In JSON data, objects are enclosed in curly braces, `{ }`, while arrays are enclosed in square brackets, `[ ]`. Each piece of data is represented as a key-value pair, where the key and value are separated by a colon, `:`. To distinguish between different key-value pairs and elements within an array, commas are used as separators.

Using Python for JSON Data

In Python, working with JSON data is made simple and efficient with the json library. This section explores how to load and parse JSON data using Python, whether the data is stored as a string or in a file.

Loading and parsing JSON data

The json library in Python provides functions to work with JSON data:

```
import json

  # JSON string
  json_string = '''
  {
    "name": "John Doe",
    "age": 30,
```

```
    "email": "john.doe@example.com",
    "address": {
      "street": "123 Main St",
      "city": "Anytown",
      "zipcode": "12345"
    },
    "phoneNumbers": [
      {"type": "home", "number": "555-555-5555"},
      {"type": "work", "number": "555-555-1234"}
    ]
  }
  '''

  # Parse the JSON string
  data = json.loads(json_string)

  # Access data from the JSON
  print(data)

  # Parse JSON string
  data = json.loads(json_string)
  print(data)
  {'name': 'John Doe', 'age': 30, 'email': 'john.doe@example.com', 'address':
  {'street': '123 Main St', 'city': 'Anytown', 'zipcode': '12345'},
  'phoneNumbers': [{'type': 'home', 'number': '555-555-5555'}, {'type': 'work',
  'number': '555-555-1234'}]}
  {'name': 'John Doe', 'age': 30, 'email': 'john.doe@example.com', 'address':
  {'street': '123 Main St', 'city': 'Anytown', 'zipcode': '12345'},
  'phoneNumbers': [{'type': 'home', 'number': '555-555-5555'}, {'type': 'work',
  'number': '555-555-1234'}]}
```

Here's how you can load JSON from a file:

```
# Load JSON data from a file
  with open('data.json', 'r') as file:
      data = json.load(file)
  print(data)
```

Extracting data from nested JSON structures

JSON data often contains nested structures, which means that values within key-value pairs can themselves be JSON objects or arrays. For example, a key can hold an entire object as its value, which can then contain more key-value pairs, or it can store an array of objects. These nested structures allow for the representation of more complex, hierarchical data, making JSON a flexible format for modeling real-world data scenarios like customer profiles, product catalogs, or API responses.

Here's how you can access nested data:

```
# Accessing nested data
  name = data['name']
  street = data['address']['street']
```

```
    work_phone = data['phoneNumbers'][1]['number']

    print(f"Name: {name}")
    print(f"Street: {street}")
    print(f"Work Phone: {work_phone}")
```

The code example produces the following output:

```
Name: John Doe
  Street: 123 Main St
  Work Phone: 555-555-1234
```

Transforming JSON data into Pandas DataFrames

Pandas provides powerful tools to transform JSON data into dataframes for easier analysis, allowing for seamless integration of hierarchical JSON data into tabular formats. With functions like `pd.json_normalize()`, Pandas can flatten nested JSON structures into columns, making it simpler to analyze complex datasets with hierarchical relationships. This transformation allows data scientists to apply Pandas rich ecosystem of data manipulation, filtering, and aggregation methods to JSON data, enabling easier exploration, visualization, and even machine learning model development. By converting JSON to dataframes, working with API responses or nested datasets becomes more intuitive and efficient:

```
import pandas as pd

  # JSON data
  json_data = '''
  [
    {"name": "John Doe", "age": 30, "email": "john.doe@example.com"},
    {"name": "Jane Smith", "age": 25, "email": "jane.smith@example.com"}
  ]
  '''
  # Parse JSON data
  data = json.loads(json_data)

  # Convert JSON to DataFrame
  df = pd.DataFrame(data)
  print(df)
```

The code example produces the following output:

```
        name  age                   email
  0    John Doe   30    john.doe@example.com
  1  Jane Smith   25  jane.smith@example.com
```

Cleaning and normalizing JSON data

For deeply nested JSON, you might need to normalize it to convert complex, hierarchical data into a flat, tabular format that is easier to work with in data analysis tools like Pandas. JSON data often contains multiple layers of nested objects or arrays, and without normalization, accessing, or manipulating, such data can become cumbersome. By using `pd.json_normalize()`, you can flatten these nested structures into columns, preserving the relationships between the nested elements while making the data more accessible for analysis. This is particularly useful when dealing with API responses or data feeds that involve deeply embedded records, such as those found in user profiles, transaction histories, or multilevel product catalogs:

```
# Example JSON data
  nested_json = '''
  {
    "store": {
      "book": [
        {"category": "fiction", "author": "John Doe", "title": "The Great
        Book", "price": 19.99},
        {"category": "science", "author": "Jane Doe", "title": "The Science
        Book", "price": 29.99}
      ]
    }
  }
  '''

  # Parse JSON data
  data = json.loads(nested_json)

  # Normalize JSON data
  books = pd.json_normalize(data['store']['book'])
  print(books)
```

The code example produces the following output:

```
category    author              title  price
  0 fiction  John Doe    The Great Book  19.99
  1 science  Jane Doe  The Science Book  29.99
```

Analyzing JSON data with Python

Once JSON data is converted into a dataframe, you can perform EDA using Pandas, Seaborn, and Matplotlib, which were covered in Chapter 4. Working with JSON data in Python involves loading and parsing JSON with the json library, transforming it into dataframes using Pandas, and then analyzing and visualizing the data. This process allows for effective handling and analysis of unstructured data, turning it into actionable insights.

Using R for JSON Data

In R, the jsonlite package provides convenient tools to load, parse, and manipulate JSON data. Whether you are working with web APIs, structured datasets, or large-scale data pipelines, jsonlite simplifies the process of converting JSON data into R objects such as dataframes and lists. This functionality makes it easier for data analysts and scientists to explore and analyze complex JSON data structures directly within the R environment.

Loading and parsing JSON data

The jsonlite package in R provides functions to work with JSON data:

```
install.packages("jsonlite")
  library(jsonlite)

  # JSON string
  json_string <- '
  {
    "name": "John Doe",
    "age": 30,
    "email": "john.doe@example.com",
    "address": {
      "street": "123 Main St",
      "city": "Anytown",
      "zipcode": "12345"
    },
    "phoneNumbers": [
      {"type": "home", "number": "555-555-5555"},
      {"type": "work", "number": "555-555-1234"}
    ]
  }
  '

  # Parse JSON string
  data <- fromJSON(json_string)
  print(data)
  Loading JSON from a File:

  # Load JSON data from a file
  data <- fromJSON("data.json")
  print(data)
```

The code example produces the following output:

```
$name
  [1] "John Doe"

  $age
  [1] 30

  $email
  [1] "john.doe@example.com"

  $address
  $address$street
  [1] "123 Main St"

  $address$city
```

```
[1] "Anytown"

$address$zipcode
[1] "12345"


$phoneNumbers
  type       number
1 home 555-555-5555
2 work 555-555-1234
```

Extracting data from nested JSON structures

Here's how you can access nested data in R:

```
library(jsonlite)

  # JSON string
  json_string <- '
  {
    "name": "John Doe",
    "age": 30,
    "email": "john.doe@example.com",
    "address": {
      "street": "123 Main St",
      "city": "Anytown",
      "zipcode": "12345"
    },
    "phoneNumbers": [
      {"type": "home", "number": "555-555-5555"},
      {"type": "work", "number": "555-555-1234"}
    ]
  }
  '

  # Parse the JSON string
  data <- fromJSON(json_string, simplifyVector = FALSE)

  # Accessing nested data
  name <- data$name
  street <- data$address$street
  work_phone <- data$phoneNumbers[[2]]$number
```

The code example produces the following output:

```
> name
  [1] "John Doe"
  > street
  [1] "123 Main St"
  > work_phone
  [1] "555-555-1234"
```

Transforming JSON data into data frames with R

The jsonlite package allows easy conversion of JSON data to a dataframe:

```
# JSON data
  json_data <- '
  [
    {"name": "John Doe", "age": 30, "email": "john.doe@example.com"},
    {"name": "Jane Smith", "age": 25, "email": "jane.smith@example.com"}
  ]
  '

  # Parse JSON data
  data <- fromJSON(json_data)

  # Convert JSON to Data Frame
  df <- as.data.frame(data)
  print(df)
```

The code example produces the following output:

```
       name age                  email
  1   John Doe  30   john.doe@example.com
  2 Jane Smith  25 jane.smith@example.com
```

Cleaning and normalizing JSON data

For deeply nested JSON, you can normalize it as this example does:

```
# Example JSON data
  nested_json <- '
  {
    "store": {
      "book": [
        {"category": "fiction", "author": "John Doe", "title": "The Great
        Book", "price": 19.99},
        {"category": "science", "author": "Jane Doe", "title": "The Science
        Book", "price": 29.99}
      ]
    }
  }
  '

  # Parse JSON data
  data <- fromJSON(nested_json, flatten = TRUE)
  print(data)
```

The code example produces the following output:

```
$store
  $store$book
    category   author            title price
  1  fiction John Doe   The Great Book 19.99
  2  science Jane Doe The Science Book 29.99
```

Analyzing JSON data with R

Once JSON data is converted into a dataframe, you can perform EDA using base R or other packages like dplyr. Use libraries like ggplot2 to visualize data insights. Working with JSON data in R involves loading and parsing JSON with the jsonlite package, transforming it into dataframes, and then analyzing and visualizing the data. This process allows for effective handling and analysis of unstructured data, turning it into actionable insights.

Social Media Data

Social media data encompasses a wide variety of information generated by users on social media platforms. This data is rich in content and diverse in format, offering a unique window into the behaviors, opinions, and interactions of individuals across the globe. Understanding the different types of social media data is important for leveraging this information for research, marketing, and various analytical purposes.

Types of Social Media Data

Social media data comes in various forms, each offering unique insights into user behavior, preferences, and interactions. One of the primary types of data is text data, which includes posts, comments, and messages shared by users on platforms like X (formerly Twitter), Facebook, and Instagram. This text data is rich in sentiment, opinions, and trends, making it valuable for sentiment analysis, opinion mining, and understanding public reactions to events or brands. Additionally, text data often includes hashtags and keywords that help identify trending topics or the most discussed subjects within a community.

Another significant type of social media data is multimedia data, which consists of images, videos, and audio files shared on platforms such as Instagram, YouTube, and TikTok. Analyzing multimedia data provides insights into visual trends, popular content types, and user engagement. For instance, by examining which types of images or videos garner the most likes, shares, or comments, and the most brands and marketers can tailor their visual content strategies to better resonate with their audience.

Engagement data, which includes likes, shares, comments, retweets, and other forms of interaction, is crucial for measuring the popularity and reach of content. It helps in understanding how users interact with different types of posts and identifying influencers within a social network. This type of data is essential for evaluating the effectiveness of marketing campaigns and determining which content strategies are most successful.

Network data—which involves the relationships and connections between users such as followers, friends, and mentions—can be analyzed to understand the structure of social networks. This type of data is particularly useful for social network analysis

(SNA), as it reveals how communities are formed and how influence flows within them. Finally, metadata—which includes information about the data itself such as timestamps, geolocation, device information, and user demographics—provides context for content creation and sharing. Analyzing metadata helps businesses understand when and where their audience is most active, making it valuable for targeted marketing efforts.

Each type of social media data offers different insights, and when combined, they provide a comprehensive view of social media behavior. This holistic understanding enables more informed decisions in areas such as marketing, customer service, and content strategy.

Posts

Posts are the core content shared on social media platforms like Facebook, Instagram, and LinkedIn. These can range from text updates and photos to videos and articles. Facebook posts, for example, can include text, images, videos, links, and user-generated tags. Instagram posts primarily focus on visual content, supplemented with captions and hashtags. LinkedIn posts often revolve around professional updates and industry-related content. Analyzing posts can reveal user preferences, engagement levels, and community interests.

X is one of the most popular platforms for generating social media data. X posts (formerly called tweets) are short messages, limited to 280 characters (except for premium users), often accompanied by hashtags, mentions, links, and multimedia content like images and videos. X posts provide real-time insights into public opinion, trending topics, and social interactions. The brief and public nature of these posts make them particularly valuable for sentiment analysis and event detection.

Reddit is a valuable data source for analyzing public discussions, opinions, and trends across various topics. As a large social media platform with numerous communities (called subreddits) focused on specific interests, Reddit provides rich, user-generated content that can be scraped and analyzed for sentiment, topic modeling, and trend analysis. The platform's diversity of opinions and real-time interactions make it particularly useful for studying consumer behavior, public sentiment, and emerging trends across different industries.

Comments

Comments are user responses in text format to posts, videos, and other content on social media platforms. They can be found on platforms like Facebook, YouTube, Instagram, and blogs. Comments often contain valuable feedback, opinions, and discussions, making them a rich source for sentiment analysis and understanding public reactions. They can also highlight community engagement and the effectiveness of content in sparking conversation.

Blogs are a rich data source for in-depth, opinionated content on a wide range of subjects, including technology, lifestyle, business, and more. As many bloggers offer expert insights, personal experiences, and detailed analyses, blog data can be valuable for sentiment analysis, trend tracking, and content marketing. The structured and thematic nature of blogs makes them suitable for extracting long-form textual data for research, competitive analysis, and understanding niche audiences.

Likes and reactions

While not textual data, likes and reactions are crucial metrics for understanding user engagement and content popularity. Platforms like Facebook offer various reactions (e.g., like, love, wow, sad, angry) that provide more nuanced insights into user sentiment than simple likes. Analyzing these interactions helps gauge public approval and emotional responses to content. Likes and reactions on social media platforms, such as Facebook's thumbs-up, heart, or angry face, are forms of engagement data that provide immediate feedback on user sentiment and content popularity. These interactions are typically stored as numerical data within databases, representing counts of each type of reaction a post receives. For instance, a post might have 200 likes, 50 hearts, and 10 angry reactions, each recorded as a separate numerical value associated with the post's ID. This data is often stored in a structured format, such as rows in a table with columns for different types of reactions, making it easy to analyze and aggregate to assess overall user engagement and sentiment toward specific content.

Shares and reposts

Shares and reposts (also formerly called retweets on Twitter) amplify content reach and indicate what users find valuable enough to disseminate within their own networks. Tracking the spread of content through shares and reposts helps in understanding virality, influence, and the spread of information across social networks. Shares and reposts on social media are forms of engagement where users distribute content created by others to their own network or followers. When a user shares or reposts a piece of content, such as an article, video, or post, it amplifies the content's reach by exposing it to a broader audience beyond the original poster's followers. Shares and reposts are typically recorded as events in the platform's database, tracking the user who shared the content, the time of the share, and any additional comments or context provided during the sharing process. This data helps measure the virality and overall impact of content across the platform.

Direct messages

Direct messages are private communications between users. They can provide deep insights into personal interactions and networks, though they are often less accessible due to privacy concerns. When available, they offer a more intimate view of user interactions and relationships.

Social media data, with its varied types, provides a multifaceted view of online user behavior and engagement. By analyzing posts, comments, likes, reactions, shares, and direct messages, researchers and businesses can gain comprehensive insights into trends, sentiments, and the impact of social media on public discourse. This data is invaluable for making informed decisions in marketing, customer service, product development, and beyond.

Using Python for Social Media Data Analysis

Python excels at handling social media data analysis. Libraries like Tweepy (for X) and Facebook's Graph API allow you to collect public posts and comments. Tweepy, a popular Python library for accessing the X API, is no longer fully free due to changes in Twitter's API access policies under the rebranding to X. Previously offering extensive free access, the new pricing model introduced by X/Twitter limits the free-tier access, requiring developers and businesses to subscribe to paid plans to utilize the full range of API features. This shift has impacted projects relying on free access for data extraction, analysis, and automation tasks, necessitating a reassessment of the costs and benefits of integrating with the platform.

Once the data is retrieved, libraries like Natural Language Toolkit (NLTK) can be used for cleaning and preprocessing the text. Powerful tools like TextBlob or Valence Aware Dictionary and sEntiment Reasoner (VADER) then come into play, leveraging machine learning to analyze the sentiment of the text data.

- NLTK is a comprehensive library for natural language processing (NLP) in Python, providing tools for tasks like tokenization, parsing, and part-of-speech tagging.
- TextBlob is a simple Python library for processing textual data, offering easy-to-use tools for sentiment analysis, translation, and text classification.
- VADER is a lexicon-based sentiment analysis tool designed specifically to analyze social media text, providing sentiment scores for positive, negative, and neutral tones.

This can reveal public perception toward a brand, product, or event on Facebook, allowing you to understand the emotional tone of the conversation and make data-driven decisions.

X data extraction using Tweepy

In the age of social media, platforms like X provide a wealth of data that can be invaluable for various forms of analysis, from sentiment analysis to trend detection. Using Python and the Tweepy library, you can easily extract this data. This section will guide you through setting up the X API, fetching X posts, and processing the data.

Setting up the X API. To extract data from X, you first need to access the X API. You will need a developer account, so visit the X Developer website and sign up for one. There is a cost to using the X API, so keep this in mind.

Once you have access to the X API, you then need to create a new app within the developer portal to generate your API keys and access tokens.

To begin, navigate to the "Apps" tab in the X Developer Portal and click "Create an App." Fill in the necessary details, such as the app name, description, and website URL. Once the app is created, you will receive essential API keys and tokens, including the API Key, API Key Secret, Access Token, and Access Token Secret. These credentials are required to authenticate and interact with the X API. Figure 10-1 depicts the X developer platform, Figure 10-2 shows the guide to setting up authentication, and Figure 10-3 shows the developer dashboard after the application is created.

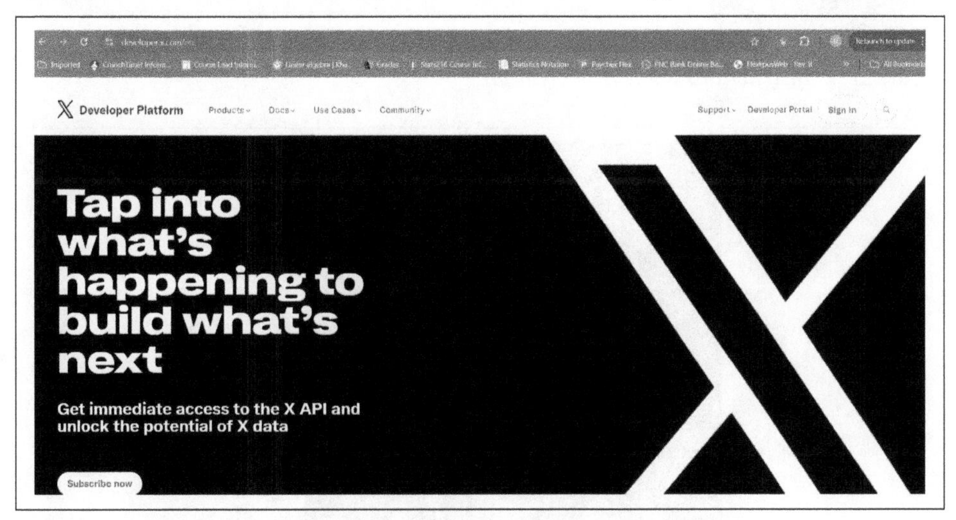

Figure 10-1. X Developer platform

API Key, API Key Secret, Access Token, and Access Token Secret are essential credentials that allow applications to authenticate with the X API. These keys grant access to X's services, enabling actions such as fetching data or posting tweets on behalf of a user. However, storing these keys and tokens directly in your code or public repositories poses significant security risks. If exposed, malicious users can exploit these credentials to access or misuse your X account or application.

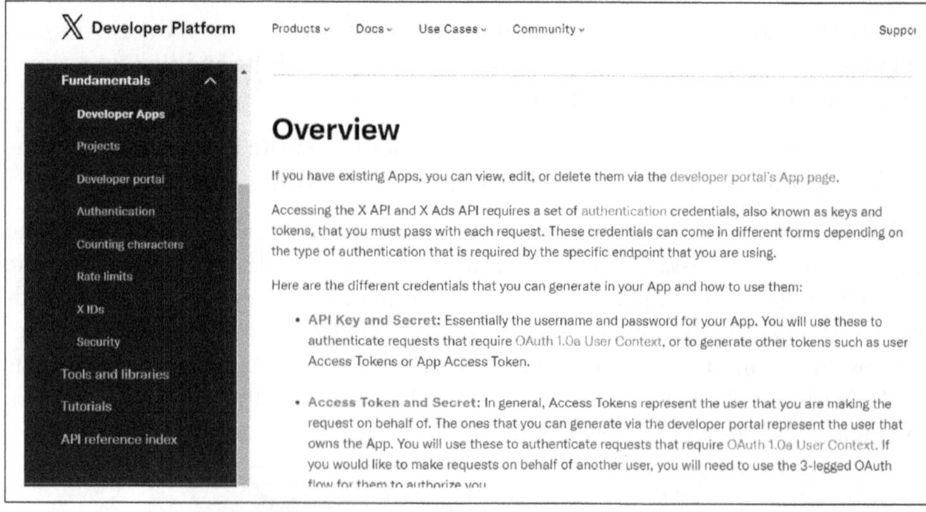

Figure 10-2. Instructions for X App creation

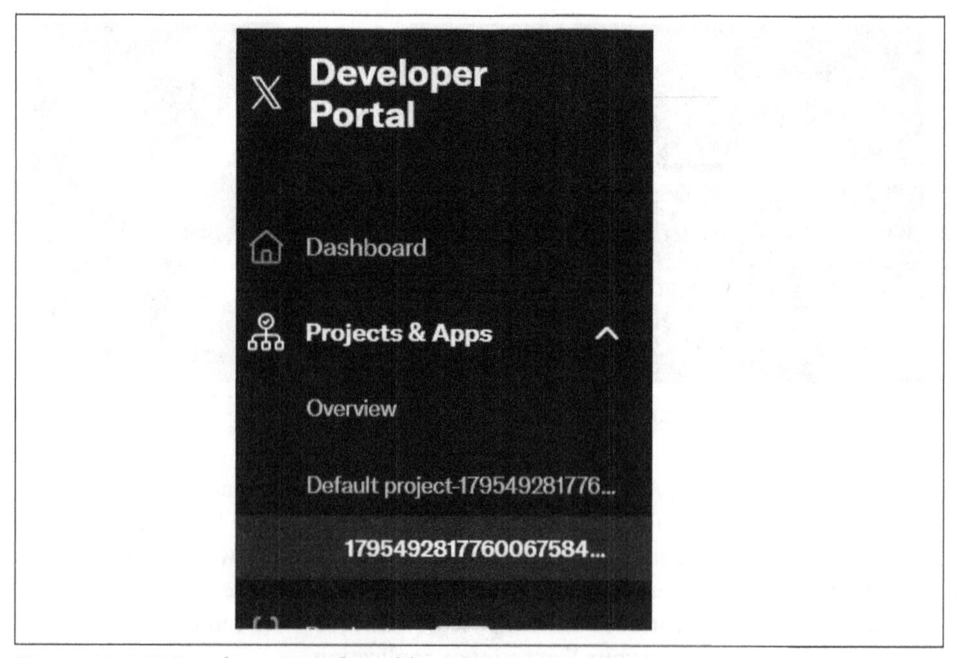

Figure 10-3. X Developer Portal Dashboard

Best practices for handling these credentials include storing them securely using environment variables or secrets management tools like AWS Secrets Manager or GitHub Secrets. This ensures that sensitive information remains protected, even if the code is shared or deployed in public repositories. Additionally, implementing role-based access control and rotating these keys regularly can further safeguard your API interactions.

Once you have the API keys and tokens, you're now ready to install the Tweepy library using pip:

```
pip install tweepy
```

Then, you can authenticate with X API by using access tokens to authenticate your requests. OAuth1 is a secure authentication protocol that allows a user to grant access to their resources without sharing their credentials directly:

```
import tweepy ❶

  # Replace these values with your own API keys and access tokens ❷
  consumer_key = 'your_API_key'
  consumer_secret = 'your_API_key_secret'
  access_token = 'your_access_token'
  access_token_secret = 'your_access_token_secret'

  # Authenticate to the X API
  auth = tweepy.OAuth1UserHandler(consumer_key, consumer_secret, access_token,
  access_token_secret) ❸
  api = tweepy.API(auth) ❹

  # Verify authentication
  try: ❺
      api.verify_credentials() ❻
      print("Authentication OK")
  except:
      print("Error during authentication")
```

❶ Import the Tweepy library, which provides easy access to the X API.

❷ Define variables to store your API keys and access tokens. These values are provided by X when you create an app on their developer platform. Replace the placeholder strings with your actual credentials.

❸ Create an authentication handler using your API keys and access tokens.

❹ Create an API object that you will use to interact with the X API.

❺ This block handles any exceptions that occur during the verification process. If the authentication is successful, it prints `"Authentication OK"`. If there is an error, it prints `"Error during authentication"`.

❻ This method attempts to verify the provided credentials. If the credentials are valid, it does nothing; otherwise, it raises an exception. When using `api.verify_credentials()` in Tweepy to verify the provided credentials, if the credentials are valid, it will return without any issues. However, if the credentials are invalid (e.g., incorrect API key, access token, etc.), it will raise an exception.

Here's an example of how you can handle the exception and log it for demonstration using `console.log()` in a Python-like context:

```
import tweepy

  # Replace these values with your own API keys and access tokens
  consumer_key = 'your_consumer_key'
  consumer_secret = 'your_consumer_secret'
  access_token = 'your_access_token'
  access_token_secret = 'your_access_token_secret'

  # Set up authentication
  auth = tweepy.OAuth1UserHandler(consumer_key, consumer_secret, access_token,
  access_token_secret)
  api = tweepy.API(auth)

  try:
      api.verify_credentials()
      print("Credentials are valid")
  except tweepy.TweepyException as e:
      print(f"Error verifying credentials: {e}")
```

This code snippet authenticates your application with the X API using Tweepy and verifies that the authentication was successful. If the credentials are valid, it prints a success message; otherwise, it prints an error message.

Fetching posts using Tweepy. Once authenticated with the X API, you can begin fetching tweets (now known as "X posts") based on specific keywords, hashtags, or search criteria. This allows you to collect and analyze relevant posts from the platform in real time. Here is an example of how to retrieve recent posts (see Figure 10-4) using a hashtag or keyword, which can be customized based on your data needs:

```
# Define the search term and the date since you want to search posts
  search_term = "#coding"
  date_since = "2024-05-01" ❶

  # Collect posts ❷
  tweets = tweepy.Cursor(api.search_tweets,
                         q=search_term,
                         lang="en",
                         since=date_since).items(100)

  # Iterate and print posts ❸
  for tweet in tweets:
      print(tweet.text)
```

```
0         Just finished a great Python tutorial! #coding
1       Loving the new features in the latest Python r...
2       Struggling with a bug in my code. Can anyone h...
3         Just finished a great Python tutorial! #coding
4       Loving the new features in the latest Python r...
                                 ...
95      Struggling with a bug in my code. Can anyone h...
96        Just finished a great Python tutorial! #coding
97      Loving the new features in the latest Python r...
98      Struggling with a bug in my code. Can anyone h...
99        Just finished a great Python tutorial! #coding
Name: text, Length: 100, dtype: object
0         Just finished a great Python tutorial! #coding
1       Loving the new features in the latest Python r...
2       Struggling with a bug in my code. Can anyone h...
3         Just finished a great Python tutorial! #coding
4       Loving the new features in the latest Python r...
                                 ...
95      Struggling with a bug in my code. Can anyone h...
96        Just finished a great Python tutorial! #coding
97      Loving the new features in the latest Python r...
98      Struggling with a bug in my code. Can anyone h...
99        Just finished a great Python tutorial! #coding
Name: text, Length: 100, dtype: object
```

Figure 10-4. Output from program: 100 X posts

❶ Define search term and date. The variable `search_term` holds the hashtag `#coding`, which the code will search for in posts. The `date_since` variable defines the start date (`2024-05-01`), so only posts from that date onward will be collected.

❷ Collect posts using Tweepy Cursor. `tweepy.Cursor` is used to fetch tweets from X. The `api.search_tweets` method searches for tweets based on the query parameters provided. `q=search_term` specifies the search query (in this case, tweets containing `#coding`). `lang="en"` restricts the search to English-language tweets. `since=date_since` ensures tweets are only retrieved from the specified date onward. `.items(100)` limits the number of tweets fetched to 100.

❸ Iterate through and print tweets. This loop iterates over each tweet fetched by the cursor, and `tweet.text` prints the content of the tweet.

This code effectively fetches and prints up to 100 English-language tweets containing the hashtag `#coding` from May 1, 2024, onward, as depicted in Figure 10-4.

Cleaning and processing X posts with Tweepy. After fetching the posts, the next step is to clean and process the data to make it suitable for analysis. This involves removing unnecessary characters, handling missing values, and potentially normalizing the text. The following code demonstrates the steps for basic X post data cleaning:

```
import re
  import pandas as pd

  # Sample list to store tweets
  tweet_list = []

  for tweet in tweets:
      tweet_list.append(tweet.text)

  # Create a DataFrame
  df = pd.DataFrame(tweet_list, columns=['tweet'])

  # Function to clean tweet text
  def clean_tweet(text):
      text = re.sub(r'http\S+', '', text)  # Remove URLs
      text = re.sub(r'@\S+', '', text)     # Remove mentions
      text = re.sub(r'#\S+', '', text)     # Remove hashtags
      text = re.sub(r'\n', '', text)       # Remove new lines
      text = re.sub(r'\W', ' ', text)      # Remove non-alphanumeric characters
      text = text.lower()                  # Convert to lowercase
      return text

  # Apply the cleaning function
  df['cleaned_tweet'] = df['tweet'].apply(clean_tweet)

  # Show cleaned tweets
  print(df.head())
```

This code removes URLs, mentions, hashtags, new lines, and non-alphanumeric characters from the tweets, and converts the text to lowercase. By following these steps, you can effectively extract, clean, and process X data using Python and Tweepy, enabling you to perform various forms of social media analysis.

The previous example removes quite a bit of content. When cleansing tweets, being less aggressive can be crucial to preserving important contextual elements that contribute to the meaning of the text. For example, instead of completely removing mentions (@usernames), hashtags, and URLs, you might want to replace them with placeholders like [USER], [HASHTAG], or [URL] to retain the structure and intent of the tweet. Additionally, allowing hashtags to keep their associated keywords (while just removing the # symbol) can preserve valuable information that might be related to topics or sentiments in the tweet. This way, you maintain more of the original content and context, making the data richer for analysis while still performing essential cleaning.

Facebook and data extraction using Facebook Graph API

Social media platforms like Facebook and Instagram are treasure troves of data that can be used for various purposes, including market research, sentiment analysis, and trend analysis. The Facebook Graph API provides a unified interface to access data from both Facebook and Instagram. By using Python, developers can automate the extraction of this data, making it easier to analyze and derive insights. This section will walk you through setting up the Facebook Graph API, extracting data from Facebook and Instagram, and processing this data for analysis.

Setting up Facebook Graph API. To set up the Facebook Graph API, begin by creating a Meta for Developers account (*https://oreil.ly/UvVnw*). Visit the Facebook for Developers (see Figure 10-5) website and sign up if you don't already have an account. Once you're logged in, navigate to the Meta for Developers Dashboard, click My Apps, and select Create App. Choose an app type based on your use case, such as For Everything Else, and provide the necessary details like your app name and contact email. After the app is created, go to the app dashboard and click "Add a Product." Select Facebook Login and follow the prompts to configure it according to your needs (e.g., for web or mobile platforms).

Next, generate access tokens by selecting Access Token Tool from the Tools section in the dashboard. Ensure that you request the correct permissions based on the data you want to access, such as `user_posts` for Facebook or `user_media` for Instagram. These tokens are essential for accessing the data via the Facebook Graph API.

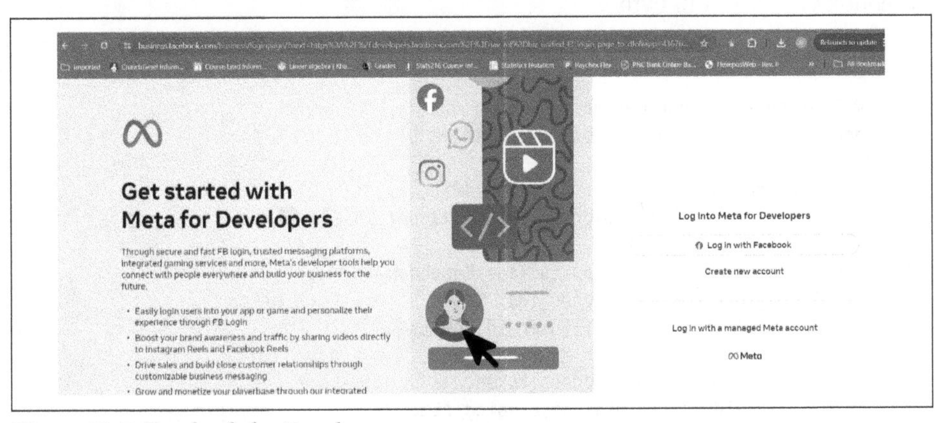

Figure 10-5. Facebook for Developers

Data extraction using Facebook Graph API. Once you have set up Facebook Graph API, you're then ready to extract Facebook or Instagram data. Before you start, make sure you have the requests library installed. You can install it using pip:

```
pip install requests
```

In the code provided, both the access token and user IDs are placeholders that need to be replaced with actual values to make the requests work.

The `access_token` is a critical security credential that you must obtain from Facebook's developer platform or Instagram's API. This token authenticates your requests to the Facebook or Instagram API and allows you to access specific user data based on the permissions granted to your app. You need to replace `'your_access_token'` with the actual token string you get when you authenticate using the API. For example:

```
access_token = 'EAAGm0PX4ZCpsBAL6ZBSpzZ...'  # Example of an access token
```

Similarly, `user_id` is the unique identifier for a Facebook or Instagram user. You need to replace `'your_facebook_user_id'` and `'your_instagram_user_id'` with the actual IDs of the users whose posts or media you want to fetch. These IDs can typically be retrieved through initial API requests or from your app's backend:

```
facebook_user_id = '123456789'
  instagram_user_id = '987654321'
```

Once these values are replaced, the code will be able to fetch posts or media from Facebook and Instagram for the specific users. The access token ensures that you have the necessary permissions to access the data, while the user IDs specify which user's content to retrieve.

Here's an example of how to extract data from Facebook and Instagram using the Facebook Graph API in Python:

```
import requests

  # Replace with your own access token
  access_token = 'your_access_token' ❶

  # Define the base URL for the Graph API
  graph_api_url = 'https://graph.facebook.com/v15.0/' ❷

  # Function to fetch user posts from Facebook
  def fetch_facebook_posts(user_id): ❸
      url = f'{graph_api_url}{user_id}/posts'
      params = {
          'access_token': access_token,
          'fields': 'id,message,created_time'
      }
      response = requests.get(url, params=params)
      data = response.json()
      return data

  # Function to fetch user media from Instagram
  def fetch_instagram_media(user_id): ❹
      url = f'{graph_api_url}{user_id}/media'
      params = {
```

```
        'access_token': access_token,
        'fields': 'id,caption,media_type,media_url,timestamp'
    }
    response = requests.get(url, params=params)
    data = response.json()
    return data

# Fetch Facebook posts
facebook_user_id = 'your_facebook_user_id'
facebook_posts = fetch_facebook_posts(facebook_user_id)
print("Facebook Posts:")
for post in facebook_posts['data']:
    print(post)

# Fetch Instagram media
instagram_user_id = 'your_instagram_user_id'
instagram_media = fetch_instagram_media(instagram_user_id)
print("\nInstagram Media:")
for media in instagram_media['data']:
    print(media) ❺
```

❶ The access token is a unique key that allows you to authenticate and access the Facebook Graph API. It should be kept secure and not shared publicly.

❷ This is the base URL for making requests to the Facebook Graph API. The version number (e.g., v15.0) should be updated according to the latest API version.

❸ The `fetch_facebook_posts` function constructs the URL for fetching posts from a user's timeline. It includes parameters such as `access_token` and fields to specify the data you want to retrieve (e.g., post ID, message, and creation time).

❹ Similarly, the `fetch_instagram_media` function constructs the URL for fetching media from a user's Instagram account. The parameters include the `access_token` and fields to specify the media data you want (e.g., media ID, caption, media type, media URL, and timestamp).

❺ The fetched data is printed to the console, where each post or media item is displayed.

This example demonstrates how to set up the Facebook Graph API and use it to extract data from both Facebook and Instagram using Python. By adjusting the parameters and endpoints, you can customize the data extraction to suit your specific needs.

Sentiment analysis on social media data in Python

Sentiment analysis is used for understanding the emotions and opinions expressed in social media data. By analyzing the sentiment of tweets, posts, and comments, you can gain insights into public opinion, brand perception, and user satisfaction. In this section, we will address how to use Python libraries such as TextBlob and NLTK for sentiment analysis and Matplotlib or Seaborn to visualize sentiment trends.

TextBlob simplifies tasks like tokenization, lemmatization, and part-of-speech tagging, making it a user-friendly tool for analyzing sentiment in text by providing polarity and subjectivity scores. NLTK, a more comprehensive library, offers a wide range of features for text processing, including tokenization, stopword removal, and stemming. Together, these tools can be used to clean and structure raw text data, enabling more accurate sentiment analysis. For instance, TextBlob can easily classify text as positive, negative, or neutral, while NLTK can help preprocess the text to remove noise and irrelevant information, improving model performance.

Performing sentiment analysis with TextBlob or NLTK libraries. TextBlob and NLTK are two popular Python libraries for NLP that can be used for sentiment analysis. The NLTK library provides a sentiment analysis tool called VADER that is specifically tuned for analyzing social media text and returns a set of scores that indicate the sentiment expressed in the text.

Before using the TextBlob and NLTK libraries, you will need to install them with pip (may need to run these independently):

```
pip install textblob
  pip install nltk
```

To then use the libraries, import them by doing the following:

```
from textblob import TextBlob
  import nltk
  from nltk.sentiment.vader import SentimentIntensityAnalyzer

  # Download NLTK data (if not already downloaded)
  nltk.download('vader_lexicon')
```

Now that the libraries have been installed and imported, you are now ready to perform sentiment analysis using TextBlob:

```
def analyze_sentiment_textblob(text):
      blob = TextBlob(text)
      return blob.sentiment.polarity

  # Example usage
  text = "I love using Python for data analysis!"
  polarity = analyze_sentiment_textblob(text)
  print(f'TextBlob Sentiment Polarity: {polarity}')
```

The code example produces the following output:

```
TextBlob Sentiment Polarity: 0.625
```

In TextBlob, sentiment analysis is done using two primary metrics: polarity and subjectivity.

When analyzing polarity, the guidelines used are:

- Positive Polarity (> 0): Indicates positive emotions, praise, satisfaction, etc.
- Negative Polarity (< 0): Indicates negative emotions, criticism, dissatisfaction, etc.
- Neutral Polarity (= 0): Indicates a neutral statement without strong positive or negative sentiments.

In the example, the results of the analysis of a sample text (could be a post or other social media type message). Suppose the content of a social media post is `"Just wrote a new script in Python! #Python #Coding"`. When TextBlob analyzes this sentence, it determines that the sentiment polarity is 0.625. This means that the analyzed text is quite positive because it's a statement of achievement and enthusiasm, though not the maximum possible (which is 1).

This score helps in quantifying the sentiment expressed in text, allowing for analysis and comparison across different texts or sets of data.

Now let's look at an example for a more nuanced sentiment analysis using NLTK's VADER tool:

```
def analyze_sentiment_nltk(text):
      sid = SentimentIntensityAnalyzer()
      sentiment_scores = sid.polarity_scores(text)
      return sentiment_scores

  # Example usage
  text = "I love using Python for data analysis!"
  sentiment = analyze_sentiment_nltk(text)
  print(f'NLTK VADESentiment Scores: {sentiment}')
```

The code example produces the following output:

```
NLTK VADESentiment Scores: {'neg': 0.0, 'neu': 0.527, 'pos': 0.473,
  'compound': 0.6696}
```

Let's look at how these results are interpreted.

When analyzing VADER sentiment scores, you use the following guidelines:

- neg: Proportion of negative sentiment in the text. Higher values indicate stronger negative sentiment.
- neu: Proportion of neutral sentiment in the text. Values close to 1.0 indicate a more neutral sentiment.

- pos: Proportion of positive sentiment in the text. Higher values indicate stronger positive sentiment.
- compound: The overall sentiment score that provides a quick summary of the overall sentiment. This is a normalized score that sums the positive, neutral, and negative scores and then scales them to be between –1 (most extreme negative) and +1 (most extreme positive).

Suppose the content of a social media post is `"Just wrote a new script in Python! #Python #Coding"`. You could interpret the results as follows:

- neg: 0.0: There is no negative sentiment in the text.
- neu: 0.527: The text is 52.7% neutral.
- pos: 0.473: The text is 47.3% positive.
- compound: 0.6696: The overall sentiment is positive, with a compound score of 0.6696, indicating a fairly strong positive sentiment.

Overall, the VADER analysis indicates that there is no negative sentiment, the text is slightly more neutral than positive, but overall, it is positive.

By using VADER Scores, you can perform detailed sentiment analysis on text data, allowing for more nuanced understanding and comparison of sentiment across different texts.

Let's look at one more example using both TextBlob and NLTK's VADER. Assume you have fetched tweets using Tweepy and stored them in a list.

Here's how you can perform sentiment analysis on these tweets:

```
import pandas as pd

  tweets = [
      "I love the new features in the latest update!",
      "The service was fine.",
      "The service has been really slow lately, very disappointed.",
      "Amazing experience, would highly recommend!",
      "Not happy with the customer service."
  ]

  # Assuming analyze_sentiment_textblob and analyze_sentiment_nltk are defined
  # functions that return sentiment scores, e.g., polarity for TextBlob and
  # compound for VADER.
  textblob_sentiments = [analyze_sentiment_textblob(tweet) for tweet in tweets]
  nltk_sentiments = [analyze_sentiment_nltk(tweet) for tweet in tweets]

  # Ensure that all columns are fully visible
  pd.set_option('display.max_colwidth', None)
```

```
# Create a DataFrame to organize the output in tabular format
df = pd.DataFrame({
    'Tweet': tweets,
    'TextBlob Sentiment': textblob_sentiments,
    'NLTK VADER Sentiment': nltk_sentiments
})

# Print the DataFrame as a table
print(df)
7]
0s
    'Tweet': tweets,
    'TextBlob Sentiment': textblob_sentiments,
    'NLTK VADER Sentiment': nltk_sentiments
})

# Print the DataFrame as a table
print(df)

                                                   Tweet  \
0                I love the new features in the latest update!
1                                    The service was fine.
2  The service has been really slow lately, very disappointed.
3                  Amazing experience, would highly recommend!
4                         Not happy with the customer service.

   TextBlob Sentiment  \
0            0.420455
1            0.416667
2           -0.637500
3            0.400000
4           -0.400000

                                             NLTK VADER Sentiment
0   {'neg': 0.0, 'neu': 0.609, 'pos': 0.391, 'compound': 0.6696}
1   {'neg': 0.0, 'neu': 0.625, 'pos': 0.375, 'compound': 0.2023}
2  {'neg': 0.298, 'neu': 0.702, 'pos': 0.0, 'compound': -0.5256}
3   {'neg': 0.0, 'neu': 0.303, 'pos': 0.697, 'compound': 0.7836}
4  {'neg': 0.375, 'neu': 0.625, 'pos': 0.0, 'compound': -0.4585}
```

Table 10-1 includes the sentiment analysis results for each tweet.

Table 10-1. Sentiment analysis results

Row number	TextBlob sentiment	NLTK VADER negative	NLTK VADER neutral	NLTK VADER positive	NLTK VADER compound
1	0.4205	0.000	0.609	0.391	0.6696
2	0.4167	0.000	0.625	0.375	0.2023
3	−0.6375	0.298	0.702	0.000	−0.5256
4	0.4000	0.000	0.303	0.697	0.7836
5	−0.4000	0.375	0.625	0.000	−0.4585

In Table 10-1, TextBlob and NLTK VADER sentiment analysis are compared. The following conclusions can be made:

Alignment in positive sentiments
: Both TextBlob and NLTK VADER generally agree on the positive sentiment. For example, the first and second rows have positive scores in both analyses, with VADER showing a higher positive component in alignment with TextBlob's relatively high positive sentiment scores.

Disagreement on negative sentiments
: In cases where TextBlob identifies negative sentiment (such as row 3 with a score of –0.6375), NLTK VADER also detects a negative sentiment with a corresponding negative compound score (–0.5256). This indicates agreement on the presence of negativity, though NLTK VADER provides more detailed analysis on how much of the sentiment is negative versus neutral.

Detailed sentiment components in VADER
: NLTK VADER provides more granular information. For example, in row 3, where TextBlob gives an overall negative score, VADER breaks it down into 29.8% negative, 70.2% neutral, and no positive sentiment. This gives more insight into the sentiment mix, whereas TextBlob offers a simpler polarity score.

TextBlob is simpler
: TextBlob offers a single polarity score ranging from –1 (negative) to +1 (positive), making it easier to interpret but less detailed than VADER, which breaks sentiment into negative, neutral, and positive components, along with a compound score summarizing the overall sentiment.

Strong positive and negative agreement
: Where TextBlob identifies very strong positive or negative sentiments (such as rows 3 and 4), VADER tends to corroborate these findings with either high positive or negative compound scores.

In summary, TextBlob provides a quick, simple assessment of overall sentiment, while NLTK VADER offers more detailed insights, especially when the sentiment is mixed or contains significant neutral content. This can be useful depending on whether you're looking for a high-level sentiment overview (TextBlob) or a more detailed breakdown (VADER).

Finally, let's look at an example using Facebook and Instagram data and TextBlob. In this example, we are starting with a JSON file that contains 1,000 records from Facebook and Instagram. JSON is the default structure when the Facebook API is used to extract data. The following is a short Python script that analyzes sentiment using TextBlob:

```
import json
from textblob import TextBlob

# Load the JSON file
with open('social_media_posts.json', 'r') as f:
    posts = json.load(f)

# Function to analyze sentiment
def analyze_sentiment(post):
    blob = TextBlob(post['content'])
    sentiment = blob.sentiment
    post['polarity'] = sentiment.polarity
    post['subjectivity'] = sentiment.subjectivity
    if sentiment.polarity > 0:
        post['sentiment'] = 'positive'
    elif sentiment.polarity < 0:
        post['sentiment'] = 'negative'
    else:
        post['sentiment'] = 'neutral'
    return post

# Apply sentiment analysis to each post
analyzed_posts = [analyze_sentiment(post) for post in posts]

# Optionally save the analyzed data to a new JSON file
with open('analyzed_social_media_posts.json', 'w') as f:
    json.dump(analyzed_posts, f, indent=4)

# Print a sample analyzed post to verify
print(json.dumps(analyzed_posts[0], indent=4))
{
   "platform": "Facebook",
   "username": "james94",
   "post_id": "8bbb3f6e-19f5-4c42-88ef-10af36d6d17b",
   "content": "Trying out some new Python packages. #Python #Coding",
   "timestamp": "2024-06-12T17:31:30.468937",
   "likes": 62,
   "comments": 433,
   "shares": 19,
   "polarity": 0.13636363636363635,
   "subjectivity": 0.45454545454545453,
   "sentiment": "positive"
}
```

This script loads the JSON file containing the generated social media posts, performs sentiment analysis on the content of each post using TextBlob, and adds the polarity, subjectivity, and overall sentiment (positive, negative, neutral) to each post. The analyzed posts are then optionally saved to a new JSON file.

Visualizing sentiment trends with Matplotlib or Seaborn libraries. Visualizing the sentiment trends over time can help you understand how opinions change. You can use libraries like Matplotlib or Seaborn for visualization. Plotly could also be used.

Before using the Matplotlib or Seaborn visualization libraries, you will need to install them with pip:

```
pip install matplotlib
  pip install seaborn
```

The following script demonstrates how to visualize the sentiment trends over time using Matplotlib and Seaborn:

```
import matplotlib.pyplot as plt
  import seaborn as sns
  import pandas as pd

  # Sample data: List of sentiment scores and corresponding dates
  dates = ["2023-01-01", "2023-01-02", "2023-01-03", "2023-01-04"]
  textblob_sentiments = [0.5, -0.2, 0.8, -0.5]
  nltk_sentiments = [0.6, -0.3, 0.7, -0.6]

  # Convert dates to a format suitable for plotting
  date_range = pd.to_datetime(dates)

  # Create a DataFrame for easier plotting
  data = pd.DataFrame({
      'Date': date_range,
      'TextBlob Sentiment': textblob_sentiments,
      'NLTK Sentiment': nltk_sentiments  # Use the raw float values directly
  })

  # Plotting the sentiment trends
  plt.figure(figsize=(12, 6))
  sns.lineplot(x='Date', y='TextBlob Sentiment', data=data, label='TextBlob
  Sentiment', marker='o')
  sns.lineplot(x='Date', y='NLTK Sentiment', data=data, label='NLTK Sentiment',
  marker='x')
  plt.xlabel('Date')
  plt.ylabel('Sentiment Score')
  plt.title('Sentiment Trends Over Time')
  plt.legend()
  plt.show()
```

The code produces the chart depicted in Figure 10-6.

By using Matplotlib and Seaborn for visualizing sentiment trends, you can gain valuable insights into the sentiment expressed in social media posts, helping you make informed decisions based on public opinion.

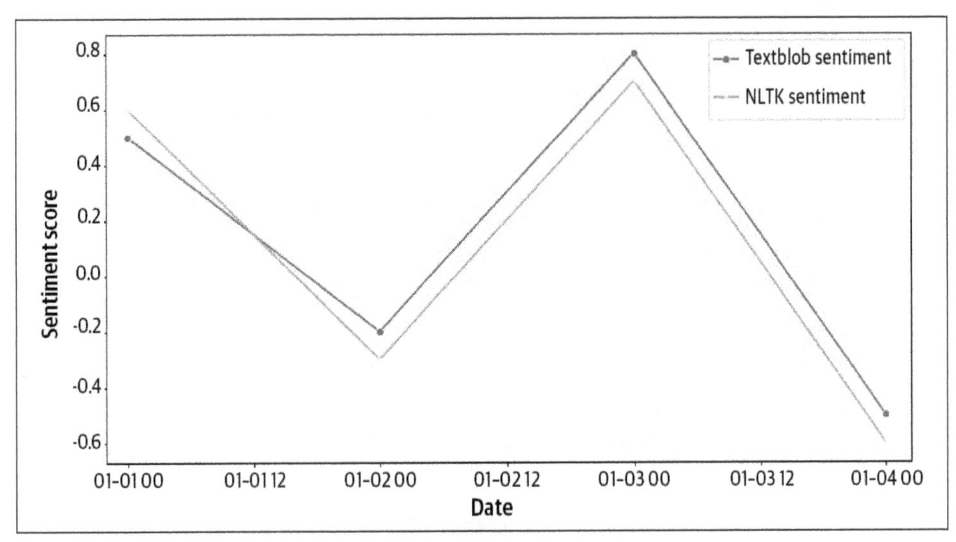

Figure 10-6. Sentiment trends over time

Next, let's look at an example that visualizes Facebook and Instagram data. To start, we will generate synthetic social media posts:

```
import random
import json
from faker import Faker

fake = Faker()

# Predefined Python-related content
python_content = [
    "Just wrote a new script in Python! #Python #Coding",
    "Learning about list comprehensions today. #Python #Coding",
    "Debugging some tricky Python code. #Python #Coding",
    "Exploring Python's standard library. #Python #Coding",
    "Just solved a problem using a Python dictionary! #Python #Coding",
    "Lambda functions are so powerful! #Python #Coding",
    "Trying out some new Python packages. #Python #Coding",
    "Understanding Python's GIL. #Python #Coding",
    "Python makes data analysis so easy! #Python #Coding",
    "Python makes data analysis hard! #Python #Coding",
    "Cannot solve a single problem with Python! #Python #Coding",
    "Just learned about Python decorators. #Python #Coding",
]

# Define a function to generate synthetic Facebook and Instagram posts
def generate_social_media_posts(num_posts, platform):
    posts = []
    for _ in range(num_posts):
        post = {
            'platform': platform,
```

```
            'username': fake.user_name(),
            'post_id': fake.uuid4(),
            'content': random.choice(python_content),
            'timestamp': fake.date_time_this_year().isoformat(),
            'likes': random.randint(0, 1000),
            'comments': random.randint(0, 500),
            'shares': random.randint(0, 200) if platform == 'Facebook' else 0
        }
        posts.append(post)
    return posts

# Generate 1,000 Facebook posts
facebook_posts = generate_social_media_posts(1000, 'Facebook')

# Generate 1,000 Instagram posts
instagram_posts = generate_social_media_posts(1000, 'Instagram')

# Combine the posts
all_posts = facebook_posts + instagram_posts

# Optionally save to a JSON file
with open('social_media_posts.json', 'w') as f:
    json.dump(all_posts, f, indent=4)

# Print a sample post to verify
print(json.dumps(all_posts[0], indent=4))
```

This code generates synthetic social media posts for Facebook and Instagram, using Python's Faker library to create random user data and timestamps, and outputs the posts in a structured format. Here are the high-level steps:

1. Import required libraries: random, json, and faker. random is used for selecting random values. json saves data as a JSON file. faker generates fake data (usernames, timestamps, UUIDs).
2. A predefined list `python_content` is created with various Python-related social media posts. The function `generate_social_media_posts(num_posts, platform)`, for example, generates a list of fake posts for the specified platform (Facebook or Instagram).
3. For each post, random user information (`username`, `post_id`) is generated. A random post content is selected from the predefined list. Timestamp, likes, comments, and shares (for Facebook only) are generated randomly.
4. Generate Facebook and Instagram posts. `generate_social_media_posts` is called to generate 1,000 posts for Facebook and 1,000 posts for Instagram.
5. Posts are combined into a single list, `all_posts`.

6. The combined posts are saved as a JSON file named *social_media_posts.json.*
7. A sample post from `all_posts` is printed in a readable format to verify the generated data. The resulting JSON file contains 2,000 synthetic social media posts with realistic user and post data.

Then sentiment analysis ratings are applied:

```
import json
from textblob import TextBlob

# Load the JSON file
with open('social_media_posts.json', 'r') as f:
    posts = json.load(f)

# Function to analyze sentiment
def analyze_sentiment(post):
    blob = TextBlob(post['content'])
    sentiment = blob.sentiment
    post['polarity'] = sentiment.polarity
    post['subjectivity'] = sentiment.subjectivity
    if sentiment.polarity > 0:
        post['sentiment'] = 'positive'
    elif sentiment.polarity < 0:
        post['sentiment'] = 'negative'
    else:
        post['sentiment'] = 'neutral'
    return post

# Apply sentiment analysis to each post
analyzed_posts = [analyze_sentiment(post) for post in posts]

# Optionally save the analyzed data to a new JSON file
with open('analyzed_social_media_posts.json', 'w') as f:
    json.dump(analyzed_posts, f, indent=4)

# Print a sample analyzed post to verify
print(json.dumps(analyzed_posts[0], indent=4))
{
    "platform": "Facebook",
    "username": "rsuarez",
    "post_id": "f7949e16-8f1d-4795-8c03-a25245a90daa",
    "content": "Python makes data analysis hard! #Python #Coding",
    "timestamp": "2024-08-24T22:30:17.679710",
    "likes": 535,
    "comments": 463,
    "shares": 24,
    "polarity": -0.36458333333333337,
    "subjectivity": 0.5416666666666666,
    "sentiment": "negative"
}
```

This code loads a JSON file of social media posts and analyzes the sentiment of each post using TextBlob:

1. The JSON file is read using *social_media_posts.json* to load the posts into memory.
2. Sentiment analysis is then applied by analyzing the content and adding polarity, subjectivity, and sentiment for each post.
3. The analyzed posts are saved to *analyzed_social_media_posts.json*, a new JSON file.
4. One sample post is printed with the sentiment analysis results for verification.

Now the sentiments will be visualized:

```
import json
  import pandas as pd
  import matplotlib.pyplot as plt

  # Load the analyzed JSON file
  with open('analyzed_social_media_posts.json', 'r') as f:
      analyzed_posts = json.load(f)

  # Convert the data to a DataFrame
  df = pd.DataFrame(analyzed_posts)

  # Plot the distribution of sentiments
  plt.figure(figsize=(10, 6))
  df['sentiment'].value_counts().plot(kind='bar', color=['green', 'red', 'blue'])
  plt.title('Sentiment Distribution of Social Media Posts')
  plt.xlabel('Sentiment')
  plt.ylabel('Number of Posts')
  plt.show()

  # Plot the average polarity and subjectivity for each sentiment
  average_sentiment = df.groupby('sentiment')[['polarity',
  'subjectivity']].mean()
  average_sentiment.plot(kind='bar', figsize=(12, 8), color=['purple',
  'orange'])
  plt.title('Average Polarity and Subjectivity by Sentiment')
  plt.xlabel('Sentiment')
  plt.ylabel('Average Value')
  plt.show()

  # Plot the number of posts over time
  df['timestamp'] = pd.to_datetime(df['timestamp'])
  df.set_index('timestamp', inplace=True)
  df['sentiment'].resample('M').count().plot(figsize=(14, 7), color='skyblue')
  plt.title('Number of Posts Over Time')
  plt.xlabel('Date')
```

```
plt.ylabel('Number of Posts')
plt.show()
```

This code example produces Figure 10-7, which contains the bar chart for sentiment distribution; Figure 10-8, which contains the bar chart for average polarity and subjectivity by sentiment; and Figure 10-9, which contains the line chart showing the number of posts over time.

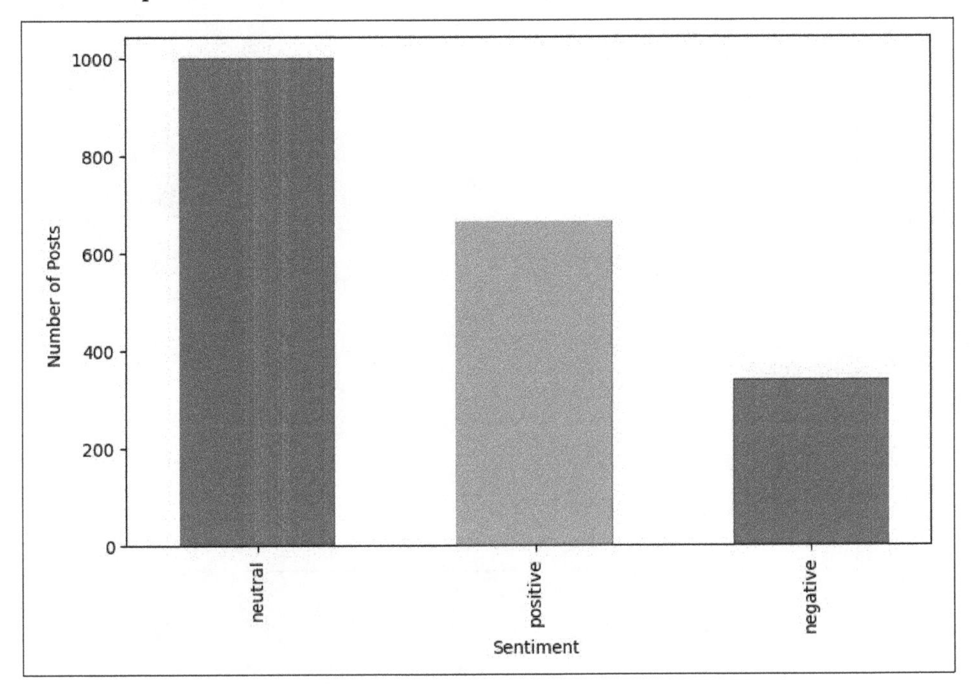

Figure 10-7. Sentiment distribution

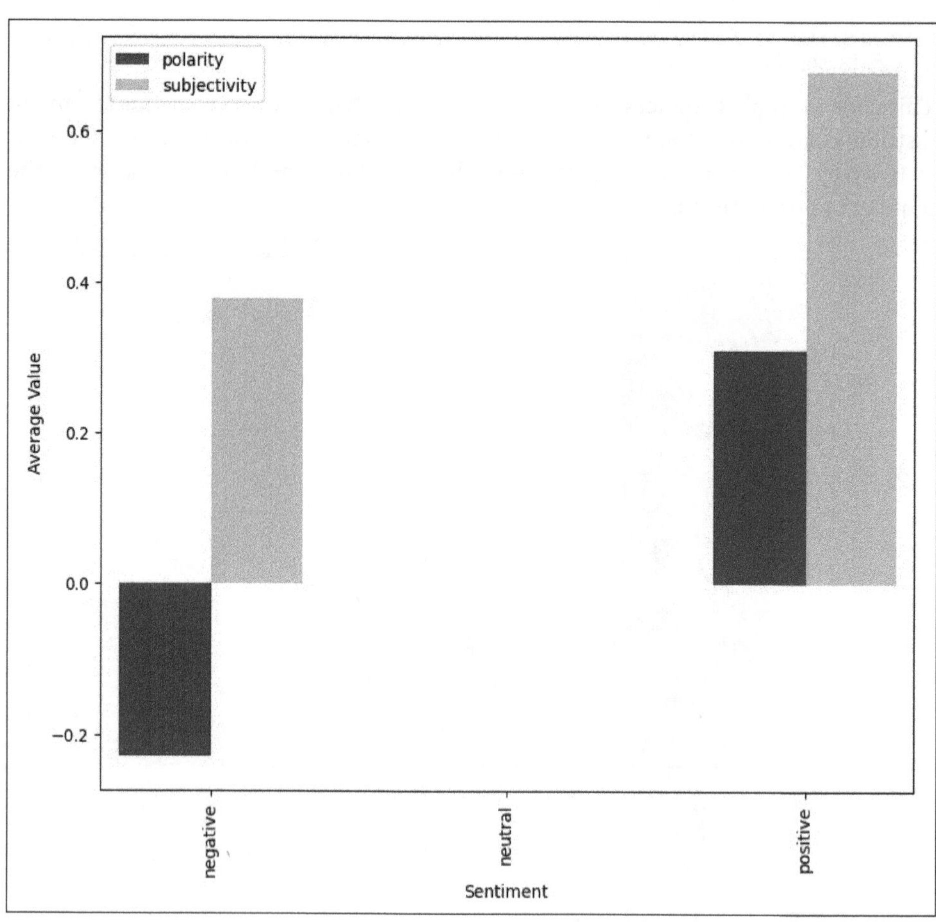

Figure 10-8. Average polarity and subjectivity by sentiment

These visualizations will help you understand the sentiment distribution, the average polarity and subjectivity for each sentiment, and the number of posts over time.

Let's explore the same social media capabilities in R.

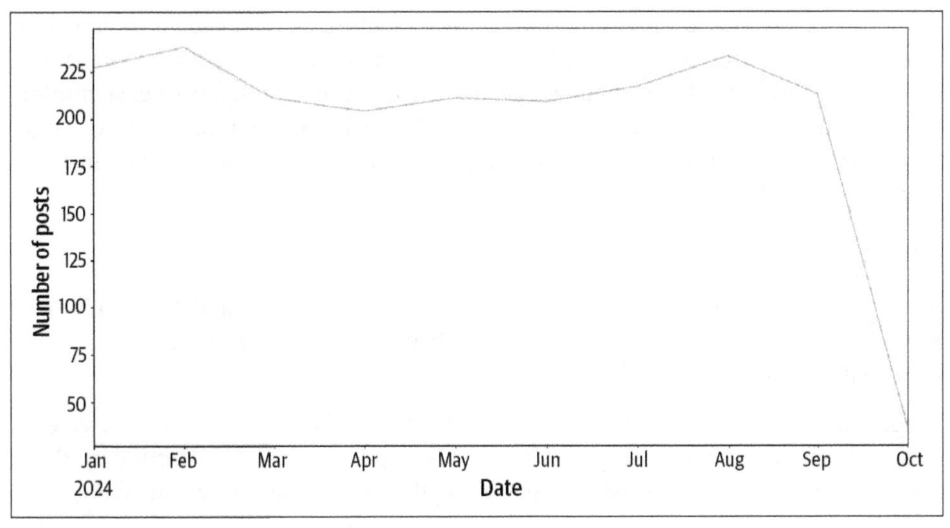

Figure 10-9. Sentiment posts over time

Using R for Social Media Data Analysis

R provides tools for social media data analysis, including Facebook data extraction and sentiment analysis. With recent updates, packages like Rfacebook have been superseded by httr and facebook.S4 for retrieving data via Facebook's Graph API, ensuring compatibility with the latest API versions. After importing data, libraries like tidytext or dplyr are instrumental in cleaning and manipulating text. Sentiment analysis can be performed using updated packages such as sentimentr or quanteda, which rely on pretrained lexicons or machine learning models to classify sentiment as positive, negative, or neutral. These analyses enable businesses to gauge public sentiment on a brand, product, or event, providing critical insights into the emotional tone of online discussions.

Httr is an R package used for making HTTP requests to web APIs. It simplifies the process of sending GET, POST, and other types of requests and handling the responses in R. It's commonly used to interact with REST APIs, including retrieving data from web services like Facebook, X, and other platforms that provide API access. Facebook.S4 is a package designed to interact with Facebook's Graph API. It allows users to retrieve data such as posts, likes, comments, and user profiles from Facebook. The package simplifies API authentication and data extraction, providing a structured way to access Facebook data for analysis and integration into R workflows.

Sentimentr is used for performing sentiment analysis on text data. It calculates sentiment scores by examining the polarity of words and their relationships within a sentence. It goes beyond simply counting positive or negative words, using context to improve accuracy. It is often applied to social media data, customer reviews, and any other text-based data to gauge emotional tone or sentiment. Quanteda is designed for quantitative text analysis. It provides a comprehensive suite of tools for managing, processing, and analyzing textual data, including functions for tokenizing, stemming, and creating document-term matrices. Quanteda is particularly well suited for large-scale text mining and NLP tasks, including sentiment analysis, topic modeling, and text classification.

X data extraction using rtweet

The rtweet package is a powerful tool for collecting and analyzing data from X. This section will walk you through setting up the rtweet package, extracting data, and processing it for analysis.

To get started with rtweet, you'll need to set up an X Developer account, create an app, and generate the necessary API keys. This step is described in "Setting up the X API" on page 369. The screenshots from the Python section apply here as well.

You will then need to install the rtweet package from CRAN and load it into your session:

```
install.packages("rtweet")
  library(rtweet)
```

Then, use your API keys to authenticate your requests. You can set up authentication using the `create_token` function:

```
# Replace with your own keys and tokens
  api_key <- "your_api_key"
  api_secret_key <- "your_api_secret_key"
  access_token <- "your_access_token"
  access_token_secret <- "your_access_token_secret"

  # Create token
  token <- create_token(
    app = "your_app_name",
    consumer_key = api_key,
    consumer_secret = api_secret_key,
    access_token = access_token,
    access_secret = access_token_secret
  )
```

Fetching posts using rtweet. Once authenticated, you can start fetching tweets. Here's an example of how to search for recent posts containing a specific hashtag (`#coding`):

```
# Search for tweets containing the hashtag #coding
  tweets <- search_tweets("#coding", n = 100, lang = "en", token = token)

  # View the structure of the fetched tweets
  str(tweets)
```

The code example produces the following output:

```
chr [1:100] "Learning about list comprehensions today. #Coding"
    "Python makes data analysis so easy! #DataScience"
```

Cleaning and processing tweet data using rtweet. After fetching the tweets, the next step is to clean and process the data to make it suitable for analysis:

```
# Load necessary library for text cleaning
  library(dplyr)

  # Clean the tweet text
  clean_tweet <- function(tweet) {
    tweet %>%
      gsub("http\\S+|www\\S+", "", .) %>% # Remove URLs
      gsub("@\\w+", "", .) %>%            # Remove mentions
      gsub("#\\w+", "", .) %>%            # Remove hashtags
      gsub("[^\\x20-\\x7E]", "", .) %>%   # Remove non-ASCII characters
      tolower()                           # Convert to lowercase
  }

  # Apply the cleaning function to the tweet text
  tweets$cleaned_text <- sapply(tweets$text, clean_tweet)

  # View cleaned tweets
  head(tweets$cleaned_text)

  [1] "just wrote a new script in python python coding"    "the service has been
    really slow lately very disappointed"
  [3] "amazing experience would highly recommend"          "not happy with the
    customer service"
```

Analyzing and visualizing post data. Once you've fetched and cleaned the tweet data, you can now perform sentiment analysis and visualize the sentiment trends. The following example demonstrates how to set up and use the rtweet package for extracting, cleaning, and analyzing data from X:

```
# Load necessary libraries for sentiment analysis and visualization
  library(tidytext)
  library(ggplot2)
  library(dplyr)

  # Ensure the tidytext library is loaded
  if (!requireNamespace("tidytext", quietly = TRUE)) {
    install.packages("tidytext")
  }
  library(tidytext)

  # Load the Bing sentiment lexicon directly from the tidytext datasets
  bing <- tidytext::get_sentiments("bing")

  # Assuming tweets_df already exists and is cleansed
  tweets_df <- data.frame(cleaned_text = c(
    "Just wrote a new script in Python!",
```

```
  "The service has been really slow lately, very disappointed.",
  "Amazing experience, would highly recommend!",
  "Not happy with the customer service."
), stringsAsFactors = FALSE)

# Tokenize the cleaned tweet text
tweet_tokens <- tweets_df %>%
  unnest_tokens(word, cleaned_text)

# Join with the Bing sentiment lexicon
tweet_sentiment <- tweet_tokens %>%
  inner_join(bing, by = "word")

# Count the number of positive and negative words
sentiment_counts <- tweet_sentiment %>%
  count(sentiment, sort = TRUE)

# Visualize the sentiment counts
ggplot(sentiment_counts, aes(x = sentiment, y = n, fill = sentiment)) +
  geom_bar(stat = "identity") +
  theme_minimal() +
  labs(title = "Sentiment Analysis of Tweets",
       x = "Sentiment",
       y = "Count")
```

Figure 10-10 shows the output.

Here is a summary of the sentiment analysis code and visualizing the sentiment trends:

- Begin by setting up your X Developer account and creating an app to get the necessary API keys. This allows you to authenticate your requests to the X API.
- Install rtweet to interact with the X API.
- Use the `create_token` function with your API keys to authenticate your session.
- Fetch tweets containing a specific hashtag with the `search_tweets` function. You can adjust the number of tweets (`n`) and the language (`lang`) as needed.
- Clean the tweet text to remove URLs, mentions, hashtags, and non-ASCII characters, and to convert the text to lowercase.
- Tokenize the cleaned tweet text and join it with the `Bing sentiment lexicon` to perform sentiment analysis. The sentiment counts are then visualized using ggplot2.

By using the rtweet package for extracting, cleaning, and analyzing data from X, it can provide you with valuable insights into public sentiment and trends.

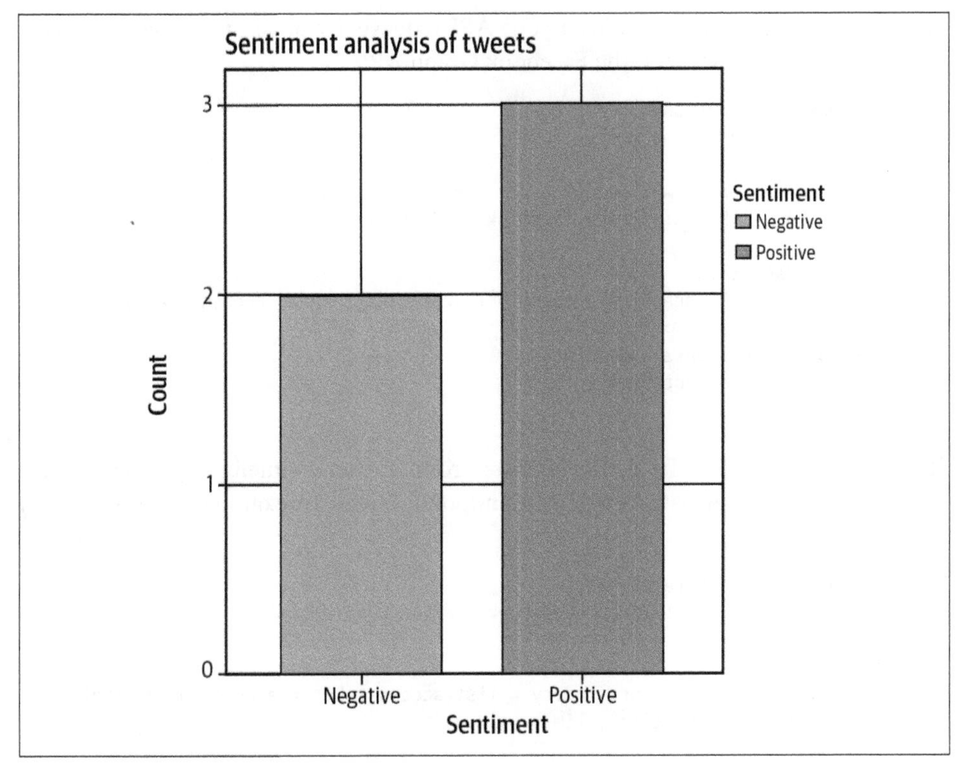

Figure 10-10. Positive and negative tweet counts

Facebook and Instagram data extraction

The facebook.S4 package is a modern tool for collecting and analyzing data from Facebook's Graph API, while httr assists with handling API requests and responses. To get started with Facebook data collection, you'll need to set up a Meta for Developers account, create an app, and generate the necessary access tokens. The steps for doing so are outlined in the Python section but apply similarly for R users.

Once you've created the app and obtained your access token, you can authenticate your requests and start fetching data from Facebook and Instagram.

Installing necessary packages. To get started, you need to install the httr and facebook.S4 packages from CRAN:

```
install.packages("httr")
  install.packages("facebook.S4")
  library(httr)
  library(facebook.S4)
```

First, you will need to authenticate your API requests by passing your access token, allowing you to interact with the Facebook Graph API:

```
# Replace with your own access token
  access_token <- 'your_access_token'

  # Define the Graph API endpoint for user information
  graph_url <- "https://graph.facebook.com/v15.0/me"

  # Fetch user info
  user_info <- GET(graph_url, query = list(access_token = access_token))

  # Parse the response and print
  user_data <- content(user_info)
  print(user_data)
```

Fetching Facebook data. To retrieve posts from a user's timeline, you can make requests to the appropriate Graph API endpoint. Here's an example of how to fetch recent posts:

```
# Define the URL to fetch posts
  posts_url <- "https://graph.facebook.com/v15.0/me/posts"

  # Fetch posts
  my_posts <- GET(posts_url, query = list(access_token = access_token, fields =
  "id,message,created_time", limit = 100))

  # Parse the response
  posts_data <- content(my_posts)
  print(posts_data)
```

The code example produces the following output:

```
[1] "Python's dynamic typing is causing bugs in my code. #Python #Coding
#Python"
[2] "Finding it hard to understand Python's scoping rules. #Python #Coding
     #Python"
[3] "Learning about list comprehensions today. #Python #Coding #DataScience"
[4] "Just wrote a new script in Python! #Python #Coding #DataScience"
[5] "Just learned about Python decorators. #Python #Coding #DataScience"
[6] "Debugging some tricky Python code. #Python #Coding #DataScience"
```

Fetching Instagram data. Fetching Instagram data is similar but requires proper permissions through your Meta for Developers account. Here's how to fetch media from an Instagram account linked to your Facebook account:

```
# Define the Instagram media endpoint
  instagram_url <- "https://graph.facebook.com/v15.0/me/media"

  # Fetch Instagram media
  instagram_media <- GET(instagram_url, query = list(access_token = access_token,
  fields = "id,caption,media_type,media_url,timestamp", limit = 100))
```

```
# Parse and print the response
instagram_data <- content(instagram_media)
print(instagram_data)
```

The code example produces the following output:

```
[1] "Just solved a problem using a Python dictionary! #Python #Coding #Python"
[2] "Exploring Python's standard library. #Python #Coding #Python"
[3] "Python's dynamic typing is causing bugs in my code. #Python #Coding
#Coding"
[4] "Python makes data analysis so easy! #Python #Coding #Python"
[5] "Debugging some tricky Python code. #Python #Coding #Coding"
[6] "Understanding Python's GIL. #Python #Coding #Python"
```

Cleaning and processing data. After fetching the data, it's essential to clean and process it to remove unnecessary characters, URLs, and special symbols for better readability and analysis:

```
# Function to clean post content
  clean_post <- function(post) {
    post <- tolower(post)  # Convert to lowercase
    post <- gsub("http\\S+\\s*", "", post)  # Remove URLs
    post <- gsub("[^[:alnum:] ]", "", post)  # Remove special characters
    post <- gsub("\\s+", " ", post)  # Remove extra whitespaces
    post <- trimws(post)  # Trim leading/trailing whitespaces
    return(post)
  }

  # Clean Facebook posts
  facebook_posts_cleaned <- sapply(posts_data$data$message, clean_post)

  # Clean Instagram media captions
  instagram_posts_cleaned <- sapply(instagram_data$data$caption, clean_post)

  # Print the cleaned posts to verify
  print(head(facebook_posts_cleaned, 10))
  print(head(instagram_posts_cleaned, 10))
```

```
platform                                           content
1 Facebook                       Lambda functions are so powerful! #Python
  #Coding #Coding
2 Facebook                       Lambda functions are so powerful! #Python
  #Coding #Coding
3 Facebook                 Just wrote a new script in Python! #Python #Coding
  #Programming
4 Facebook          Getting errors I can't debug in Python. #Python #Coding
  #Programming
5 Facebook                     Python makes data analysis so easy! #Python
  #Coding #Coding
6 Facebook                    Trying out some new Python packages. #Python
  #Coding #Coding
7 Facebook  Just solved a problem using a Python dictionary! #Python #Coding
```

```
  #Programming
8 Facebook              Python makes data analysis so easy! #Python #Coding
  #DataScience
9 Facebook                 Just learned about Python decorators. #Python
  #Coding #Coding
10 Facebook Finding it hard to understand Python's scoping rules. #Python
  #Coding #Coding
                                                          cleaned_content
1                        lambda functions are so powerful python coding coding
2                        lambda functions are so powerful python coding coding
3                       just wrote a new script in python python coding coding
4             getting errors i cant debug in python python coding programming
5                      python makes data analysis so easy python coding coding
6                     trying out some new python packages python coding coding
7  just solved a problem using a python dictionary python coding programming
8                 python makes data analysis so easy python coding datascience
9                      just learned about python decorators python coding coding
10  finding it hard to understand pythons scoping rules python coding coding

platform                                      content
1 Instagram                       Lambda functions are so powerful! #Python
  #Coding #Python
2 Instagram            Learning about list comprehensions today. #Python
  #Coding #Programming
3 Instagram                    Just wrote a new script in Python! #Python
  #Coding #Programming
4 Instagram                      Trying out some new Python packages. #Python
  #Coding #Coding
5 Instagram Wish Python's performance was better for heavy computations. #Python
  #Coding #Python
6 Instagram            Struggling with Python's indentation rules. #Python
  #Coding #Programming
7 Instagram            Struggling with Python's indentation rules. #Python
  #Coding #DataScience
8 Instagram                  Exploring Python's standard library. #Python
  #Coding #DataScience
9 Instagram                     Just learned about Python decorators. #Python
  #Coding #Coding
10 Instagram                   Lambda functions are so powerful! #Python
  #Coding #DataScience
                                                          cleaned_content
1                        lambda functions are so powerful python coding coding
2                        lambda functions are so powerful python coding coding
3                       just wrote a new script in python python coding coding
4             getting errors i cant debug in python python coding programming
5                      python makes data analysis so easy python coding coding
6                     trying out some new python packages python coding coding
7  just solved a problem using a python dictionary python coding programming
8                 python makes data analysis so easy python coding datascience
9                      just learned about python decorators python coding coding
10  finding it hard to understand pythons scoping rules python coding coding
```

You can now use the cleaned data to perform sentiment analysis and visualize the sentiment trends.

Sentiment analysis on social media data in R. In this section, we explore how to preprocess and analyze text data from social media platforms like Facebook and Instagram using R. Tokenization serves as a foundational step in this process by breaking down text into individual words or tokens, enabling deeper insights through the application of NLP techniques. Using the `unnest_tokens()` function from the tidytext package, we can efficiently split cleaned text into tokens for further analysis.

Once the text is tokenized, we perform sentiment analysis by mapping words to predefined lexicons like Bing, which categorizes words as positive or negative. This allows us to quantify the overall sentiment expressed in the posts, helping businesses or analysts understand public opinion more effectively.

Here's an example of performing sentiment analysis:

```
# Load necessary libraries for sentiment analysis and visualization
  library(tidytext)
  library(ggplot2)

  # Perform sentiment analysis using the Bing lexicon
  bing <- get_sentiments("bing") ❶

  # Tokenize the cleaned Facebook post text ❷
  facebook_tokens <- facebook_posts %>%
    unnest_tokens(word, cleaned_content)

  # Tokenize the cleaned Instagram post text
  instagram_tokens <- instagram_posts %>%
    unnest_tokens(word, cleaned_content)

  # Join with the Bing sentiment lexicon for Facebook ❸
  facebook_sentiment <- facebook_tokens %>%
    inner_join(bing, by = "word")

  # Join with the Bing sentiment lexicon for Instagram
  instagram_sentiment <- instagram_tokens %>%
    inner_join(bing, by = "word")

  # Count the number of positive and negative words for Facebook ❹
  facebook_sentiment_counts <- facebook_sentiment %>%
    count(sentiment, sort = TRUE)

  # Count the number of positive and negative words for Instagram
  instagram_sentiment_counts <- instagram_sentiment %>%
    count(sentiment, sort = TRUE)

  # Visualize the sentiment counts for Facebook
  ggplot(facebook_sentiment_counts, aes(x = sentiment, y = n, fill =
```

```
sentiment)) +
  geom_bar(stat = "identity") +
  theme_minimal() +
  labs(title = "Sentiment Analysis of Facebook Posts",
       x = "Sentiment",
       y = "Count")

# Visualize the sentiment counts for Instagram
ggplot(instagram_sentiment_counts, aes(x = sentiment, y = n, fill =
sentiment)) +
  geom_bar(stat = "identity") +
  theme_minimal() +
  labs(title = "Sentiment Analysis of Instagram Posts",
       x = "Sentiment",
       y = "Count")
```

Figures 10-11 and 10-12 show the output.

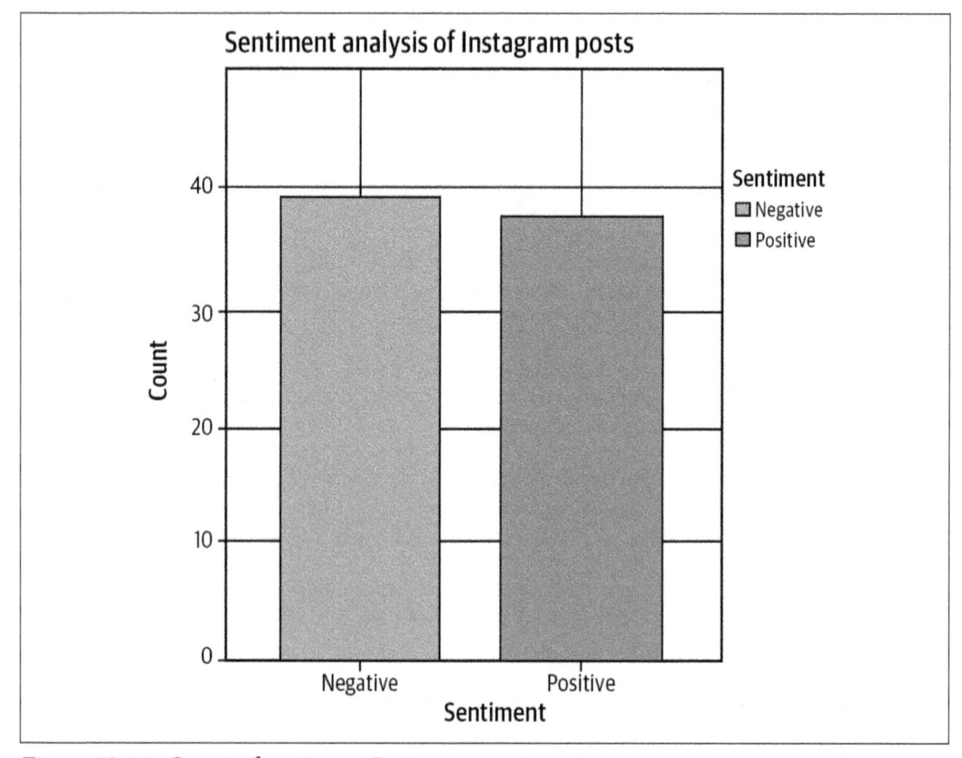

Figure 10-11. Count of sentiment Instagram posts

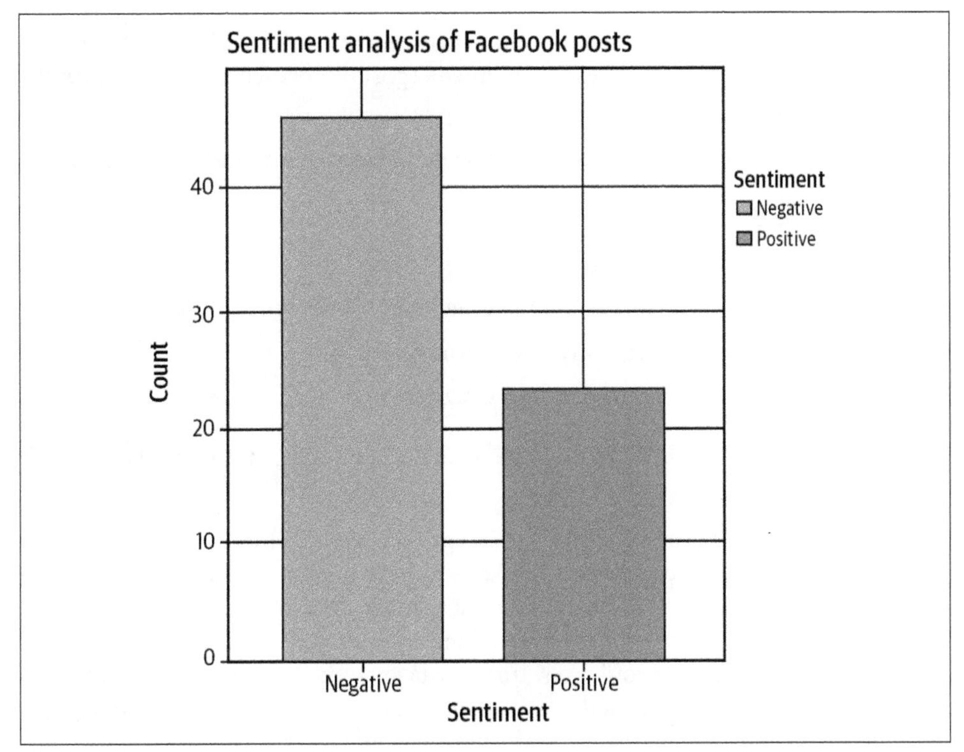

Figure 10-12. Count of sentiment Facebook posts

The following is an explanation of the code provided for performing sentiment analysis using the Bing lexicon, followed by tokenizing, joining, counting, and visualizing sentiment for Facebook and Instagram posts:

❶ This line loads the `Bing sentiment lexicon` from the tidytext package. The Bing lexicon classifies words as either positive or negative.

❷ Tokenization is the process of splitting text into individual words or tokens. The `unnest_tokens` function from the tidytext package splits the `cleaned_content` column into individual words, creating a new row for each word. The resulting tokens are stored in the word column.

❸ This step matches each tokenized word with its corresponding sentiment (positive or negative) from the Bing lexicon. The `inner_join` function merges the tokenized words with the Bing lexicon, joining on the word column. Only words that appear in both the tokenized data and the Bing lexicon are kept.

❹ This step tallies the number of positive and negative words. The `count` function counts the occurrences of each sentiment (positive or negative) and sorts the counts in descending order.

This code performs sentiment analysis on cleaned Facebook and Instagram posts by:

- Loading the Bing sentiment lexicon
- Tokenizing the cleaned text into individual words
- Joining the tokenized words with the Bing lexicon to assign sentiments
- Counting the number of positive and negative words
- Visualizing the sentiment counts using bar plots

These steps help in understanding the overall sentiment of the posts on both platforms.

Transitioning from sentiment analysis to image data analysis highlights a key shift in the types of data being processed. While sentiment analysis focuses on extracting emotions, opinions, and intent from text data, image data analysis deals with visual content. Both tasks are crucial for understanding user-generated content, but they require different techniques and tools. Image data analysis involves working with pixels, patterns, and features extracted from images, utilizing advanced techniques such as computer vision, deep learning, and image recognition algorithms. Just as sentiment analysis helps interpret text, image data analysis can reveal valuable insights from visual media, such as recognizing objects, detecting emotions in faces, or identifying trends in visual content. Now, we turn our attention to the methodologies and tools used to analyze image data.

Image Data

Image data analysis has become a fundamental part of many fields, from healthcare to social media, and Python and R offer powerful tools for performing these tasks. Both languages provide a range of libraries that simplify complex processes such as image processing, object detection, and classification. Python, with libraries like Open Source Computer Vision Library (OpenCV), is renowned for its flexibility in handling image data. Similarly, R leverages packages such as magick for image manipulation. This section will explore how these tools work, focusing on basic image processing workflows in Python and R.

Image Processing in Python

Python offers a robust ecosystem for analyzing image data, with a variety of libraries that simplify tasks like image processing, object detection, and image classification.

Some popular libraries include OpenCV, Pillow (PIL), and TensorFlow/Keras for deep learning-based image analysis.

OpenCV is one of the most widely used libraries for image processing and computer vision. It provides functions to perform tasks such as reading, resizing, and manipulating images. Figure 10-13 is an image that will be used for the example.

Figure 10-13. Sample image

This code is performing basic image processing operations using OpenCV and Matplotlib in Python and produces Figure 10-14:

```
import cv2 ❶
  import matplotlib.pyplot as plt

  # Read an image from file ❷
  image = cv2.imread('sample_image.jpg')

  # Resize the image ❸
  resized_image = cv2.resize(image, (300, 300))

  # Convert the image from BGR (OpenCV format) to RGB for proper visualization
```

```
# in Matplotlib ❹
resized_image_rgb = cv2.cvtColor(resized_image, cv2.COLOR_BGR2RGB)

# Display the resized image ❺
plt.imshow(resized_image_rgb)
plt.title('Resized Image')
plt.show()
```

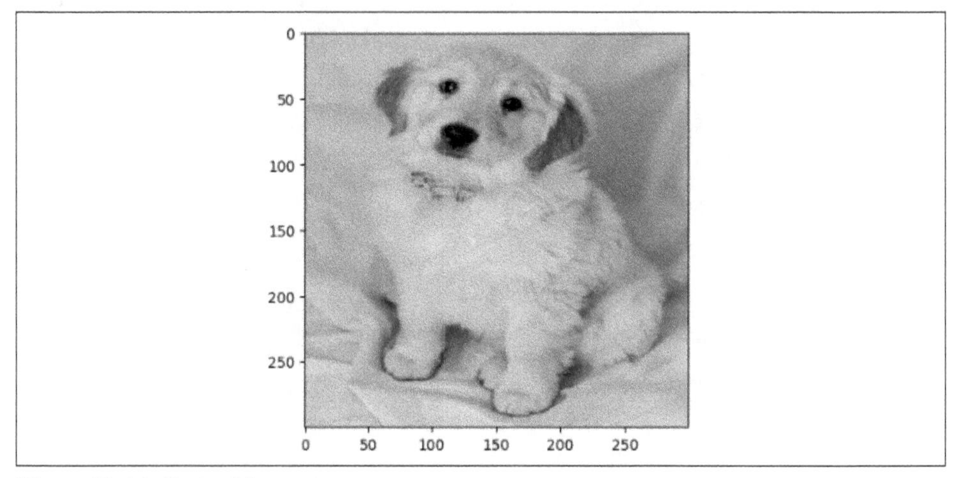

Figure 10-14. Resized image

❶ cv2 is the OpenCV library, commonly used for image processing tasks. `matplotlib.pyplot` is part of the Matplotlib library used for plotting and visualizing data, including images.

❷ The `cv2.imread()` function is used to read the image from the file. OpenCV reads images in BGR (Blue-Green-Red) format by default, unlike the common RGB format used by most image viewers.

❸ The `cv2.resize()` function resizes the image to a 300 × 300 pixel dimension. This helps adjust the size of the image for display or other processing.

❹ OpenCV reads images in BGR format, but Matplotlib (used for displaying the image) expects the image in RGB format. The `cv2.cvtColor()` function is used to convert the image from BGR to RGB format, ensuring correct color rendering when displayed. BGR (Blue, Green, Red) and RGB (Red, Green, Blue) are two different ways to represent color in images. In BGR, the channels are ordered with blue first, followed by green and red, while in RGB, red is first, followed by green and blue.

❺ `plt.imshow()` displays the image using Matplotlib. `plt.title()` adds a title to the image, in this case, `'Resized Image'`. `plt.show()` displays the image in a window.

The code reads an image, resizes it to 300 × 300 pixels, converts it from BGR to RGB format, and then displays it using Matplotlib. This is a basic image-processing workflow to load, resize, convert, and visualize an image.

Image Processing in R

In R, image data analysis is possible through packages like magick for image manipulation and keras for deep learning–based tasks. These libraries enable users to process and analyze image data efficiently.

The magick package provides a simple interface to the ImageMagick library, which allows for reading, writing, and manipulating images. This code produces Figure 10-15:

```
# Install and load the magick package ❶
  install.packages("magick")
  library(magick)

  # Read an image ❷
  image <- image_read('sample_image.jpg')

  # Resize the image ❸
  resized_image <- image_resize(image, "300x300")

  # Display the resized image ❹
  print(resized_image)
```

Figure 10-15. Resized image in Posit Viewer

This code performs basic image manipulation using the magick package in R:

❶ `install.packages("magick")` installs the magick package, and `library(magick)` loads it into the current R session, enabling the use of its image manipulation functions.

❷ `image_read('sample_image.jpg')` reads the image file into the R environment. The image is stored in the image variable.

❸ The `image_resize()` function resizes the image to 300 × 300 pixels. The resized version is stored in the `resized_image` variable.

❹ `print(resized_image)` displays the resized image in the R output (likely in the RStudio viewer or similar).

This process involves reading, resizing, and displaying an image in R using the magick package.

Python and R offer robust solutions for image data analysis through their respective libraries, OpenCV and magick. Both languages provide intuitive workflows for reading, resizing, and manipulating images, enabling users to efficiently handle image data.

Video Data

Video data for analytics can come from a variety of sources and be categorized into different types based on its origin and purpose. Common sources include surveillance cameras, which provide continuous footage for security and monitoring purposes, and social media platforms, where user-generated content offers rich insights into trends and consumer behavior. Broadcast media, such as television news and live events, provide structured video data for real-time analysis and content extraction. Additionally, commercial applications, such as retail store cameras and traffic monitoring systems, generate video data to optimize operations and improve customer experiences. These videos can be analyzed for various purposes, including motion detection, facial recognition, object tracking, and behavioral analysis.

The diverse origins and types of video data, from real-time surveillance to user-generated content, provide a wealth of information that can be leveraged for enhancing security, marketing strategies, operational efficiency, and overall decision-making processes. The types of video data include:

Movies and TV shows
: Used for entertainment analysis and content recommendation.

Surveillance footage
: Used for security and monitoring.

User-generated content
: Videos from platforms like YouTube, Instagram, and TikTok.

Scientific and medical videos
: Used in research and diagnostics.

Sports footage
: Used for performance analysis and strategy planning.

Using Python for Video Data

OpenCV is a powerful tool for video processing, allowing you to extract frames from a video file easily. By reading the video file with `cv2.VideoCapture()` and iterating through it frame by frame, you can save specific frames as images for further analysis or processing. This is useful in tasks such as motion detection, object tracking, and frame-based image recognition from video:

```
pip install opencv-python
```

Using a sample video that shows a forest from a window, this example shows how the video file can be read:

```
import cv2

  # Open a video file
  video_path = 'TestMovie.MOV'
  cap = cv2.VideoCapture(video_path)

  # Check if video opened successfully
  if not cap.isOpened():
      print("Error: Could not open video.")
  else:
      print("Video opened successfully.")

  # Release the video capture object
  cap.release()
  Video opened successfully.
```

Extracting frames and saving them as images

Analyzing video content often begins with breaking the video into individual frames, which allows for more detailed examination and manipulation. By applying basic image processing techniques to these frames using tools like OpenCV, you can extract insights from visual data and save them as images for further processing:

```
import os
```

```
# Create a directory to save frames
output_dir = 'frames'
os.makedirs(output_dir, exist_ok=True)

# Open video file
cap = cv2.VideoCapture(video_path)

frame_count = 0
while cap.isOpened():
    ret, frame = cap.read()
    if not ret:
        break

    # Save frame as image
    frame_filename = os.path.join(output_dir, f'frame_{frame_count:04d}.jpg')
    cv2.imwrite(frame_filename, frame)
    frame_count += 1

cap.release()
print(f"Extracted {frame_count} frames.")
Extracted 107 frames.
```

Analyzing video content

After extracting the frames, it is now possible to apply different processing techniques to gain insight.

Basic image processing. Once you have extracted frames, you can apply basic image processing techniques using OpenCV. Converting video frames to grayscale is a common image processing technique that simplifies the analysis of visual data by reducing the computational complexity and noise often present in color images. Grayscale images contain only intensity information, which can be particularly useful when the color of the objects in the frame is not important for the task, such as in object detection, edge detection, or basic visual analysis. By eliminating the color data, we can focus on essential features like contrast and brightness, making grayscale an efficient approach for many computer vision tasks. The following example demonstrates how to convert a video frame to grayscale using OpenCV, and then the video frame is saved and displayed for further analysis:

```
# Convert a frame to grayscale
frame = cv2.imread('frames/frame_0000.jpg') ❶
gray_frame = cv2.cvtColor(frame, cv2.COLOR_BGR2GRAY) ❷

# Save the grayscale frame
cv2.imwrite('frames/gray_frame_0000.jpg', gray_frame) ❸
```

❶ The function `cv2.imread('frames/frame_0000.jpg')` reads an image from the specified file (`frame_0000.jpg`) and loads it into the variable `frame`. This image is assumed to be in BGR (Blue-Green-Red) format, as OpenCV reads images in that format by default.

❷ The function cv2.cvtColor(frame, cv2.COLOR_BGR2GRAY) converts the frame from its original BGR color format to grayscale. The resulting grayscale image is stored in gray_frame.

❸ The function cv2.imwrite('frames/gray_frame_0000.jpg', gray_frame) saves the grayscale version of the frame to a new file called *gray_frame_0000.jpg* in the specified directory.

The following code uses Matplotlib to display the grayscale image:

```
# Display the grayscale image using matplotlib
  plt.imshow(gray_frame, cmap='gray')
  plt.title('Grayscale Image')
  plt.axis('off')  # Turn off the axis labels
  plt.show()
```

The code produces the grayscale image from the video (Figure 10-16).

Figure 10-16. Grayscale image from sample video

Object detection using pretrained models. You can use pretrained models like YOLO, SSD, or Haar Cascades for object detection on video frames. Pretrained models like YOLO (You Only Look Once), SSD (Single Shot Multibox Detector), and Haar Cascades are popular object detection techniques that allow you to identify objects in images or video frames. These models are trained on large datasets and can detect various objects such as cars, people, animals, and more.

YOLO is a fast and highly accurate object detection model. It treats object detection as a single regression problem, predicting both the class of objects and their bounding boxes simultaneously in one pass over the image. YOLO's speed and efficiency make it well suited for real-time object detection in video streams. SSD divides the image into a grid and detects objects within each grid cell by applying predefined bounding boxes of different sizes and aspect ratios. It balances speed and accuracy, making it a popular choice for detecting multiple objects in a single image. Haar Cascades are older, lightweight object detection models that are particularly good at detecting faces, eyes, and other objects with specific shapes. Although not as robust or versatile as modern models like YOLO or SSD, Haar Cascades are fast and efficient for simpler detection tasks.

These models can be applied to video frames by processing each frame individually, detecting objects, and drawing bounding boxes around the detected objects. They are useful in a wide range of applications, from surveillance to autonomous driving.

In video processing, you can use these models to analyze each video frame, detect objects like cars or pedestrians, and highlight them with bounding boxes. This makes it possible to track objects across frames in real time, which is useful for monitoring traffic, detecting people in security footage, or even identifying products in a shopping video.

Storing and visualizing video data insights

Processed data, such as images or video frames with detections or modifications, can be saved to disk for later use or further analysis. This is especially useful when you work with large datasets, need to review outputs at a later time, or want to compare the results of different detection algorithms or models. For instance, in object detection tasks, after you process a video or image with bounding boxes around detected objects, saving those processed frames can help with performance analysis or model debugging.

By saving the processed frames, you ensure that the results of your model's inference are easily accessible and can be reused without needing to reprocess the raw data every time, which is both time-saving and efficient:

```
# Assuming 'frame' is the processed image with object detections (e.g.,
    # bounding boxes drawn)
  cv2.imwrite('processed_frames/detected_frame_0000.jpg', frame)
```

Creating visual summaries of video analysis. Creating visual summaries of video analysis involves extracting key information or events from a video and presenting them in a more concise, meaningful way. This could include generating shorter video clips that focus on moments where objects are detected, highlighting specific frames where significant events occur, or summarizing the overall analysis using visualizations such as charts or annotated frames. These summaries help users quickly understand the critical aspects of a video without needing to watch the entire footage, making it ideal for tasks like surveillance, sports analysis, or even content creation.

For example, in object detection scenarios, you can highlight moments where objects of interest (e.g., vehicles, people, or animals) appear in the video by cropping or annotating those frames. You can also combine frames or segments where detections occur into a new video clip, which allows stakeholders or analysts to focus on the most relevant sections.

There are several use cases where video analysis and visual summaries can be highly beneficial. In surveillance, for instance, video analysis can detect intruders or unusual activities and compile clips of those moments for quick review. In sports, it can be used to extract and analyze key moments such as goals, fouls, or specific player movements. Similarly, for video content review, it helps to summarize critical parts of long videos, such as customer interactions or tutorials, allowing users to focus on the most relevant sections without watching the entire footage. These applications help streamline the process of identifying important events, improving efficiency and decision making:

```
import numpy as np

  # Create a video summary from processed frames
  output_video_path = 'video_summary.avi'
  fourcc = cv2.VideoWriter_fourcc(*'XVID')
  out = cv2.VideoWriter(output_video_path, fourcc, 20.0, (width, height))

  for i in range(frame_count):
      frame_filename = os.path.join('processed_frames',
      f'detected_frame_{i:04d}.jpg')
      frame = cv2.imread(frame_filename)
      out.write(frame)

  out.release()
  print(f"Video summary saved to {output_video_path}")
```

Working with video data in Python involves reading and processing video files using OpenCV, extracting frames, analyzing video content through image processing and object detection, and storing and visualizing the results. This workflow allows for effective handling and analysis of video data, turning it into actionable insights.

Using R for Video Data

When working with video data in R, one of the most fundamental steps is extracting individual frames for analysis. This process allows you to treat each frame as a separate image, enabling you to apply a range of image processing techniques or machine learning models on the extracted data. The opencv package in R provides a convenient interface to OpenCV functions, allowing you to handle video files, extract frames, and perform various transformations.

Extracting frames from video

To work with video data in R, you can use the opencv package, which provides an interface to OpenCV functions. Before you do anything, you need to extract frames from video:

```
install.packages("opencv")
  library(opencv)

  # Open a video file
  video_path <- "TestMovie.MOV"
  cap <- ocv_video(video_path)

  # Check if video opened successfully
  if (is.null(cap)) {
    print("Error: Could not open video.")
  } else {
    print("Video opened successfully.")
  }

  # Release the video capture object
  ocv_release(cap)
```

Extracting frames and saving them as images

You can extract frames from the video and save them as images for further processing. Here is an example:

```
library(opencv)

  # Create a directory to save frames
  output_dir <- "frames"
  if (!dir.exists(output_dir)) {
    dir.create(output_dir)
  }

  # Open video file
  cap <- ocv_video(video_path)
  frame_count <- 0

  while (TRUE) {
    frame <- ocv_read(cap)
    if (is.null(frame)) {
      break
    }

    # Save frame as image
    frame_filename <- file.path(output_dir, sprintf("frame_%04d.jpg",
    frame_count))
    ocv_write(frame_filename, frame)
    frame_count <- frame_count + 1
  }
```

```
  ocv_release(cap)
  print(paste("Extracted", frame_count, "frames."))
```

Analyzing video content

Once you have extracted frames, you can apply basic image processing techniques using OpenCV. Here is an example:

```
# Convert a frame to grayscale
  frame <- ocv_read(file.path(output_dir, "frame_0000.jpg"))
  gray_frame <- ocv_gray(frame)

  # Save the grayscale frame
  ocv_write(file.path(output_dir, "gray_frame_0000.jpg"), gray_frame)
```

Storing and visualizing video data insights

Processed data can be saved for later use or further analysis. Here is an example:

```
# Save processed frame with detections
  ocv_write(file.path(output_dir, "detected_frame_0000.jpg"), frame)
```

Creating visual summaries of video analysis

You can create visual summaries such as video clips of detected objects or highlight specific frames. Here is an example:

```
library(av)

  # Create a video summary from processed frames
  output_video_path <- "video_summary.mp4"
  av_encode_video(input = list.files(output_dir, pattern =
  "detected_frame_\\d{4}.jpg", full.names = TRUE), output = output_video_path,
  framerate = 20)
  print(paste("Video summary saved to", output_video_path))
```

The OpenCV package in R may not support direct video handling like OpenCV in Python, so the workaround using `ffmpeg` is a reliable method to extract frames and then process them using `opencv` functions in R. The R package is mainly designed for working with image transformations.

An alternative to extracting frames from a video is to convert the video to a series of images externally (using `ffmpeg`, for example) and process them using `opencv` in R. Alternatively, you can call `ffmpeg` from R to process the video into frames.

Here's an example using `ffmpeg` from within R to convert a video to images and then processing the images with OpenCV:

```
# Ensure you have ffmpeg installed on your system
  # Example command for splitting video into frames:
  system("ffmpeg -i TestMovie.MOV -vf fps=1 frames/frame_%04d.jpg") ❶
```

```
# Now process the frames using the opencv package
library(opencv) ❷

# Directory to save frames
output_dir <- "frames"
frame_files <- list.files(output_dir, pattern = "frame_\\d{4}.jpg",
full.names = TRUE)

# Process each frame
for (frame_file in frame_files) {
  frame <- ocv_read(frame_file)

  # Convert to grayscale
  gray_frame <- ocv_gray(frame) ❸

  # Save the grayscale frame
  gray_frame_file <- sub("frame", "gray_frame", frame_file)
  ocv_write(gray_frame_file, gray_frame)

  # Print progress
  print(paste("Processed", frame_file))
}
```

❶ This command extracts frames from the video using `ffmpeg`, saving one frame per second.

❷ After extracting the frames, opencv in R can read, process, and save each image frame. The code processes each frame by converting it to grayscale and saving the result.

❸ The `ocv_gray()` function converts each frame to grayscale, and the `ocv_write()` function saves the processed frame.

Working with video data in R involves reading and processing video files using the OpenCV package, extracting frames, analyzing video content through image processing, and storing and visualizing the results. This workflow allows for effective handling and analysis of video data, turning it into actionable insights.

Summary

In the current data-driven era, businesses encounter various data types beyond traditional structured formats, including semistructured data like JSON, social media data, image data, and video data. Managing these diverse data sources is crucial for gaining comprehensive insights and making informed decisions. This chapter highlighted the importance of integrating various modern data types in business analytics and explored the tools and techniques for handling them effectively.

CHAPTER 11

Measuring Business Value from Analytics and the Role of AI

Business analytics has become an indispensable tool for organizations seeking to gain a strategic edge. Several examples have been explored where value is attained, such as a global retailer leveraging business analytics to optimize its supply chain operations. By analyzing sales data, inventory levels, and market trends, the retailer can make data-driven decisions that reduce stockouts and excess inventory, leading to cost savings and increased revenue. The integration of AI with business analytics further amplifies these benefits, enabling predictive and prescriptive analytics that offer deeper insights and more accurate forecasts, such as predicting future demand based on historical data and external factors like weather or market shifts.

Measuring business value involves evaluating metrics such as increased revenue, cost savings, and improved customer satisfaction. In this retailer's case, AI-driven analytics could automate the reordering process, ensuring that high-demand items are always in stock, while also offering personalized product recommendations to customers based on their purchase history. This combination of traditional analytics with AI leads to greater operational efficiency and effectiveness, resulting in a higher return on investment. By aligning data initiatives with organizational goals, businesses can achieve measurable improvements in performance and profitability.

What Is Business Value in Analytics?

In the context of analytics, business value refers to the tangible and intangible benefits that an organization gains from leveraging data insights to drive decision making and strategic initiatives. It encompasses a wide range of outcomes, from direct financial gains to nonfinancial improvements. The fundamental idea is that by systematically analyzing data, organizations can unlock insights that inform better decisions, leading to improved performance and competitive advantage.

Strategic Impact

Analytics provides both direct and indirect business value by enabling organizations to make informed, data-driven decisions that support long-term strategic goals. Direct impacts include the ability to identify emerging trends and customer preferences, which can guide product development and market positioning. For instance, a retail company might use analytics to forecast future demand, allowing it to capitalize on upcoming market opportunities. A well-known example is Netflix, which leverages analytics to determine which shows and movies to produce based on viewer data. This data-driven content strategy directly aligns production with audience preferences, enabling Netflix to maintain its competitive edge in the streaming industry while expanding its subscriber base.

Indirectly, the insights gained from analytics can help businesses strengthen strategic partnerships and improve organizational agility. For example, understanding shifting consumer preferences through analytics may enable a company to negotiate better supplier agreements, securing materials ahead of demand spikes, further reinforcing its market position.

Operational Efficiency

Analytics contributes to operational efficiency both directly and indirectly by optimizing processes and improving resource allocation. Direct impacts are seen when organizations use predictive analytics to anticipate equipment failures and plan maintenance schedules, reducing downtime and enhancing productivity. A manufacturing company might rely on analytics to identify inefficiencies in production, leading to process improvements and cost savings. UPS provides a notable example of operational efficiency, using route optimization algorithms to cut fuel consumption and reduce delivery times. These direct improvements lead to immediate cost reductions and increased productivity.

Operational efficiency achieved through analytics has indirect value through enhancing an organization's capacity to scale and respond quickly to market changes. Streamlined operations not only reduce costs but also make the business more agile, enabling faster adaptation to fluctuations in demand and competitive pressures.

Customer Satisfaction and Loyalty

Analytics also plays a crucial role in enhancing customer satisfaction and loyalty, both directly and indirectly. Directly, businesses use analytics to personalize offerings and improve customer experiences. For example, Amazon's recommendation system, driven by analytics, increases customer satisfaction by suggesting products based on purchase history and browsing behavior, leading to repeat sales and higher customer retention.

By analyzing customer feedback and interactions, companies can implement targeted service improvements that foster long-term customer loyalty. For instance, airlines may use analytics to monitor passenger feedback and adjust services to meet evolving customer expectations. High levels of customer satisfaction lead to increased brand loyalty, positive word of mouth, and long-term profitability.

Business value in analytics is multifaceted, encompassing strategic impact, operational efficiency, and customer satisfaction and loyalty. Organizations can achieve significant improvements across various dimensions of their business through the use of business analytics. These enhancements not only drive financial performance but also contribute to sustained competitive advantage in an increasingly data-driven world. Let's explore how business value for analytics can be measured.

Metrics and KPIs for Measuring Business Value

Measuring business value derived from analytics initiatives involves the use of various metrics and key performance indicators (KPIs) that quantify the impact of these initiatives across different dimensions of the organization. These metrics are essential for evaluating the effectiveness of data-driven strategies and ensuring they align with the company's overarching goals. Key metrics typically fall into three categories: financial metrics, operational metrics, and customer metrics.

Financial Metrics

Financial metrics are crucial for quantifying the direct economic impact of analytics initiatives. Return on investment (ROI) is a primary metric used to assess the profitability of analytics projects by comparing the financial gains against the costs incurred. For instance, a retail company might implement a predictive analytics system to optimize inventory management. By reducing overstock and stockouts, the company can save significant costs and improve revenue, yielding a high ROI. However, it's important to recognize that implementation, development, monitoring, and ongoing refinement of predictive analytics systems come with associated costs. These may include software, infrastructure, personnel, and training expenses. Despite these up-front and operational costs, the goal is to ensure a net positive ROI by driving substantial gains in efficiency, cost savings, and revenue growth.

For example, a utility company using data analytics to predict equipment maintenance needs can prevent costly breakdowns and reduce maintenance costs, resulting in overall financial benefits. Additionally, revenue growth can be tracked by examining how analytics initiatives contribute to increased sales, such as through targeted marketing campaigns that boost customer acquisition and retention. Net positive

ROI is achieved when the benefits of analytics initiatives, including cost savings and revenue growth, outweigh the total costs of implementation and maintenance.

Operational Metrics

Operational metrics focus on the efficiency and effectiveness of business processes, highlighting how analytics can streamline operations and enhance productivity. One of the most critical metrics in analytical projects is full-time equivalent (FTE) savings, which measures the reduction in labor required to complete tasks through automation or improved processes. For example, a company implementing an analytics-driven customer service chatbot may significantly reduce the need for human intervention, saving numerous FTEs and reallocating those resources to higher-value tasks.

Efficiency gains can also be measured by evaluating improvements in process speed, accuracy, and resource utilization. For instance, a manufacturing firm using machine learning algorithms to predict production bottlenecks can optimize workflows, resulting in faster production times and lower operational costs. Process optimization is another key metric, where analytics can identify and eliminate inefficiencies. A logistics company might use route optimization software to reduce delivery times and fuel consumption, further demonstrating operational improvements. By tracking FTE savings alongside these other metrics, organizations can quantify the operational benefits of their analytics initiatives, justifying investments and making informed decisions to enhance business processes.

Customer Metrics

Customer metrics are vital for understanding the impact of analytics on customer satisfaction and loyalty. Retention rates measure the percentage of customers who continue to engage with a company over a specific period, providing insight into customer loyalty. For example, a telecommunications company using predictive analytics to identify at-risk customers and implement retention strategies, such as personalized offers, can see an increase in retention rates. Satisfaction scores, obtained through customer surveys and feedback, gauge the overall satisfaction level of customers with the company's products or services. A hospitality business using sentiment analysis on customer reviews to improve service quality can see a rise in satisfaction scores. These metrics help organizations understand how analytics initiatives enhance the customer experience, leading to higher customer loyalty and lifetime value.

Aligning Metrics with Organizational Goals and Objectives

To effectively measure business value, it is essential to align metrics with the organization's strategic goals and objectives. This alignment ensures that analytics initiatives support the broader mission of the company and contribute to its success. For instance, if a company's objective is to expand its market share, analytics projects should focus on metrics like customer acquisition cost (CAC) and market penetration rates. By aligning metrics with goals, organizations can prioritize analytics initiatives that have the most significant impact and allocate resources more efficiently. This alignment also facilitates better communication with stakeholders, demonstrating how analytics contributes to achieving key business objectives.

Leveraging Metrics to Demonstrate Value

Analytics initiatives can leverage these metrics to demonstrate value by providing clear, data-driven evidence of their impact. For example, a company might present a dashboard showing the ROI of various analytics projects, highlighting the financial benefits realized over time. Operational improvements can be showcased through before-and-after comparisons of process efficiency metrics, such as reduced cycle times or lower error rates. Customer metrics can be visualized through trend analyses, illustrating how satisfaction scores and retention rates have improved following the implementation of analytics-driven strategies. By regularly tracking and reporting these metrics, organizations can continuously assess the effectiveness of their analytics initiatives and make data-informed adjustments to optimize outcomes.

Metrics and KPIs in Practice

Consider a healthcare provider that uses analytics to improve patient care and operational efficiency. Financial metrics might include cost savings from optimized scheduling and reduced readmission rates, demonstrating a positive ROI. Operational metrics could highlight efficiency gains from streamlined administrative processes and enhanced resource allocation, such as better staff utilization. Customer metrics might show improved patient satisfaction scores and higher retention rates due to personalized care plans informed by data analytics. By aligning these metrics with the healthcare provider's goals of improving patient outcomes and reducing operational costs, the organization can clearly demonstrate the value of its analytics initiatives to stakeholders, reinforcing the importance of data-driven decision making in achieving its mission.

In summary, by defining and tracking key metrics and KPIs, organizations can effectively measure and demonstrate the business value of their analytics initiatives. Aligning these metrics with strategic goals ensures that analytics projects support the broader objectives of the company, driving meaningful and sustainable improvements across financial, operational, and customer dimensions.

Business Case Examples of Value for Analytics

Let's explore a detailed example of how analytics value is measured in a business scenario. We'll use a retail company that wants to improve its overall performance by leveraging data analytics. The key areas of focus will be inventory management, customer engagement, and sales performance.

Consider the following scenario, where a retail company wants to improve performance with data analytics. The company operates a chain of retail stores and an online ecommerce platform. Facing challenges with inventory management, customer engagement, and optimizing sales strategies, the company decides to implement data analytics to address these issues and measure the impact on business value.

Step 1: Problem Definition and Setting Measurable Outcomes

First, the retail company defines its key challenges: reducing stockouts, enhancing customer engagement, and improving sales strategies. The problem is broken down into measurable outcomes, such as improving inventory turnover, increasing customer retention, and boosting sales. By defining these outcomes up front, the company establishes a clear focus for its analytics initiatives.

Step 2: Identifying Metrics to Measure Success and Failure

Next, the company identifies specific metrics to measure both the success and potential failures of its analytics efforts. Financial metrics you might consider measuring include:

Return on investment (ROI)
: Calculate the profitability of analytics initiatives.

Cost savings
: Identify reductions in operational costs, such as inventory holding costs.

Revenue growth
: Measure the increase in sales revenue due to targeted marketing and optimized inventory.

Operational metrics you might consider measuring include:

Inventory turnover
: Track the rate at which inventory is sold and replaced over time.

Stockout rate
: Measure how frequently items are out of stock.

Process efficiency
: Evaluate improvements in supply chain and logistics operations.

Customer metrics you might consider measuring include:

Customer retention rate
Measure the percentage of returning customers.

Customer satisfaction scores
Gather feedback to gauge customer satisfaction.

Customer lifetime value (CLV)
Estimate the total revenue expected from a customer over the relationship.

Step 3: Implementing Analytics Solutions

Implementing analytics solutions driving efficiency, personalization, and informed decision-making across key business areas:

Inventory management
The company implements predictive analytics to forecast demand more accurately. Machine learning models analyze historical sales data, seasonality, and market trends to predict future inventory needs, aiming to optimize stock levels and reduce stockouts.

Customer engagement
Customer analytics is used to segment the customer base and personalize marketing efforts. Purchase history, browsing behavior, and preferences are analyzed to create targeted campaigns aimed at improving retention and satisfaction.

Sales performance
Sales analytics identifies high-performing and underperforming products. This insight allows the company to adjust its product mix, pricing strategies, and promotions to enhance sales performance.

Step 4: Measuring and Demonstrating Value

Financial impact might include:

ROI calculation
The company invests $500,000 in analytics technology, leading to a revenue increase of $1,200,000 and cost savings of $300,000 within the first year.

Cost savings
Predictive inventory management reduces excess stock by 15%, saving $200,000 annually.

Operational improvements might include:

Inventory turnover
: The turnover rate increases from six to eight times per year, improving stock management.

Stockout rate
: Stockouts decrease by 25%, boosting customer satisfaction and minimizing lost sales.

Process efficiency
: Optimized logistics reduce delivery times by 20%, improving overall efficiency.

Customer outcomes might include:

Customer retention rate
: Retention increases from 65% to 75%, driven by targeted campaigns.

Customer satisfaction
: A 10% improvement in satisfaction scores, attributed to personalized experiences and better product availability.

Customer lifetime value (CLV)
: Engagement strategies raise CLV by 20%, reflecting higher long-term revenue potential.

Step 5: Reporting and Continuous Improvement

Dashboards and reports visualize the metrics and communicate the success of the analytics initiatives to stakeholders. Regular monitoring helps identify further areas for improvement, ensuring the company continuously leverages data analytics effectively.

This scenario illustrates how data analytics can drive significant business value, improve inventory management, enhance customer engagement, and optimize sales. By systematically measuring financial, operational, and customer metrics, the company can justify investments and guide future strategies.

AI and Generative AI in Business Analytics

AI has transformed business analytics by offering advanced tools and techniques that enhance decision-making processes through data-driven insights. One of the most impactful areas is predictive analytics powered by machine learning. Machine learning algorithms analyze historical data to forecast trends, identify patterns, and aid in decision making. For instance, retailers leverage predictive analytics to forecast demand, optimize inventory levels, and tailor marketing strategies. This enables businesses to anticipate customer needs, reduce waste, and enhance operational efficiency, driving growth and profitability.

Natural language processing (NLP) plays a pivotal role in analyzing unstructured text data. NLP enables companies to extract valuable insights from sources like customer reviews, social media posts, and support tickets. Sentiment analysis and text classification are common use cases where NLP helps companies understand customer emotions, preferences, and pain points. For instance, financial institutions use NLP to analyze customer feedback to detect service issues and improve customer satisfaction. Similarly, ecommerce platforms employ NLP to enhance product recommendations and customer support by analyzing real-time customer queries, improving customer engagement and loyalty.

Computer vision (CV), another critical AI subfield, focuses on extracting insights from visual data, such as images and videos. Manufacturing companies use computer vision for quality control, detecting defects in real time during production, which reduces product recalls and ensures consistent quality. In retail, computer vision is applied to monitor customer behavior through video analytics, optimizing store layouts based on foot traffic and customer interactions with products. Healthcare providers use CV to analyze medical images, enabling faster and more accurate diagnostics, leading to better patient outcomes.

Additionally, generative AI has introduced new possibilities in content creation and product design. For example, fashion brands use generative AI to create personalized clothing recommendations based on a user's style preferences. In marketing, generative AI tools automate the creation of engaging content, such as social media posts and product descriptions, that are tailored to individual customer segments, enhancing the effectiveness of campaigns and improving brand visibility.

Overall, AI, NLP, and CV bring value to business analytics by optimizing processes, deepening customer understanding, and enhancing operational efficiencies across industries. These technologies drive innovation and deliver measurable outcomes in a variety of business contexts, from improving customer experiences to streamlining production and marketing operations.

Introduction to Generative AI

Generative AI represents a cutting-edge advancement in the field of AI, focusing on the creation of new, original content. Unlike traditional AI, which typically analyzes and interprets existing data, generative AI can produce creative outputs such as text, images, music, and more. This capability is driven by sophisticated models like generative adversarial networks (GANs) and variational autoencoders (VAEs). These models learn patterns from vast datasets and generate new data that is remarkably similar to the original, making them powerful tools for creative content generation.

Applications in Product Design

In product design, generative AI can be a game-changer. Designers can use AI algorithms to generate multiple design prototypes based on specific parameters and constraints. For example, automotive companies can use generative AI to create new car models that optimize for aerodynamics and aesthetics. This not only accelerates the design process but also introduces innovative design solutions that might not have been conceived through traditional methods. Additionally, generative AI can help in customizing products to meet individual customer preferences, enhancing the overall user experience.

Applications in Content Creation

Generative AI has also revolutionized content creation across various media. In the field of writing, AI models like GPT-4 can generate articles, stories, and even poetry that mimic human creativity. This can be particularly useful for businesses that require large volumes of content for marketing, blogging, or social media. By automating content generation, businesses can maintain a consistent and engaging online presence without overwhelming their creative teams. Furthermore, AI-generated content can be tailored to match the brand's tone and style, ensuring coherence across different platforms.

Applications in Marketing

Marketing is another area where generative AI is making a significant impact. AI can generate personalized marketing messages, advertisements, and even entire campaigns. By analyzing customer data, generative AI can create targeted content that resonates with individual preferences and behaviors. For instance, AI can generate personalized email campaigns that address the specific interests and needs of each recipient, leading to higher engagement and conversion rates. Additionally, generative AI can be used to create dynamic advertisements that adjust in real time based on user interactions, making marketing efforts more responsive and effective.

Enhancing Customer Experience

The integration of AI and generative AI in business analytics significantly enhances customer experience. By leveraging AI for predictive analytics, companies can anticipate customer needs and preferences, offering personalized recommendations and services. Generative AI takes this a step further by creating customized content and products that align with individual customer tastes. For example, an online retailer can use generative AI to recommend unique fashion items based on a customer's browsing history and style preferences. This level of personalization not only boosts customer satisfaction but also fosters loyalty and long-term engagement.

Improving Operational Efficiency

AI and generative AI contribute to improving operational efficiency across various business functions. Predictive maintenance powered by AI can foresee equipment failures and schedule timely interventions, reducing downtime and maintenance costs. Generative AI can also optimize workforce scheduling by analyzing demand patterns and employee availability to create efficient shift plans. For example, a large retail chain could leverage AI to generate optimized staffing schedules during peak shopping hours, ensuring sufficient customer service while minimizing labor costs during slower periods. These enhancements in efficiency translate to cost savings and improved productivity, giving businesses a competitive edge.

Future Prospects and Challenges

The future of AI and generative AI in business analytics holds immense potential, but it also presents challenges. As AI technologies continue to evolve, their applications will become more sophisticated, offering even greater value to businesses. However, issues such as data privacy, ethical considerations, and the need for significant computational resources must be addressed. Ensuring that AI systems are transparent, fair, and secure will be crucial for gaining and maintaining public trust. Businesses must also invest in continuous learning and development to keep pace with the rapid advancements in AI technology and fully leverage its potential.

Use Cases for AI and Generative AI in Business Analytics

In today's rapidly evolving business landscape, AI and generative AI have become integral to driving innovation and growth across various industries. By harnessing the power of these technologies, businesses can enhance their analytics capabilities, streamline operations, and create personalized experiences that cater to the unique needs of their customers. This section explores specific use cases where AI and generative AI have made significant impacts, illustrating how these technologies incorporate business analytics to deliver measurable value. Whether it's improving customer engagement, optimizing supply chains, or enhancing decision-making processes, AI is transforming the way organizations operate and compete in the market.

Use Case 1: AI-Driven Customer Insights and Recommendations

AI-driven customer insights and personalized recommendations are among the most effective applications of machine learning in business analytics. By analyzing vast amounts of customer data—such as purchase history, browsing behavior, and demographic details—AI can generate tailored recommendations that improve customer engagement. For example, companies like Amazon and Netflix use AI to provide real-time, data-driven suggestions that enhance the user experience and increase customer loyalty. By leveraging analytics to understand customer behavior, businesses can optimize marketing efforts, improve retention, and boost overall satisfaction.

Use Case 2: Generative AI in Content Creation

Generative AI has revolutionized content creation in industries such as marketing, entertainment, and design. AI models like OpenAI's GPT-4 can generate high-quality marketing copy, social media posts, and even long-form content, reducing the time and resources needed for content creation. In marketing, generative AI can produce consistent brand messaging across platforms, while in the creative arts, it can generate artwork, music, and product designs. For instance, small businesses can use generative AI to create engaging marketing materials without a large design team, ensuring that they maintain a professional image while keeping costs down. This integration of AI into business analytics enables data-driven content strategies that resonate with target audiences.

Use Case 3: AI-Powered Supply Chain Optimization

AI's ability to improve supply chain operations through predictive analytics and optimization algorithms has become a critical factor in business success. By analyzing historical sales data, market conditions, and seasonal trends, AI can predict demand more accurately and help businesses manage their inventory more efficiently. For example, a retailer can use AI to forecast demand during holiday seasons, ensuring optimal stock levels to meet customer needs without overstocking. This use of business analytics to drive supply chain efficiency results in cost savings, reduced waste, and improved customer satisfaction.

Use Case 4: Enhancing Decision Making with AI

AI empowers organizations to make smarter, data-driven decisions in real time. In finance, AI-driven analytics can process vast datasets to uncover market trends, assess risks, and identify investment opportunities. In human resources, AI can optimize the recruitment process by analyzing candidate profiles to find the best fit for a position, ultimately improving employee retention and reducing hiring costs. These applications demonstrate how AI, combined with advanced business analytics, supports more informed decision-making processes that minimize risk and maximize efficiency.

Use Case 5: AI in Healthcare Analytics

AI is transforming healthcare by enabling predictive analytics that improves patient outcomes and optimizes hospital operations. For example, AI can analyze patient data to identify individuals at high risk of developing chronic conditions, allowing healthcare providers to intervene early and provide preventive care. Additionally,

AI-driven insights help hospitals optimize staffing schedules, manage medical supplies, and streamline patient care, reducing operational costs and improving overall efficiency. Business analytics plays a pivotal role in ensuring that AI applications in healthcare are aligned with organizational goals, ultimately enhancing the quality of care delivered to patients.

Use Case 6: Generative AI for Personalized Customer Experiences

Generative AI is pushing the boundaries of personalized customer experiences by creating content tailored to individual preferences. In the fashion industry, for instance, AI can generate customized clothing designs based on a user's style preferences and trends, offering virtual try-ons that enhance the shopping experience. Similarly, entertainment platforms use AI to create personalized playlists or recommend movies that align with users' specific interests. By incorporating business analytics, companies can continuously refine these AI-driven recommendations to boost customer engagement and satisfaction.

Use Case 7: AI in Retail Analytics

In retail, AI's ability to analyze customer behavior through foot traffic patterns and optimize store layouts is reshaping the shopping experience. Retailers can strategically place products to maximize visibility and sales based on real-time data, improving both operational efficiency and revenue generation. Additionally, AI-powered chatbots provide instant customer support and product recommendations, enhancing customer service and loyalty. By integrating AI with retail analytics, businesses can make data-driven decisions that improve the overall customer experience and drive profitability.

Addressing Factual Inconsistencies and Human-AI Collaboration

While AI and generative AI bring immense benefits to business analytics, their implementation comes with critical challenges, particularly the risk of factual inconsistencies. In customer service chatbots, for instance, inaccurate or misleading information can erode trust and lead to significant reputational damage. Similarly, in marketing, relying on AI-generated content that misrepresents facts or misaligns with the brand's message could harm customer relationships and brand credibility. These risks highlight the importance of human oversight and cocreation in AI workflows. Ensuring that AI systems work alongside human experts is crucial for maintaining accuracy, especially in customer-facing roles.

In addition, data privacy and security remain paramount concerns, as AI systems often require access to vast amounts of sensitive information. Businesses must comply with regulations, implement stringent security measures, and uphold ethical standards to prevent biases in AI algorithms. A collaborative, human-AI approach—one that blends human judgment with AI efficiency—will help mitigate these risks, ensuring that AI-driven initiatives not only remain accurate but also aligned with ethical practices and organizational goals.

Future Prospects

The future of AI and generative AI in business analytics holds immense potential. As AI technologies continue to evolve, their applications will become more sophisticated, offering even greater value to businesses. Advances in NLP, computer vision, and machine learning will enable more nuanced and comprehensive analyses, driving further innovation. Businesses that invest in AI and generative AI now will be well positioned to leverage these advancements, gaining a competitive edge and driving sustained growth in an increasingly data-driven world.

Challenges and Considerations

The integration of AI and generative AI into business analytics presents transformative opportunities for organizations, enabling sophisticated data-driven decision making and innovative solutions. However, alongside these benefits, these advanced technologies bring challenges such as data privacy, integration issues, bias mitigation, and resource management. Navigating complex privacy laws like GDPR and CCPA, ensuring ethical data use, and addressing compatibility with legacy systems are crucial to maintaining customer trust and regulatory compliance.

Additionally, businesses must overcome technical and organizational hurdles to successfully deploy AI solutions. Cultivating a data-driven culture, securing executive buy-in, and providing necessary training are essential to align AI projects with business goals. Rigorous cost-benefit analyses are needed to justify substantial investments in AI technologies. Future-proofing AI investments through continuous learning and adaptation ensures that businesses remain competitive and compliant in a rapidly evolving technological landscape. By addressing these challenges, organizations can harness AI's full potential to drive innovation and achieve sustainable growth.

Integration Challenges and Scalability in Deploying AI Solutions

Integrating AI solutions into existing business processes and systems poses significant challenges. Many organizations struggle with legacy systems that are not compatible with modern AI technologies, leading to integration issues that require substantial time and resources to resolve. Furthermore, scalability is a major concern. While AI solutions might perform well in pilot projects or limited deployments, scaling these solutions across the entire organization can be difficult. This involves ensuring that the infrastructure can handle increased data loads and that AI models maintain their performance and accuracy as they are applied to larger datasets and more complex scenarios.

Mitigating Biases and Ensuring Fairness in AI-Driven Decisions

Bias in AI algorithms is a critical issue that can lead to unfair and discriminatory outcomes. AI systems are only as good as the data they are trained on, and if this data contains biases, the AI will likely perpetuate these biases in its decisions. For example, biased hiring algorithms can disadvantage certain groups of candidates, and biased customer service bots can provide unequal assistance based on demographic factors. Ensuring fairness requires meticulous efforts to identify and mitigate biases in training data and algorithms. This involves implementing practices such as diverse data collection, bias testing, and the use of fairness metrics to evaluate and adjust AI models continually.

Technical and Organizational Challenges in AI Deployment

Deploying AI solutions is not just a technical challenge but also an organizational one. Companies need to cultivate a data-driven culture where employees at all levels understand and embrace the use of AI. This requires training and upskilling staff to work effectively with AI tools and analytics. Moreover, aligning AI projects with business goals and securing executive buy-in are essential for successful deployment. Resistance to change and lack of clear strategic direction can hinder the adoption of AI, making it crucial for leadership to communicate the benefits and potential of AI clearly and consistently.

Cost and Resource Considerations

The implementation of AI technologies can be costly, involving significant investments in software, hardware, and skilled personnel. Developing and maintaining AI models require specialized knowledge and expertise, which can be expensive to acquire and retain. Additionally, continuous monitoring and updating of AI systems are necessary to ensure they remain effective and relevant, further adding to the costs. Organizations need to conduct thorough cost-benefit analyses to ensure that the potential benefits of AI deployment justify the investment. This includes considering the long-term savings and efficiencies gained through AI, as well as the initial outlays.

Future-Proofing AI Investments

To future-proof AI investments in a rapidly evolving landscape, businesses must remain agile and ready to adapt. AI technologies advance quickly, and today's solutions may soon be outdated. Staying updated on new AI developments, regulations, ethical standards, and best practices is essential. This requires building flexible AI systems, investing in continuous learning, and fostering a culture of innovation. By doing so, businesses can ensure their AI investments remain relevant and effective over time.

While AI offers significant potential, challenges such as data privacy, ethical considerations, scalability, bias mitigation, and cost management must be addressed. Tackling these issues and maintaining a mindset of continuous improvement will allow organizations to fully harness AI's potential for innovation and long-term growth.

Summary

This chapter explored the crucial role of business analytics in driving organizational growth, efficiency, and profitability. By integrating AI, businesses can enhance their analytics capabilities, leading to better decision making, improved inventory management, and optimized customer experiences. The chapter emphasizes the importance of aligning analytics initiatives with organizational goals to achieve measurable success, using KPIs like ROI, cost savings, and customer retention.

AI's transformative impact is demonstrated through use cases in predictive analytics, NLP, and supply chain optimization. Generative AI is highlighted for its ability to revolutionize content creation and product design, enabling businesses to engage customers more effectively.

The chapter also addresses challenges such as data privacy, AI biases, and scalability, emphasizing the importance of human-AI collaboration and continuous adaptation to ensure ethical, effective, and future-proof AI investments.

Index

A

B

C

D

E

H

I

J

K

L

M

N

O

P

Q

R

S

T

U

V

W

X

Y

Z

About the Author

Deanne Larson, Ph.D., is a data science practitioner and academic whose passion is helping others be successful in applying analytics to achieve business value. Her research has focused on implementing an enterprise data strategy, applying agile analytics, and data science best practices. Dr. Larson is passionate about teaching and applying analytics. Deanne attended executive training at the Harvard Business School focusing on IT leadership, Stanford University focusing on data science, MIT focusing on AI, and New York University focusing on business analytics. She has presented at multiple conferences including TDWI, TDWI Europe, IRM UK, PMI, and other academic conferences. She is Principal Faculty, has consulted for several Fortune 500 companies, and has authored multiple research articles on data science methodology and best practices. She holds Project Management Professional (PMP), Project Management Agile Certified Practitioner (PMI-ACP), Certified Business Intelligence Professional (CBIP), and Six Sigma certifications.

Colophon

The animal on the cover of *Modern Business Analytics* is a Rajah Brooke's birdwing butterfly (*Trogonoptera brookiana*). This species is the national butterfly of Malaysia and can be found in rainforests throughout Southeast Asia.

Rajah Brooke's birdwings have large black wings with tooth-shaped, green markings and a red patch by the head. The males sport more vibrant colors than the females. Their diet is part defense mechanism; it includes plants that are toxic to other species, including birds.

Rajah Brooke's birdwing butterfly has an IUCN conservation status of Least Concern, but they are protected by the Convention on International Trade in Endangered Species of Wild Fauna and Flora. Many of the animals on O'Reilly covers are endangered; all of them are important to the world.

The cover illustration is by Karen Montgomery, based on an antique line engraving from *Insects Abroad*. The series design is by Edie Freedman, Ellie Volckhausen, and Karen Montgomery. The cover fonts are Gilroy Semibold and Guardian Sans. The text font is Adobe Minion Pro; the heading font is Adobe Myriad Condensed; and the code font is Dalton Maag's Ubuntu Mono.

www.ingramcontent.com/pod-product-compliance
Lightning Source LLC
Jackson TN
JSHW052015151224
75332JS00005B/2

9781098140717